ORGANIC CHEMISTRY
AN INTRODUCTORY TEXT

C. W. WOOD, M.Sc., M.Ed., LL.B., F.R.I.C.
Formerly Senior Science Master
St. Bees School

A. K. HOLLIDAY, Ph.D., D.Sc., F.R.I.C.
Reader in Inorganic Chemistry
The University of Liverpool

and

R. J. S. BEER, M.A., D.Phil.
Reader in Organic Chemistry
The University of Liverpool

LONDON
BUTTERWORTHS

ENGLAND:	BUTTERWORTH & CO (PUBLISHERS) LTD LONDON: 88 Kingsway, WC2B 6AB
AUSTRALIA:	BUTTERWORTH & CO (AUSTRALIA) LTD SYDNEY: 586 Pacific Highway, Chatswood, NSW 2067 MELBOURNE: 343 Little Collins Street, 3000 BRISBANE: 240 Queen Street, 4000
CANADA:	BUTTERWORTH & CO (CANADA) LTD TORONTO: 14 Curity Avenue, 374
NEW ZEALAND:	BUTTERWORTH & CO (NEW ZEALAND) LTD WELLINGTON: 26-28 Waring Taylor Street, 1
SOUTH AFRICA:	BUTTERWORTH & CO (SOUTH AFRICA) LTD DURBAN: 152-154 Gale Street

First Published 1960
Second impression 1961
Third impression 1962

Second Edition 1963
Second impression 1964
Third impression 1965
Fourth impression 1967

Third Edition 1968
Second impression 1970
Third impression 1972

Suggested U.D.C. Number: 547
ISBN 0 408 61401 3

PRINTED AND BOUND IN GREAT BRITAIN BY
HAZELL WATSON AND VINEY LTD
AYLESBURY, BUCKS

B/2/9 22 £1.50

This book is to be returned on or before
the last date stamped below.

A

1977 8 APR 1948
Y 1977
977

ORGANIC CHEMISTRY

CONTENTS

PREFACE TO THIRD EDITION

Our aim in preparing this extensively revised third edition has been to present the reactions of organic chemistry in terms of simple concepts of structure and mechanism, so that their interrelationship is understood from an early stage. Hence, although the factual basis of the book is preserved, and the various topics are treated in essentially the same order as before, the reader is now introduced to the idea of 'mechanism' at the outset. Various aspects of electronic theory are discussed at appropriate places in the text, with certain more advanced theoretical concepts reserved for Chapter 26.

Methods of preparation have been largely omitted from the earlier chapters. We believe it is disconcerting for the beginner in organic chemistry to be confronted with sets of reactions which are only related in the sense that they lead to the same type of product. Reactions capable of being used as preparative methods appear logically as the subject is developed and are subsequently collected together in a separate chapter, which thus serves as a revision chapter for much of the foregoing factual material.

The changes mentioned above have necessitated the complete re-writing of large parts of the book; elsewhere, many minor alterations have been made and the opportunity has been taken to correct a number of errors. Some of the examination questions which appeared in the second edition have been omitted and new questions, taken mainly from first-year examinations at the University of Liverpool and intended to test understanding rather than memory, have been included. We are again grateful to the University and to the various examining boards for permission to reproduce questions; those originally set by the Joint Matriculation Board are now designated J.M.B.

A.K.H.
R.J.S.B.

PREFACE TO SECOND EDITION

In this new edition of our three books we have been concerned primarily with additions to cover new developments, and alterations where a change of emphasis was needed; there are no major changes in the subject-matter. In response to many requests, we have included questions from examination papers at the ends of most of the chapters.

In this organic chemistry book, we have further emphasized the importance of functional groups, e.g. by giving the appropriate group formulae at the head of each chapter. New reagents and methods of synthesis have been added—in particular, the powerful new synthetic method of hydroboration has been treated in an elementary way. The continued growth in the use of petroleum-based substances in organic chemistry on the industrial scale has required the inclusion of brief accounts of the ways in which many aliphatic and aromatic compounds are obtained from petroleum. The section on polymers has been expanded, and the polymerization of simple substances such as acetylene has been more fully treated. A short treatment of proteins and amino acids has been included. More emphasis has also been placed on the shape and stereochemistry of simple molecules.

We are grateful to the following examination boards for permission to reproduce questions set in recent years in 'A' level (A) and scholarship (S) papers: Associated Examining Board (A), Cambridge Local Examination Syndicate (C), London University (L), Northern Universities Joint Matriculation Board (N), Oxford and Cambridge Schools Examination Board (O and C), Oxford Local Examinations (O), Southern Universities Joint Board (S), Welsh Joint Education Committee (W); and to the University of Liverpool for questions from various examination papers. We are also grateful to all those who have helped us with comments and criticisms, and especially to Mr. D. R. Lewis, M.Sc., whose sharp eye has enabled us to remove many obscurities and imperfections from the text.

C.W.W.
A.K.H.

bibliotheca SelfCheck System

Customer name: Jana Aljohwani
Customer iD: L001 80789

Items that you have checked out

Title: Organic chemistry :
ID: a167861 01
Due: 05 March 2025

Total items: 1
Account balance: 0.00 EUR
Checked out: 1
Overdue: 0
Hold requests: 0
Ready for collection: 0
19/02/2025 16:48

Thank you for using the bibliotheca SelfCheck
System

PREFACE TO FIRST EDITION

This is one of three books which together meet the requirements of the G.C.E. Advanced and Scholarship level examinations of the various examining bodies. These books should be of value also to university students taking a first year chemistry course leading either to a degree in Chemistry or to a degree in some other science subject, and to students at colleges of technology taking courses for H.N.C., Grad. R.I.C., and similar examinations. A knowledge of ' O ' level chemistry is assumed and therefore detailed accounts of, for example, the preparation of the common gases, studied at that level, are not given.

The separation of the various branches of chemistry is unfortunately necessitated by the great accumulation of knowledge in each branch; but we believe that no one branch should be studied in isolation. It is true that organic chemistry can still claim some individuality, though the modern studies of, for example, organometallic compounds are making the dividing line between organic and inorganic chemistry a very indistinct one. But both inorganic and organic chemistry today demand some previous knowledge of basic physical chemistry, of valency theory and of molecular structure. In this trilogy we have devoted a complete book to each of the three branches, physical, inorganic, and organic chemistry. It must be emphasized, however, that each book is complete in itself and may certainly be used independently of the others; thus, the inorganic and organic books contain sufficient about atomic structure and valency to understand the subject-matter, but these topics are considered in greater detail in the physical chemistry book.

The historical approach has not been neglected in these books; the aspiring scientist can gain inspiration by looking backwards as well as forwards. But looking forward has been encouraged by the inclusion of some of the newer developments in chemistry, e.g., ion-exchange resins, high polymers, ' clathrate ' and ' sandwich ' compounds, and the story of the new elements beyond uranium. It is hoped that these will stimulate an interest in the ever-expanding domain of chemical knowledge.

Organic chemistry is usually encountered at a later stage of learning than is inorganic and physical chemistry. In the first chapter of this book, we have endeavoured to show how the historical separation of the chemistry of carbon compounds came about, and how organic compounds differ from inorganic. Thereafter, organic compounds are grouped together for convenience of study; emphasis is placed, however, upon the reactions of groups rather than on individual compounds. The reactions of each group are summarized at the end of the appropriate chapter. The use of newer reagents such as lithium aluminium hydride and the borohydrides has been included where applicable.

Many references are made to the large-scale uses of the compounds considered, and to important syntheses of industrial importance, e.g. those involving water-gas and Fischer–Tropsch. Some of these are described more fully where the underlying chemistry is within the comprehension of the readers for whom the book is intended. It is hoped that this will emphasize that organic chemistry is not just another laboratory subject.

Late in the book, a chapter considers the application of valency theory to problems of organic structure and reactivity. This chapter goes somewhat beyond the present needs of the candidate for Advanced level chemistry, but should be useful for those sitting the Scholarship papers and also for degree students. In it an attempt has been made to explain some of the facts of organic chemistry, using theory which is equally applicable to other branches of chemistry. Hence, the use of too much essentially 'organic' symbolism has been avoided, and the resonance concept has not been over-emphasized.

Contrary to common practice, methods of separation and purification, determination of molecular and constitutional formulae are treated fully at the end of the book, although frequent references to these subjects have been made earlier in the text. Methods of determining molecular weights are not included since these are usually described in detail in books of physical chemistry. It has been found by experience that to start a course of organic chemistry by first describing these methods is unsatisfactory, for without some knowledge of organic compounds, such sections are not readily appreciated. For the same reason, there is a final chapter on the different kinds of isomerism, for with the facts of earlier pages, a full treatment of this most important phenomenon is possible.

Our grateful thanks are due to Dr. J. S. E. Holker who has read the book through and made many invaluable criticisms and suggestions for improvement, and to those who have assisted us in the reading of the proofs.

C.W.W.
A.K.H.

INTRODUCTION

WHAT we now call inorganic chemistry began with the extraction of metals from their ores which were found on or under the surface of the earth. Nevertheless, certain processes of a chemical nature had been carried out from earlier times with the body substances of plants as the raw materials. Beer is known to have been made as early as 7000 B.C. in the valleys of the Tigris and Euphrates where barley grew wild, the method being one of fermentation which was essentially the same in principle as that in use today. At this early period also, in the countries where the vine was grown, wine was prepared by a similar fermentation of the sugar in the grape. The fact that wine turned sour on keeping in air was also known (although it was not until the end of the eighteenth century that Lavoisier showed that the souring was due to the oxidation of the alcohol of the wine to acetic acid by the oxygen of the air). Acetic acid was thus perhaps the first acid to be identified. Some of its properties were known at an early date, for we find the effervescing action of sour wine on chalk described two centuries before Christ.

Coloured substances were obtained from plants and animals and used for dyeing cloth. The famous Tyrian purple dye was extracted from a shellfish found in the Mediterranean as far back as 1500 B.C.

For some centuries after the birth of Christ, however, the investigation of the chemical compounds making up the bodies of plants and animals lagged far behind the study of compounds found in the earth's crust or derived from them, although there were many opportunities for such investigation. Distillation—a process much more commonly used to study the chemical compounds in living organisms than those on or in the earth—was known and described in great detail in the fifth century. Accounts can be read of it being actually used in the eighth century by the Arabs to concentrate the alcohol in fermented liquors and the acetic acid in sour wine.

Isolated discoveries were made at much later dates. A mixture of sulphuric acid and alcohol was distilled in the thirteenth century. The distillate, no doubt, consisted of a mixture of alcohol and ether. Paracelsus (1493–1541), the famous Swiss-born doctor-chemist (or Iatro-chemist, to give such men their proper name) used this distillate in medicine. Boyle (1627–91) isolated methyl alcohol from the products of the destructive distillation of wood. He prepared some salts of acetic acid and distilled some of them, notably those of lead and calcium, to obtain the substance acetone.

In the seventeenth century, for the first time, pure specimens of acetic acid and alcohol were prepared, as distinct from concentrated solutions obtained by the distillation of sour wine and alcoholic liquor respectively. Acetic acid (m.p. 17°C) was obtained by the distillation of verdigris (a basic copper acetate).

However, it was not until the eighteenth century that any real progress in the study of the chemical compounds in living organisms was made. There were two causes for this. First, there were the wasted centuries of alchemy—the fruitless search for the philosopher's stone. After Boyle's discrediting of alchemy in the seventeenth century, there followed a further misfortune. The German chemist Becher (1635–82) and later, his pupil Stahl (1660–1734) put forward the erroneous Phlogiston Theory of Combustion, which influenced chemistry for over a hundred years. Most of the compounds already isolated from living organisms were combustible, and thus, according to the theory, were rich in phlogiston. At a time when some progress might have been made in the further study of these compounds, it was unfortunate that a false theory caused confusion as to their nature.

During the eighteenth century, the German chemist Scheele (1742–86), working in Sweden, succeeded in isolating many important vegetable and animal compounds. He prepared oxalic acid by oxidizing cane sugar by means of nitric acid. He extracted this same acid from the sorrel plant. He obtained citric acid from lemons, malic acid from apples, lactic acid from sour milk, uric acid from urine, and glycerine from olive oil—to mention a few only. Scheele enunciated no principle nor formulated any great theory, but his contributions to chemistry were nevertheless great, for, by preparing a big variety of compounds for the first time, he provided the data on which principles and theories could be built up in succeeding years.

In the eighteenth century, the German chemist Marggraf obtained formic acid by distilling red ants. In 1773, urea, the excretion product of many animals, was isolated from urine.

The scientists of this period were unable to understand the existence of so many compounds closely related to each other in composition and yet differing widely in chemical behaviour. It was suggested, for example, that there was only one vegetable acid, viz. acetic acid, and all those others which Scheele had obtained, were some ' combined forms ' of it. Had they known, as we know to-day, that both acetic acid and lactic acid have the same empirical formula of CH_2O, the confusion would have been greater than ever; however, at that time quantitative chemistry was in its infancy.

There had been growing too, during this century, the erroneous idea that there was something fundamentally different between the chemical compounds of mineral origin and those derived from living organisms. Public recognition of this was first made by the Swedish chemist Berzelius in 1808. To the chemistry of those compounds of living origin—or as he said from ' organized nature ' —Berzelius gave the name of *organic chemistry*, while a study of those of mineral origin constituted *inorganic chemistry*.

Lavoisier was the first to attempt to analyse organic compounds. He found that most of those obtained from plants consisted of the three elements carbon, hydrogen and oxygen, while many of those present in the animal body contained nitrogen also. The chemists of this period had been used to two or three elements combining together to form one compound only, and the multiplicity of compounds of the three elements carbon, hydrogen and oxygen strengthened the belief in the new classification of chemistry into inorganic and organic. It was even suggested that the compounds of organic chemistry did not obey the recently discovered laws of definite and multiple proportions, on which the new Atomic Theory of Dalton had been formulated. Berzelius, who was a believer in this idea, attempted to explain it by suggesting that the compounds of ' organized nature ' had been built up in the living body by some strange ' vital force ' of unknown character. Up to this date no chemist had succeeded in synthesizing an organic compound, i.e. those which were known at the time had been obtained from raw materials of living origin in which the ' force ' was supposed to be already present.

Two events soon happened to show that the classification of substances into the two groups was based on an unsound foundation.

First, during the first quarter of the nineteenth century, the methods of analysis started by Lavoisier were greatly improved. Many famous European chemists contributed to this, notably Berzelius, Gay-Lussac, Thénard, Dumas and especially Liebig. They proved conclusively that the simpler organic compounds, at least, consisted of elements united in simple atomic proportions according to the laws governing the composition of inorganic compounds. It was from this analytical work at the beginning of the nineteenth century that organic chemistry as a branch of science proper can be said to date.

Secondly, in 1828, the German chemist Wöhler prepared by accident in the laboratory a white crystalline substance identical with the urea first isolated from urine fifty years previously*. Ordinary methods of laboratory practice had been used. Reporting this discovery to his former teacher Berzelius, Wöhler wrote that he had

---

* Urea was probably first synthesized in 1812 by John Davy (brother of Humphry Davy). Davy had prepared phosgene, $COCl_2$, by the direct addition of chlorine to carbon monoxide in sunlight. On treatment of phosgene with ammonia, he obtained a white crystalline solid which did not give carbon dioxide on addition of an acid, thus showing that it was not ammonium carbonate. The solid was undoubtedly urea but Davy did not identify it with the compound obtained from urine in the previous century.

prepared urea ' without requiring a kidney or an animal, either man or dog '. He rightly claimed to have disproved the ' vital force ' theory of Berzelius and to have shown that no fundamental distinction between organic and inorganic chemistry existed.

In 1822, Wöhler (who was primarily an inorganic chemist—he was the first to isolate aluminium) prepared cyanic acid (of formula HOCN), and some of its salts, e.g. silver, lead and potassium cyanates. On attempting to make ammonium cyanate, by evaporating the solution, crystals of urea separated. In his original preparation of cyanic acid, Wöhler had used cyanogen which had been obtained from compounds of animal origin. Some claimed therefore, that such compounds possessed the ' vital force ' already and this contributed to the formation of urea. Shortly afterwards, however, Kolbe synthesized acetic acid from raw materials which were of purely mineral origin. Other similar syntheses soon followed.

A new definition of organic chemistry was now needed. It was given by the German chemist Gmelin in 1848 as *the chemistry of the compounds of carbon*—a definition which still stands today. The oxides of carbon and carbonates are thus, by definition, organic compounds; it is, however, usual to study these substances in a course of inorganic chemistry. There is, in fact, no theoretical reason why organic chemistry should be studied in a separate course —this is merely done on grounds of convenience, some of which are given on p. xxiii.

An interesting discovery which subsequently proved of immense importance to the study of organic chemistry was made in 1823. In 1822, Wöhler had prepared silver cyanate, a compound which we now know to have the formula AgOCN. Another compound of silver, oxygen, carbon and nitrogen called silver fulminate, had been prepared and in 1823, Liebig showed that this compound had a percentage composition identical with that of Wöhler's silver cyanate.

When a few years later, Wöhler obtained urea in an attempt to prepare ammonium cyanate, the idea grew up that the properties of a compound may depend not only on the number and kind of atoms in each molecule, but on the *way in which the atoms are arranged*. Thus, silver cyanate differed from silver fulminate probably because the atoms in the respective molecules were arranged differently. Other similar pairs of substances were discovered, and to this phenomenon of the *occurrence of two substances having the same molecular formula but different chemical and physical properties*, the name of *isomerism* was given by Berzelius in 1832.

With the more accurate methods of quantitative analysis available and with the idea of the existence of molecules first put forward by Avogadro and elaborated by Cannizzaro, distinctions between empirical formulae and molecular formulae were being realized. Dalton had isolated two hydrocarbons of empirical formula CH_2; these were ethylene and butylene of molecular formulae C_2H_4 and C_4H_8 respectively. The hydrocarbon benzene of molecular formula C_6H_6 (first isolated by Michael Faraday) and the hydrocarbon acetylene of molecular formula C_2H_2 constituted another similar pair. Here were different compounds *possessing the same empirical formula, with the molecular formula of one some simple multiple of that of the other.* To this hitherto unknown phenomenon, Berzelius gave the name of *polymerism.* For reasons we shall discuss later, isomerism and polymerism are not so well-known in the realm of inorganic chemistry.

The next logical step in the solution of these problems was to attempt to find the reason for the joining together of atoms in chemical combination and the manner of their arrangement within the molecule.

Various suggestions were made by a number of chemists. It was left to the English chemist Frankland (1825–99) in 1852 to combine these and formulate a *Theory of Valency.* Frankland maintained that each of the atoms of the different elements possessed a definite *combining capacity* or *valency,* which could be expressed as a whole number. Hydrogen was to be the standard with a valency of one. In the simplest hydrocarbon, methane, of molecular formula CH_4, one atom of carbon is combined with four atoms of hydrogen, and the valency of carbon is therefore four. Representing a single valency as a dash (—), we get a picture, represented in one plane, of a methane molecule as

This formula shows the number and kind of atoms in a molecule, but gives an imperfect ' picture ' of the molecule because all the atoms are shown in one plane. It is often called a *graphical, constitutional* or *structural* formula; we shall use the last term.

Frankland's theory was first applied to the problems of the structure of organic molecules in 1858 by Kekulé and the young Scots chemist Couper, working independently. They maintained that organic chemistry was based on two fundamental principles:

(1) *That carbon has a valency of 4, i.e. one atom of it can combine with four univalent atoms or groups of atoms.* (It was necessary to include ' groups of atoms ', for as early as 1832, Wöhler and Liebig

had shown that associated groups of atoms, which were residues of compounds (called *radicals*) could behave in chemical reactions like single atoms. Two simple examples are CH_3— and C_2H_5—, called *methyl* and *ethyl* respectively. Many others will be met later.)

(2) Carbon atoms can unite with themselves apparently indefinitely to form chains which may be straight, or branched, or closed rings by joining up the ends, e.g.

$$
\begin{array}{cc}
& \text{H} \quad \text{H} \\
| \quad | & | \quad | \\
-\text{C}-\text{C}-, \text{ e.g. in } \textit{ethane}, & \text{H}-\text{C}-\text{C}-\text{H} \\
| \quad | & | \quad | \\
& \text{H} \quad \text{H}
\end{array}
$$

$$
\begin{array}{cc}
| \quad | \quad | \quad | & \\
-\text{C}-\text{C}-\text{C}-\text{C}- & \\
| \quad | \quad | \quad | & \\
-\text{C}- \quad \text{or} & \\
| & \\
-\text{C}- & \\
|
\end{array}
$$

Great progress in the study of organic chemistry was now made. Structural formulae of compounds already discovered were worked out by methods which will be described later. But, in addition— and this is important—the existence of hitherto unknown compounds was forecast and the later discovery of some of these served to confirm the basic principles of Kekulé and Couper.

Until the end of the nineteenth century, chemists could know little of the structure of either atoms or the molecules made up from them. At the beginning of the present century, investigations, mainly due to physicists such as J. J. Thomson, Rutherford and Planck, had made it possible to understand something of the nature of the atom. It is now recognized that any atom is made up of a positively charged nucleus surrounded by negatively charged particles of almost negligible mass, called *electrons*. The nucleus is composed of *neutrons* (having no charge) and positively charged particles called *protons*; the charge on the proton is equal and opposite to that on an electron, and every atom contains equal numbers of protons (in the nucleus) and electrons (outside it). The number of electrons (or protons) is the *atomic number* of the element concerned. For example, the atomic number of hydrogen is 1 (one proton + one electron), and the atomic number of sodium is 11. Different numbers of neutrons in atoms of the same element give rise to *isotopes*, i.e. atoms of the same atomic number but differing in mass. It is, however, the light extranuclear electrons which are important in determining the chemical behaviour of the element; and when the atoms of two or more elements come together, the possibility of union to form a compound must depend upon the interaction of the electrons.

The Danish physicist Bohr suggested that the electrons in an atom were situated in certain definite *energy levels*, visualized by Bohr as orbits. Thus, hydrogen was thought of as a central proton with one electron ' revolving ' round it. These orbits (n) were numbered outwards from the nucleus ($n = 1, 2, 3$, etc.). It is now known that Bohr's theory was inadequate; but his idea of energy levels is correct, and his orbits are now called *shells*.

If we start at hydrogen and proceed along the first and second period elements of the periodic classification, through helium to neon, the atomic number increases from 1 to 10. In helium we have two electrons and they fill the first ' shell ', corresponding to the original orbit of Bohr, and now more commonly called the K shell. After helium, the next shell L (corresponding to the old orbit 2), fills up and is full at neon, i.e. the L shell is filled by eight electrons. Spectroscopic studies indicate that this shell can be divided into four 'sub-shells', each capable of containing two electrons and having, under certain circumstances, slightly different energies. In fact, in all atoms—and in molecules also—there can be many different energy levels, but each can contain two, and not more than two, electrons.

A few years after Bohr had put forward his theory of atomic structure, Kossel, Lewis and others began to develop an electronic theory of valency. It was recognized that the atoms of the noble gases were particularly stable and inactive, and so it was suggested that when one atom of an element combines with another, the electrons in their outermost shells are rearranged in some way to give each atom a noble gas structure. Since the outer shell of electrons in all the noble gases (except helium) contains eight electrons, this suggestion usually implied that each element attained an outer stable *octet* of electrons. Now this stable octet might be attained either (*a*) by the giving or receiving of electrons, or (*b*) by the sharing of electrons.

(a) The giving or receiving of electrons

Consider the sodium atom with two electrons in the first (K) shell, eight in the L shell and one in the M shell, i.e. having the electronic configuration 2.8.1, with one electron more than neon (2.8). The fluorine atom has the configuration 2.7, i.e. one electron less than neon. Clearly, if one electron is transferred from the sodium to the fluorine, then both atoms have the neon structure, thus:

$$\underset{2.8.1}{\text{Na}\cdot} \; + \; \underset{2.7}{{}^{\times}\overset{\times\times}{\underset{\times\times}{\text{F}}}{}^{\times}} \longrightarrow \underset{2.8}{\text{Na}^{+}} \; \underset{2.8}{\left[{}^{\times}_{\times}\overset{\times\times}{\underset{\times\times}{\text{F}}}{}^{\times}_{\times} \right]^{-}}$$

(The dots and crosses are used to depict the electrons on sodium and fluorine atoms for convenience, and there is, of course, no distinction between the electrons on the different atoms.) When this transfer occurs, not only do the two atoms attain the noble gas structure (associated with stability) but they become charged, i.e. we obtain a sodium *ion*, Na^+, and a fluoride ion, F^-. Between these two ions, electrostatic attraction exists; we have an *ionic* or *electrovalent bond*, and each element has an electrovalency of one. It is now known that the crystal of sodium fluoride is made up of sodium ions and fluoride ions only. Due to the work of Arrhenius, it had been realized previously that ions existed in solution; but it is now clear that ions exist already in most solid salts. Moreover, in a crystal of, say, sodium fluoride, the electrostatic attraction between the oppositely charged ions is not confined to particular pairs, i.e. is not localized, but extends over the whole crystal, since the latter contains positive and negative ions arranged alternately in a three-dimensional lattice. There is, then, no such thing as a sodium fluoride molecule.

An ionic crystal is not easily broken up by heating, although it may easily break up in solution because of the reduced electrostatic attraction.

(b) *The sharing of electrons*

The simplest example of sharing of electrons between two atoms is in the formation of the hydrogen molecule, H_2, thus:

$$\text{H}\cdot + {}^{\times}\text{H} \longrightarrow \text{H}{}^{\bullet}_{\times}\text{H}$$

Here the two electrons are shared, and give to each hydrogen atom the configuration of helium. Another example is the formation of methane, from carbon with four electrons in its outer shell, and four atoms of hydrogen, thus:

$$\underset{\times}{\overset{\times}{\times}\,\text{C}\,\times} \; + \; 4\text{H}\cdot \longrightarrow \text{H}{}^{\bullet}_{\times}\overset{\overset{\textstyle\text{H}}{\bullet\times}}{\underset{\underset{\textstyle\text{H}}{\times\bullet}}{\text{C}}}{}^{\bullet}_{\times}\text{H}$$

Here, the carbon attains the stable neon configuration. Moreover, it is clear that each shared electron pair is equivalent to the line used to depict valency in structural formulae (p. xvii). There is apparently no separation of charge, and no ions are formed. Shared electron-pair bonds are called *covalent bonds*. They have certain characteristics. First, they are *localized* between particular atoms; commonly, we find a relatively small number of atoms bound together to form a molecule, but the compound as we know it—be it solid, liquid or gas—is composed of these molecules held together by relatively weak intermolecular forces (p. 292), i.e. the covalent bonds are localized inside each molecule. Hence although covalent bonds may be quite strong, most covalent compounds are easily melted or vaporized when the temperature is raised, because the molecules are easily driven apart, and no covalent bonds need be broken. Only rarely do we find covalent bonds extending throughout a crystal; when this happens (e.g. in the diamond, with strong co-valent bonds between the carbon atoms) we obtain a highly stable covalent lattice which cannot easily be broken up either by heat or in solution (cf. the ionic lattice).

Secondly, covalent bonds are *directional*. The discovery of a further type of isomerism (p. 332) proved that the picture of the methane molecule as shown in one plane on p. xvii was inadequate. In 1874, Van't Hoff and le Bel put forward the theory of the *tetrahedral carbon atom*, in which the carbon atom was represented at the centre of a regular tetrahedron, with its valency bonds directed towards the four corners. The methane molecule is thus three-dimensional, and may be represented by any one of the following diagrams:

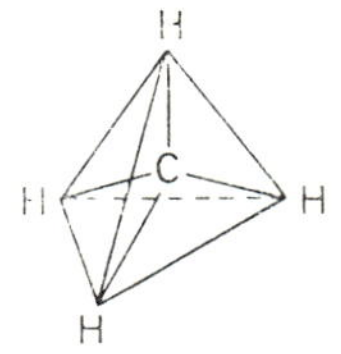

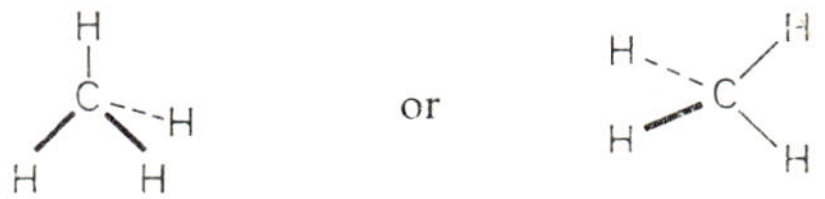

In the two lower diagrams, bonds represented by ordinary lines are supposed to be
in the plane of the paper, bonds shown as
heavy lines project towards the viewer, and
bonds shown as broken lines recede behind
the plane of the paper.

(A simple molecular model shows the shape better than any of the above diagrams.) For many simple organic compounds, structural formulae in one plane are adequate representations; but it must always be remembered that organic molecules in general are three-dimensional structures.

In some covalent bonds, both electrons come from one atom and none from the other. For example, in the formation of ammonium chloride from ammonia and hydrogen chloride, the nitrogen provides both electrons to give the covalent N:H bond in the ammonium ion, NH_4^+:

$$
\begin{array}{c}
H \\
H \!-\!\! \overset{\displaystyle}{N} : + \; H\overset{x}{}Cl \longrightarrow \left[\begin{array}{c} H \\ H\!-\!\!N:H \\ H \end{array} \right]^{+} + \;\; {}_{x}^{x}Cl^{-} \\
H
\end{array}
$$

A bond of this kind is often called a *co-ordinate link*, but notice that all the four N—H bonds in the ammonium ion are identical, covalent bonds. Notice also that two originally covalent molecules react to form a new ionic compound, ammonium chloride. This type of reaction is often important, as, for example, when hydrogen chloride (covalent) dissolves in water (covalent), when we have the reaction

$$
H\!-\!Cl + \begin{array}{c} H \\ \diagdown \\ \diagup \\ H \end{array}\!\!O \longrightarrow H_3O^+ + Cl^-
$$

and the formation of the hydroxonium ion H_3O^+ is responsible for the acidic behaviour of an aqueous solution of hydrogen chloride.

Wöhler's ammonium cyanate is seen to be *ionic* (like many inorganic compounds), made up of ammonium ions, NH_4^+, and cyanate ions, OCN^-. But when heated it changed into the *covalent* compound urea (and most organic compounds are essentially covalent) thus explaining the isomerism of these two compounds:

$$
[NH_4]^+OCN^- \;\rightleftharpoons\; O\!=\!C \begin{array}{c} \diagup N \diagdown \!\!{}^{H}_{H} \\ \diagdown N \diagdown \!\!{}^{H}_{H} \end{array}
$$

Here, two ions combine to give a covalent molecule—the opposite situation to that just discussed.

Ionic compounds are frequently soluble in water; in the process of solution, the ions are separated (i.e. *dissociated*) but often retain their identities. Thus if sodium chloride is dissolved in water, sodium ions, Na^+, and chloride ions, Cl^-, exist in the solution, and in an aqueous solution of silver nitrate there are silver ions, Ag^+, and nitrate ions, NO_3^-. When one of these solutions is added to the other, there is a white precipitate of silver chloride due to the association of silver and chloride ions, thus:

$$AgNO_3 + NaCl \longrightarrow AgCl \downarrow + NaNO_3$$

i.e. $$\underset{white}{Ag^+ + Cl^- \longrightarrow AgCl \downarrow}$$

Here, we are concerned with a simple 'change of partners' in solution—a *metathetic* reaction. Such reactions are usually very *rapid*, go to *completion*, and are often *quantitative*, i.e. there are no *side-reactions*. For such reactions we can write complete, balanced equations, which are said to be *stoichiometric*.

Some covalent compounds form ions when dissolved in water, e.g. hydrogen chloride; but most organic covalent compounds are only sparingly soluble in water, and remain unchanged in solution, i.e. do not break up into ions. Thus ethyl chloride, C_2H_5Cl (with a molecule like that of ethane, p. xviii, in which one hydrogen is replaced by chlorine) is covalent, contains no chloride ions, and hence gives no immediate precipitate with silver nitrate solution. Any reaction in which ethyl chloride is to take part will necessitate the *breaking of a covalent bond*; a simple example is (p. 12):

$$C_2H_5Cl + OH^- \longrightarrow C_2H_5OH + Cl^-$$

in which the C—Cl bond must break*. Such a reaction is often *slow*, does not necessarily go to completion, and side-reactions (due to other forms of attack by the reagent added) can often occur. It is therefore often impossible to write down a stoichiometric equation for an organic reaction; more commonly, the main reaction must be represented (e.g.) thus:

$$A \xrightarrow{reagent} B + C$$

The two rather typical reactions just discussed, one inorganic and the other at least partly organic, show some striking differences. However, it is important not to create a general distinction between inorganic and organic reactions (as has sometimes been done). Some inorganic reactions involve the breaking of covalent bonds, e.g. the hydrolysis of (covalent) boron trichloride to give boric acid:

$$BCl_3 + 6H_2O \longrightarrow B(OH)_3 + 3H_3O^+ + 3Cl^-$$

---

* This reaction is also an example of a *hydrolysis* reaction. This word, although meaning, strictly, 'splitting by water' is often used to describe reactions in which the attacking species is the hydroxide ion, since the latter anion is derived from water. In fact, the 'true' hydrolysis reaction $C_2H_5Cl + 2H_2O \longrightarrow C_2H_5OH] + H_3O^+ + Cl^-$ does proceed very slowly in the absence of alkali.

whilst in many organic reactions, *ionic intermediates* are known to be involved (see, for example, the discussion of the reaction between ethylene and bromine on p. 21). Moreover, according to modern views, there is no clear-cut distinction between an ionic and a covalent bond. Consider the two atoms A and B; the ionic and covalent representations for a bond between the two would be

$$A^+ B^- \quad \text{and} \quad A \overset{\times}{\bullet} B$$

with the two electrons equally shared between A and B in the covalent representation. Very often, when an actual covalent bond is formed, the two electrons are not *equally* shared. Thus, if B has a bigger share of the bonding pair than A, the actual bond will be 'in-between' the completely unequal form A^+B^- and the equally-shared form $A^{\times}B$. In such a case, the covalent bond between A and B is said to possess *polarity*, and, as will be seen, this property is very important when we consider the reactivity of organic molecules.

The number of organic compounds known today is probably over a million, and thousands are being discovered each year. How are we to set about the study of such a vast number of compounds? In attempting to answer this question, it is instructive to consider some work of Dumas from 1827 onwards. He prepared a number of derivatives of ethyl alcohol and, later, corresponding derivatives of methyl alcohol. Certain similarities in the behaviour of these corresponding substances were noticed and this was explained by the presence within the molecules of the same groups of atoms. This led eventually to the principle we accept today that *the properties of an organic compound may be regarded as the sum of the properties of the individual groups within its molecule.* Anyone with only an elementary training in organic chemistry can look at the constitutional formula of a simple organic compound unknown to him, identify groups with which he is familiar, make allowances for conflicting properties of certain groups and finally arrive at a fairly accurate forecast of the properties of the unknown substance. Usually, in a simple compound, one particular group is most important in deciding the properties—*the functional group* (see below).

It is also possible to arrange organic compounds possessing similar groups, and therefore similar properties, into families called *homologous series.* To become familiar with the members of a family, all that is necessary is to study in detail the behaviour of one particular member. From such knowledge, and bearing in mind certain rules governing homologous series, the properties of other members of the family can be given with reasonable accuracy. This is the method adopted in the following pages.

The characteristics of a homologous series may be summarized as follows:

(a) *All the members can be represented by a general formula.* The first homologous series to be studied is a class of hydrocarbons (i.e. compounds of hydrogen and carbon only) known as the *paraffins* or *alkanes*, the first member of which is methane, CH_4. The general formula is C_nH_{2n+2}. Substituting $n = 1$, the formula of methane is obtained. The second member, ethane, where $n = 2$, has the formula of C_2H_6. Subsequent members, in order, have formulae of C_3H_8, C_4H_{10}, C_5H_{12}, etc.

Thus to pass from one member to the next in order of increasing molecular weight in this and all other homologous series, the group CH_2 is added.

(b) *The chemical properties of the different members are similar.* A homologous series is usually characterized by the presence of a functional group, which largely determines the properties of each member of the series. Thus the alcohols (Chapter 6) all contain the —OH group, which gives them similar properties. Sometimes the functional groups may be a particular kind of *bonding*—thus the functional group of the olefins (Chapter 2) is the *double-bonded* pair of carbon atoms $\diagup C{=}C\diagdown$ —it is this that gives the olefins their characteristic properties. It must be noted, however, that as n and therefore the molecular weight of the member increases, the chemical reactivity decreases. It may conceivably happen that a member of low molecular weight (i.e. where n is small) readily exhibits certain chemical behaviour, whereas a member of high molecular weight, when subjected to the same treatment, is relatively unreactive.

(c) *There is a definite gradation of physical properties among the members.* As molecular weight increases, i.e. as the value of n increases, the melting point, boiling point and density rise. Methane, the first alkane, is a gas. In general, those alkanes where $n > 4$ are liquids at room temperature; those where $n > 16$ are solids. (See p. 293.)

We have seen that while *there is no fundamental difference between organic and inorganic compounds*, the two classes are usually studied in separate courses. Some of the reasons why this is so are given below.

(1) The enormous number of compounds of carbon and the similarities which exist among them make it more convenient to study them separately from the compounds of other elements. The existence of so many compounds is due to facts already stated, viz.:

(a) the *quadrivalency of carbon*

(b) the existence of the *strong linkage*, and hence,

(c) the ability to form *chains of carbon atoms* of indefinite length,* or *rings* of various sizes.

(2) Many organic compounds, particularly those found in living organisms, have huge molecules made up of a thousand atoms or over. For example, one of the simpler proteins (which are important naturally-occurring organic compounds) is believed to have a molecular formula of $C_{662}H_{1020}N_{193}O_{201}S_4$. Most proteins have even bigger molecules. Few compounds having molecules of comparable size are, as yet, definitely known in inorganic chemistry. With such large molecules, the possibilities of isomerism are obviously very great. Even with fairly small numbers of atoms, many isomers can be formed. The chemical behaviour of these isomers depends on the actual arrangement within each molecule in a way which is less commonly encountered amongst inorganic compounds.

(3) Many inorganic compounds are, as already mentioned, made up of ions, and are often very stable to heat (ammonium cyanate is an exception here). By contrast, most organic compounds are unstable to heat. Many are decomposed at temperatures as low as 300°C and few are stable at 1000°C. The number of inorganic compounds stable at 1000°C is quite large.

Because most organic compounds are covalent, many are liquids at room temperature and many which are solids, melt at relatively low temperatures. Only a relatively small number of inorganic compounds possess these properties. Because of these differences, different laboratory techniques are used in the study of organic compounds. For example, distillation is a process of purification much more commonly used in organic than in inorganic chemistry; determination of a melting point and/or boiling point provide quick and easy methods of determining the purity of a specimen of an organic compound. Such methods are rarely used in inorganic chemistry.

(4) If soluble in water (and relatively few are), aqueous solutions of organic compounds are, in general, *non-conductors of electricity*. This follows from the covalent character of most organic compounds,

---

* Silicon, the next element below carbon in Group IV of the periodic classification of the elements, has a valency of four also, and can form compounds such as silane, SiH_4, analogous to methane, CH_4. But the $-\overset{|}{\underset{|}{Si}}-\overset{|}{\underset{|}{Si}}-$ linkage is much weaker than the $-\overset{|}{\underset{|}{C}}-\overset{|}{\underset{|}{C}}-$ and so only a few compounds contain silicon chains, and even these are easily broken.

contrasting with the ionic compounds of inorganic chemistry, aqueous solutions of which are electrolytes.

Reference has been made (p. xxiii) to the covalent character of ethyl chloride, C_2H_5Cl, and the fact that it gives no immediate precipitate with aqueous silver nitrate. Methods of analysis to identify chlorine (and other elements) in organic compounds differ from those used in inorganic chemistry.

(5) Although we no longer believe in the ' vital force ' of Berzelius, one must always bear in mind the very intimate connexion between organic chemistry and living organisms. All terrestrial life-forms, at any rate, are based on carbon compounds and their transformations. In this sense, carbon is uniquely important amongst the elements.

PARAFFINS OR ALKANES

HYDROCARBONS—compounds of carbon and hydrogen—are among the most important organic compounds. Many different homologous series are known. In some kinds of natural petroleum, hydrocarbons of eight different homologous series have been identified. The molecules of some of these compounds contain chains of carbon atoms, which may be straight or branched: in others the carbon atoms are joined to form rings of varying numbers of such atoms.

Some of the hydrocarbons occurring in nature are simple and some very complex. Two simple ones, methane and ethylene, of molecular formulae CH_4 and C_2H_4 respectively, are constituents of the gases trapped above the oil in most oil wells. The characteristic yellow colour of carrots, butter, and yolk of egg is due to a hydrocarbon of molecular formula $C_{40}H_{56}$. It is understandable that many isomers (p. xvi) having this molecular formula are known, and another isomer of the same hydrocarbon gives the red colour to tomatoes, rose hips and the berries of the hawthorn. The smooth glossy appearance of the leaves of members of the cabbage family is due to the presence of hydrocarbon waxes containing 29 and 31 atoms of carbon in their molecules. Natural rubber is a hydrocarbon with long chain-like molecules of formula $(C_5H_8)_n$, where n is several thousand.

The structural formulae of the two simple hydrocarbons mentioned above and that of a third one, acetylene, of molecular formula C_2H_2, are as follows:

methane
(tetrahedral)

ethylene
(planar)

acetylene
(linear)

In the molecule of methane, each of the four valencies of the carbon atom is joined to a hydrogen atom. The four valencies of the carbon atom are thus being used to their maximum combining capacity.

Methane is the first member of a homologous series of hydro-carbons of which the second is ethane, of molecular formula C_2H_6, and structural formula:

(1)   (2)   or   (3)

Formulae 2 and 3 attempt to show the three-dimensional nature of the ethane molecule.

Here again, all valencies are used to the greatest combining power. Such compounds are said to be *saturated*.

In the molecule of ethylene, two valency bonds link together the two carbon atoms and in the molecule of acetylene, the corresponding link consists of three bonds. In each case one bond would have been sufficient. The ethylene molecule would appear to be capable of combining with two more hydrogen atoms and that of acetylene with a further four, to give in each case a molecule of ethane, C_2H_6. We therefore say that ethylene and acetylene are *unsaturated*. Notice that in ethylene and acetylene, the bonds from the carbon atoms are no longer tetrahedral. Further reference to the bonds in the molecules of these two hydrocarbons will be made later.

Table 1.1.—*The paraffins or alkanes*

Name of compound(s)	Formula	State	No. of isomers
Methane	CH_4		1
Ethane	C_2H_6		1
Propane	C_3H_8	Gas	1
Butanes	C_4H_{10}		2
Pentanes	C_5H_{12}		3
Hexanes	C_6H_{14}		5
Heptanes	C_7H_{16}		9
Octanes	C_8H_{18}	Liquid	18
	$C_{17}H_{36}$		
	⋮		
	$C_{30}H_{62}$ etc.	Solid	>4 million

The three hydrocarbons methane, ethylene and acetylene are the first members of three homologous series commonly called *paraffins*, *olefins* and *acetylenes* respectively, with the general formulae and alternative names given as follows:

SATURATED HYDROCARBONS:

(1) *Paraffins* or *Alkanes** C_nH_{2n+2}

The names of some of these are shown in *Table 1.1.*

UNSATURATED HYDROCARBONS :

(1) *Olefins* or *Alkenes* C_nH_{2n}

(2) *Acetylenes* or *Alkynes* C_nH_{2n-2}

Structurally, hydrocarbons of these types are the parents of many other organic compounds. Thus the compound *ethyl chloride*, C_2H_5Cl, mentioned in the Introduction (p. xxiii), is related to ethane in the sense that a hydrogen atom is replaced with a chlorine atom, i.e. C_2H_6 becomes C_2H_5Cl. Ethyl chloride is the second member of a homologous series of organic chlorides, the first being CH_3Cl, derived from methane and called *methyl chloride*.

Ethyl alcohol, or *ethanol*, C_2H_5OH, is also related structurally to ethane, but differs from ethyl chloride in that it contains the hydroxyl group (OH) instead of a chlorine atom. Ethyl alcohol and ethyl chloride are quite distinct compounds with different chemical and physical properties; they both contain the hydrocarbon residue C_2H_5—, but have different functional groups —OH and —Cl.

Methane, CH_4

Methane is a colourless gas with practically no smell. It is almost insoluble in water, and is a product of the decomposition of vegetable and animal matter in the absence of air. It is thus found in coal mines (and hence its name *fire-damp*), in marshy districts where it is a product of the bacterial decay of vegetable matter (e.g. cellulose), under water (and thus also called *marsh gas*), under the sea (*natural gas*), and where oil is found. A typical natural gas contains nearly 80 per cent methane; in some samples, the percentage of methane is as high as 98.8 per cent. Methane is present in the human intestine and in the gut of most herbivorous animals, being formed by the breaking down of the cellulose contained in the plant body.

Large amounts of methane are set free in the purification of sewage (p. 13).

---

* With the enormous number of closely related organic compounds some systematic nomenclature, based on characteristic or functional (to use an alternative term) groups present in the molecules, must be adopted. Without this, an extensive study of organic chemistry would be extremely difficult. International conferences to build up such a systematic nomenclature are held periodically, The terms alkane, alkene and alkyne are general systematic names which are now universally accepted.

Determination of the molecular formula of methane

Some methane is collected over mercury in a eudiometer as shown in *Figure 1.1*. The volume and mercury level are read. An amount of oxygen, in excess of that required for complete combustion*, is then introduced. While holding the eudiometer on a rubber pad, a spark is passed, when combustion occurs. When the pad is removed mercury enters the eudiometer which now contains the carbon dioxide formed during combustion and the excess amount of oxygen. Water is also formed and condenses to a liquid of negligible volume.

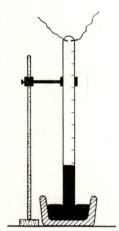

Figure 1.1. Determination of the molecular formula of methane.

By means of a bent pipette, a little caustic soda solution is introduced into the eudiometer to find the volume of carbon dioxide present and thus the volume of unused oxygen. The latter volume is needed in order to be able to calculate the volume of oxygen actually used in the combustion.

At each stage of the experiment, the volumes and mercury levels are read. Provided that after the explosion some interval be allowed for cooling, the temperature of all volume readings will be the same, and the laboratory temperature need not be taken.

After all volumes are converted to the same pressure, it will be found that:—

1 vol. of methane + 2 vol. of oxygen ——⟶ 1 vol. of carbon dioxide

Let the molecular formula for methane be C_xH_y.

* Preliminary experiments may have to be done in order to find suitable volumes of methane and oxygen to use in the experiment.

Since by Avogadro's Law, equal volumes of different gases at the same temperature and pressure contain the same number of molecules, it follows that:

1 molecule of methane combines with 2 molecules of oxygen to form 1 of carbon dioxide, thus:

$$C_xH_y + 2O_2 \longrightarrow CO_2$$

Hence, $x = 1$

Since the single carbon atom of the methane molecule needs 2 atoms of oxygen for combustion, there are 2 (of the total 4) atoms of oxygen left for the hydrogen in the methane molecule. Since in water, the combining ratio of hydrogen and oxygen atoms is 2:1, there are therefore in the methane molecule 4 atoms of hydrogen. Hence the molecular formula of methane is CH_4.

This eudiometric method is available for the determination of the molecular formula of any other hydrocarbon, no matter to which homologous series it belongs, provided that, at room temperature, it exists as a gas.

Bonding in methane

The C—H bonds in methane are covalent bonds (Introduction p. xvii) in which the electron pairs are almost equally shared between carbon and hydrogen atoms. This is not true of all covalent bonds; in general, electron sharing between two atoms A and B is such that one atom (A) has a somewhat greater share than the other (B). We represent this inequality as

$$\overset{\delta-}{A}\text{——}\overset{\delta+}{B}$$ (where $\delta+$ and $\delta-$ are small fractional charges)

or as $A\text{——}{\leftarrow}B$

Note the direction of the arrow-head which indicates the tendency of the electrons to be rather towards one end of the bond than towards the other.*

The absence of any electrical inequality in simple hydrocarbons like methane means that these compounds are *stable* towards ionic reagents, for example caustic alkalis and strong mineral acids. In contrast, organic halides like methyl and ethyl chlorides are attacked by reagents which contain the hydroxide ion OH^-:

$$OH^- + C_2H_5Cl \xrightarrow[\text{(solvent)}]{} C_2H_5OH + Cl^-$$

This type of reaction is discussed in detail later (Chapter 5), but here it is instructive to note that the reactivity of ethyl chloride towards alkalis and the complete lack of reactivity of saturated

* A more detailed and sophisticated discussion of covalent bonds is given in Chapter 26.

hydrocarbons are consequences of the types of covalent bond present in these molecules. In ethyl chloride the electronegative chlorine atom acquires more than an equal share of the two electrons of the C—Cl bond and the molecule can be represented*:

$$
\underset{\substack{|\\H}}{\overset{\substack{H\\|}}{H-C}}-\underset{\substack{|\\H}}{\overset{\substack{H\\|}}{C}}\!\!\overset{\delta+}{\rule{0pt}{0pt}}\!\!-Cl^{\delta-} \quad \text{or} \quad \underset{\substack{|\\H}}{\overset{\substack{H\\|}}{H-C}}-\underset{\substack{|\\H}}{\overset{\substack{H\\|}}{C}}\!\rightarrow\! Cl
$$

The fractional positive charge on the carbon atom to which chlorine is attached makes possible attack by (i.e. reaction with) the negatively charged hydroxide ion (p. 12):

$$
HO^- + \underset{CH_3}{\overset{H\diagdown\ \diagup H}{C}} \rightarrow Cl \longrightarrow HO-\underset{CH_3}{\overset{H\diagdown\ \diagup H}{C}} + Cl^-
$$

ethyl chloride   ethyl alcohol

In the case of methane and similar alkanes, there are no positively charged atoms at which attack can occur, and these compounds are consequently inert towards alkalis. Since many of the common laboratory reagents are ionic in character, it is not surprising that saturated hydrocarbons are often referred to as unreactive substances, but as shown later, this statement requires some qualification.

GENERAL CHEMISTRY OF THE ALKANES

(1) Nomenclature

The name of an alkane, and of other hydrocarbons also, is composed of two parts:

(a) a part to indicate the number of carbon atoms present in the molecule:

i.e.	meth-	1 carbon atom
	eth-	2 carbon atoms
	prop-	3 carbon atoms
	but-	4 carbon atoms
	pent-	5 carbon atoms
	hex-	6 carbon atoms

* Diagrams like these are used for convenience and simplicity, but it must always be remembered that the structures are really three-dimensional ones in which the carbon is approximately tetrahedral.

(*b*) a part to indicate the series to which the hydrocarbon belongs. For the alkanes this is *-ane*, thus:

methane	CH_4
ethane	C_2H_6
propane	C_3H_8
butanes	C_4H_{10}
pentanes	C_5H_{12}
hexanes	C_6H_{14}

Isomerism is possible with all alkanes containing four or more carbon atoms in the molecule. To differentiate between the possible isomers, the ' straight chain ' isomer* is usually referred to as the *normal* compound, abbreviated to the prefix ' n ' in naming the compound, and the branched chain one as the ' iso ' compound, thus:

$$CH_3 \cdot CH_2 \cdot CH_2 \cdot CH_3 \qquad (CH_3)_2CH \cdot CH_3$$
n-butane   isobutane

(Note the simplified structural formulae; the dots are used between composite groups like CH_2)

This system of nomenclature breaks down with the pentanes, for here in addition to the normal isomer, there are two branched chain isomers, thus:

$$CH_3 \cdot CH_2 \cdot CH_2 \cdot CH_2 \cdot CH_3 \qquad (CH_3)_2CH \cdot CH_2 \cdot CH_3 \qquad (CH_3)_4C$$
n-pentane

To distinguish between the two branched chain isomers, a more precise naming is necessary. The fully systematic nomenclature is beyond the scope of this book, but $(CH_3)_2CH \cdot CH_2 \cdot CH_3$ is called 2-methylbutane and $(CH_3)_4C$, 2,2-dimethylpropane.

(2) *Isomerism*

With the general formula for the series as C_nH_{2n+2}, one member differs from its neighbour in series by the group CH_2. Thus in

---

* The nomenclature of the alkanes was probably due to Hofmann. The first four were named because of their respective structural similarity to compounds already known. Thus *meth*ane, CH_4, is related to the inflammable compound of formula CH_3OH originally obtained by the distillation of wood and called ' wood spirit ', but later *methyl* alcohol (from Greek *methu* = wine (spirit), and *hule* = wood). *Eth*ane, C_2H_6, was similarly related to $(C_2H_5)_2O$, *ether* (meaning *to burn* or *glow*). *Prop*ane, C_3H_8, contains the grouping of the acid of formula $C_3H_7 \cdot COOH$, thought (wrongly) to be the first of the fatty (carboxylic) acids (*pro* = first, *pion* = fat). Similarly, butane, C_4H_{10}, is related to $C_4H_9 \cdot COOH$, the acid isolated after hydrolysis of butter.

The suffix *-ane* has probably no significance, being merely a common suffix in our language. For the series of hydrocarbons called the *alkenes* or *olefins* (p. 15), the suffix *-ene* is used; hence an alternative better name for ethylene, C_2H_4, is *ethene*.

The series of *alkynes* (p. 31) might have been given the suffix *-ine*. In fact, *-yne* has been used.

passing from methane to ethane and in turn to propane, one atom of hydrogen in the molecule is replaced by the CH_3 group:

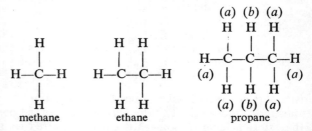

methane	ethane	propane

Now in the molecule of methane, each hydrogen atom is similarly situated. In building from it the formula for ethane, it is immaterial which particular hydrogen atom is replaced by the CH_3 group, and the same obtains in passing from ethane to propane.

In the molecule of propane, however, there are two different groupings of hydrogen atoms. Each of the hydrogen atoms labelled (a) is similar, for each is attached to a carbon atom which in turn carries two other hydrogen atoms and the C_2H_5 group. Each of the (b) atoms is attached to a carbon atom which carries one other hydrogen atom and two CH_3 groups. Thus two different structural formulae for the hydrocarbon butane can be obtained according to whether a hydrogen atom (a) or one of (b) is replaced.

$$
\begin{array}{cccc}
H & H & H & H \\
| & | & | & | \\
H{-}C{-}C{-}C{-}C{-}H \\
| & | & | & | \\
H & H & H & H
\end{array}
\qquad
\begin{array}{ccc}
H & H & H \\
| & | & | \\
H{-}C{-}C{-}C{-}H \\
| & | & | \\
H & C & H \\
& H &
\end{array}
$$

(replacing H(a))   (replacing H(b))
n-butane   isobutane

These two structural formulae represent the two isomers of butane.

For convenience, these formulae are more usually written as:

$CH_3{\cdot}CH_2{\cdot}CH_2{\cdot}CH_3$ $CH_3{\cdot}CH{\cdot}CH_3$
    n-butane | or $(CH_3)_2CH{\cdot}CH_3$
                                                              CH_3
                                                           isobutane

It must again be emphasized that because the graphical formulae given above are drawn in one plane, they are not strictly representative of the molecules. They should all be pictured with each carbon atom surrounded in space by four other atoms positioned at the corners of a tetrahedron.

(3) *Physical properties*

As is general in a homologous series, the boiling point rises as *n* increases. If the boiling points of the normal alkanes are plotted against the number of carbon atoms in their molecules, a regular smooth curve is obtained. The rise in boiling point is due to the increased attraction between the molecules as the number of atoms increases (p. 293).

Branched chain isomers boil at temperatures lower than those of the corresponding normal isomer. Here the molecules are more compact and in consequence the attractive forces between them are smaller.

If the melting points of the normal alkanes are plotted against the number of carbon atoms in their molecules, the graph obtained has a zig-zag path. Two smooth curves are, however, obtained if the points for (*a*) those alkanes containing an odd number of carbon atoms, and (*b*) those containing an even number are joined up.

(4) *Chemical properties*

The lack of reactivity of alkanes towards some common reagents has already been mentioned. This, in fact, is the reason for the term *paraffin* (derived from Latin *parum* = little, and *affinis*). They are not affected to any appreciable extent by acids (p. 5), alkalis or oxidizing agents except (in the latter case), under very drastic conditions.

(*a*) Lower members in particular are stable, although at very high temperatures even the lower members undergo decomposition. At red heat, propane decomposes into methane and ethylene:

$$C_3H_8 \xrightarrow[\text{heat}]{\text{red}} CH_4 + C_2H_4$$

This thermal decomposition is called *pyrolysis* or *cracking* and, as we shall see later (Chapter 4), is extremely important in the petroleum industry.

(*b*) In a plentiful supply of air or oxygen, alkanes burn to form carbon dioxide and water, e.g., $CH_4 + 2O_2 \longrightarrow CO_2 + 2H_2O$.
In an insufficient supply, water and carbon monoxide or even carbon are obtained. The gaseous and volatile members form explosive mixtures with air or oxygen, a most useful property which provides the energy of the internal combustion engine.

As the number of carbon atoms in the molecule increases, the flame becomes more luminous.

(c) Alkanes react with the halogens (except iodine), forming *substitution products*. An apparently simple example is the reaction of methane with chlorine, brought about by the presence of light. In this reaction or, more correctly, series of reactions, hydrogen atoms of the methane molecule are successively replaced by chlorine atoms (hence the word ' substitution '):

$$CH_4 + Cl_2 \longrightarrow CH_3Cl + HCl$$
methyl chloride

$$CH_3Cl + Cl_2 \longrightarrow CH_2Cl_2 + HCl$$
methylene chloride

$$CH_2Cl_2 + Cl_2 \longrightarrow CHCl_3 + HCl$$
chloroform

$$CHCl_3 + Cl_2 \longrightarrow CCl_4 + HCl$$
carbon tetrachloride

Essentially similar reactions occur with ethane and higher members of the homologous series.

Such reactions are now quite well understood and are believed to differ in a fundamental way from reactions like the previously mentioned one between ethyl chloride and hydroxide ions, which involve ionic reagents or intermediates. Although alkanes, with their electrically neutral atoms, are inert to attack by hydroxide ions, they can undergo what are known as *free radical reactions*. In these, attack is due to the presence of *free radicals*, which are atoms or groups of atoms that have one unpaired electron. In the chlorination of methane, the reaction is initiated by the formation of chlorine atoms; in daylight (or at elevated temperatures) these are formed by the dissociation of some chlorine molecules:

$$Cl—Cl \rightleftharpoons Cl \cdot + \cdot Cl$$

The covalent bond linking the atoms splits so that each atom takes one electron (shown as a dot above—do not confuse this with the dot used in simplified structural formulae, p. 7). The resulting atoms are highly reactive and can recombine to form chlorine molecules; but in the presence of methane molecules an alternative reaction is the stripping of a hydrogen atom from its linkage to carbon:

$$\begin{array}{ccc} H & & H \\ | & & | \\ H—C—H + \cdot Cl \longrightarrow & H—C \cdot & + \quad H—Cl \\ | & & | \\ H & & H \end{array}$$

The immediate products are a stable molecule of hydrogen chloride and another highly reactive free radical, the *methyl radical* CH_3·; the dot again representing the unpaired electron. The methyl radical is sufficiently reactive to attack undissociated chlorine molecules:

$$\begin{array}{ccc} & H & & & & H \\ & | & & & & | \\ H-C\cdot & + & Cl-Cl & \longrightarrow & H-C-Cl & + & \cdot Cl \\ & | & & & & | \\ & H & & & & H \end{array}$$

The products are methyl chloride and another chlorine atom, which can attack another molecule of methane, giving methyl chloride and a new methyl radical, or it can attack a molecule of methyl chloride similarly, giving a different kind of free radical:

$$\begin{array}{ccc} & H & & & & H \\ & | & & & & | \\ Cl-C-H & + & \cdot Cl & \longrightarrow & Cl-C\cdot & + & HCl \\ & | & & & & | \\ & H & & & & H \end{array}$$

This new radical, in turn, is capable of attacking a chlorine molecule:

$$\begin{array}{ccc} & H & & & & H \\ & | & & & & | \\ Cl-C\cdot & + & Cl-Cl & \longrightarrow & Cl-C-Cl & + & \cdot Cl \\ & | & & & & | \\ & H & & & & H \end{array}$$

It should now be clear that once a chlorine atom has been generated in this system, a *chain of reactions* becomes possible, leading to the observed substitution products of methane, viz. methyl chloride, methylene chloride, chloroform and carbon tetrachloride. For this reason, such reactions are called *radical chain reactions*. The chain of reactions can obviously be broken by the recombination of two radicals or by exhaustion of the materials.

Other reactions which the alkanes undergo, for example cracking (p. 9) and combustion (p. 9) also involve free radicals.

One further reaction of saturated hydrocarbons should be mentioned. Under ordinary conditions, sulphuric and nitric acids have little action on alkanes, but with nitric acid in the vapour state at 400°C and under 10 atm pressure a substitution reaction occurs, leading to aliphatic nitro-compounds like *nitromethane*, CH_3NO_2. Here again the reaction probably involves free radicals since

' nitration ' of ethane gives not only nitroethane, $C_2H_5NO_2$, but also nitromethane. Nitroalkanes are of industrial importance as solvents and are also important in organic synthesis (p. 192).

Comparison of free radical and ionic reactions

The chlorination of methane discussed above is a fairly typical ' free radical reaction '. On the other hand, the reaction of ethyl chloride with alkali, giving ethyl alcohol, is a typical organic ionic reaction.*

It is important to understand the difference between these two types of reaction.

Consider a hypothetical molecule A—B with a normal covalent bond between A and B. This molecule could split in several ways, as illustrated below:

$$(i) \quad A\text{---}B \longrightarrow A \cdot + \cdot B$$

$$(ii) \quad A\text{---}B \longrightarrow A \colon + B$$

$$(iii) \quad A\text{---}B \longrightarrow A + \colon B$$

(The dots represent the electrons of the covalent bond between A and B)

In process (i), the bond splits evenly, each residue A and B taking one electron. In other words A· and B· are free radicals and any reaction in which this happened would be classified as a *free radical reaction*. This type of fission of a bond is also known as *homolysis* (i.e. equal splitting). Processes (ii) and (iii) are essentially similar to one another; the *bonding pair of electrons* goes either to A or to B. If A—B is electrically neutral, then in process (ii) the product A: will be negatively charged, whilst B will be positively charged. In process (iii) the positions are reversed, but in both cases, ions are formed and the bonding pair of electrons remains intact. Processes (ii) and (iii) are therefore *ionic reactions* and the fission is often described as *heterolysis* (i.e. unequal splitting).

The classification of the methane chlorination reaction as a free radical reaction should be quite clear. The ethyl chloride-hydroxide ion reaction, which we have classified as ionic, requires some further discussion. This reaction has already been represented as:

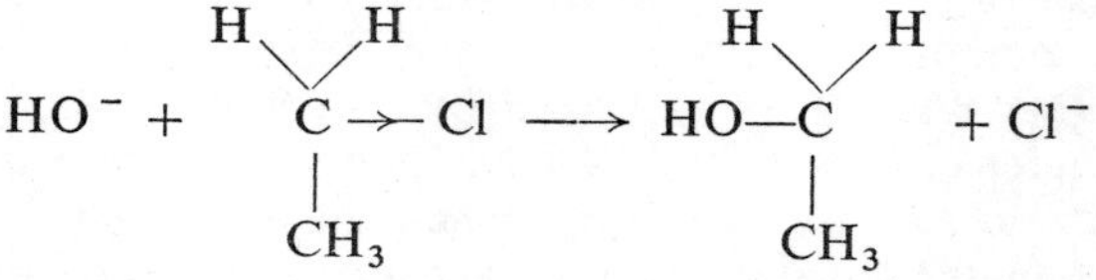

$$HO^- + \quad \begin{matrix} H & H \\ \diagdown & \diagup \\ C \rightarrow Cl \end{matrix} \longrightarrow HO\text{---}C \quad \begin{matrix} H & H \\ \diagdown & \diagup \\ \quad \end{matrix} + Cl^-$$
$$\qquad\qquad \underset{CH_3}{\mid} \qquad\qquad\qquad \underset{CH_3}{\mid}$$

---

* The term ' ionic reaction ' as used in organic chemistry, does not refer to the metathetic reactions of the type familiar in inorganic chemistry (see Introduction, p. xxiii).

In the hydroxide anion, the oxygen atom has the normal octet of electrons. During the reaction, *two* of these electrons are shared with the carbon atom, forming the new C—O bond. At the same time, the C—Cl bond undergoes ionic fission (heterolysis), the bonding pair of electrons going wholly to the chlorine atom, which thus becomes a chlorine anion. Such a reaction is commonly represented as follows:

$$HO^- \quad CH_2\!-\!Cl \longrightarrow HO-CH_2 + Cl^-$$
$$\qquad\quad |\qquad\qquad\qquad\quad |$$
$$\qquad\quad CH_3\qquad\qquad\qquad CH_3$$

where the *curved* arrows imply the movement of *pairs* of electrons. Shifts of electron pairs in this way are characteristic of ionic reactions.

We are now able to understand a useful generalization about the chemistry of alkanes.

Alkanes are generally inert to ionic reagents under normal conditions. They can, however, undergo free radical reactions which usually require some special initiation process (e.g. the generation of radicals through the action of light or heat).

(5) *Uses*

The alkanes are perhaps the most extensively used of all organic compounds. They are used directly as fuels to provide heat or power. They also constitute the important raw materials from which a whole range of other organic compounds can be manufactured. In America, methane—an important constituent of the natural gas of the oil well—is pumped to the cities for combustion. Until methane from natural gas deposits under the North Sea and north east England becomes available, liquefied methane (which occupies much less space than in the gaseous state) is being brought here from oil wells abroad in specially-constructed ships, and is added to the gas supply. Attempts are being made to remove the methane present in certain coal mines (and thereby reduce the risk of explosions). For removal, the methane is passed along with steam through retorts at a temperature of about 450°C and containing nickel as a catalyst, when it is converted into carbon monoxide and hydrogen, which are fed direct into the gas mains:

$$CH_4 + H_2O \xrightarrow[\text{Ni}]{450°C} CO + 3H_2$$

In some sewage works, the sewage is delivered into large tanks which are maintained at a temperature of between 80 and 90°F. The sewage is broken down by the bacteria present and a gas consisting of 70 per cent methane and 30 per cent carbon dioxide, and a little hydrogen sulphide is evolved and collected. The hydrogen sulphide is removed and also about 20 per cent of the carbon dioxide. The remaining gas is compressed and used as a motor fuel for the corporation vehicles. Successful experiments in the use of methane as a fuel were made in Germany some years ago.

Calor gas is a mixture of n-butane and isobutane, with a small proportion of pentane.

Higher alkanes are used as solvents in the manufacture of lacquers and varnishes, as fuels for internal combustion engines, for heating kilns and furnaces, as lubricating oils, for the manufacture of fats, soaps, and as raw materials for the plastic industries and for many other purposes. Reference to some of these uses will be made in later chapters.

SUMMARY

(1) *Hydrocarbons* contain only carbon and hydrogen.

(2) Paraffins or *alkanes* are hydrocarbons with the general formula C_nH_{2n+2}. They are *saturated*, i.e. they contain no double or triple bonds between the carbon atoms.

(3) Alkanes are *unreactive* towards ionic reagents (e.g. caustic alkalis and mineral acids) under normal conditions.

(4) Under special conditions, alkanes undergo *free radical reactions*. The chlorination of methane to give CH_3Cl, CH_2Cl_2, $CHCl_3$ and CCl_4 is an example; it is also an example of a *substitution* reaction.

(5) Most organic reactions can be classified as either
Free radical reactions, which involve atoms or groups with an unpaired electron,
or
Ionic reactions, which involve movement or redistribution of pairs of electrons.

(6) *Fission of bonds:*

Homolysis A:B $\longrightarrow$ A• + •B (a free radical process)

Heterolysis A:B $\Bigl\langle$ A: + B (ionic process—if A—B is
A + :B neutral, ions are formed)

2

OLEFINS OR ALKENES: POLYMERS

Functional Group: $\diagup C = C \diagdown$

THE general formula of this homologous series of unsaturated hydrocarbons is C_nH_{2n}. The 'hydrocarbon' of formula CH_2 (i.e. methylene where $n = 1$) has been obtained as a highly reactive intermediate in certain reactions, but the first stable member is *ethylene* or *ethene* of formula C_2H_4. The old name for this hydrocarbon was *olefiant* gas and from this the general name of *olefins* was derived.

The names of the members of the series end in *-ylene* or *-ene* (according to the nomenclature used), in place of the ending *-ane* in the alkane series. Thus:

Ethylene or ethene	C_2H_4	
Propylene or propene	C_3H_6	gases
Butylenes or butenes	C_4H_8	

Only one structure is possible for propylene, but there are several butylene isomers.

Ethylene or ethene, C_2H_4

The gas was first discovered by Becher. Small amounts of it are present in coal gas and in the natural gas of oil-bearing districts. Much more ethylene is, however, obtained in the refining of petroleum during the cracking operations and this is being used increasingly in various industrial processes.

Dalton discovered that in ethylene the weight of carbon combined with unit weight of hydrogen is exactly twice the amount in methane. This discovery led him to the Law of Multiple Proportions.

The formula of ethylene

The molecular formula is obtained by the eudiometric method given for methane on p. 4).

The accepted structural formula of ethylene is that of a symmetrical planar molecule, i.e. all six atoms are in one plane:

$$\begin{matrix} H & & H \\ \diagdown & & \diagup \\ & C = C & \\ \diagup & & \diagdown \\ H & & H \end{matrix}$$

15

The presence of a double bond between the carbon atoms is the reason for the reactivity of ethylene. Although this double bond is actually shorter and mechanically stronger than a single bond, it is chemically more reactive.

Properties of ethylene

Ethylene is a colourless gas with a sweetish smell. It is an anaesthetic but it is now rarely, if ever, used. It is almost insoluble in water but is appreciably soluble in alcohol and ether.

In contrast with the chemical inertness of the alkanes under normal conditions, ethylene and other alkenes are very reactive. The activity is due to the unsaturated character of these hydrocarbons, i.e. the presence of the double bond in the molecule of each of them.

(1) Ethylene burns in air or oxygen and forms an explosive mixture with them. This provides a means of testing to see if all air is displaced from an ethylene generator before jars of the gas are collected.

$$C_2H_4 + 3O_2 \longrightarrow 2CO_2 + 2H_2O$$

Unlike the flame of burning methane, that of ethylene is very luminous and sometimes smoky. This is due to the high carbon content.

(2) Ethylene burns in chlorine, as do many hydrocarbons, depositing carbon:

$$C_2H_4 + 2Cl_2 \xrightarrow{\text{combustion}} 4HCl + 2C$$

(3) Ethylene shows *additive* properties. An additive or *addition* reaction is one where two substances combine together to form a single compound only. The additive properties of ethylene are due to the fact that the carbon atoms in its molecule are not exerting their maximum combining power, i.e. are due to the presence of the *double bond*. Similar double bonds occur in many other compounds, not merely in hydrocarbons, and in general such compounds possess additive power. The following addition reactions are thus characteristic of the double bond $\diagup\!\!\!\overset{\textstyle \diagdown}{C}\!\!=\!\!\overset{\textstyle \diagup}{C}\!\!\diagdown$ and are properties of ethylene by virtue of its containing such a bond.

(a) *Addition of hydrogen:*—Ethane is formed. Catalysts such as nickel and platinum black make the process possible:

$$\underset{H}{\overset{H}{\diagdown}}C\!\!=\!\!C\underset{H}{\overset{H}{\diagup}} + H_2 \xrightarrow{\text{catalyst}} C_2H_6$$

[This saturation of a double bond by hydrogen is used extensively in a variety of industries. Many of the fatty oils occurring in nature, e.g. olive oil, coconut and groundnut oils, whale oil, etc., consist of compounds containing a double bond. They are heated to about 180°C with hydrogen under pressure, when they are converted to solid fats which are used in the manufacture of soaps, candles, margarine and detergents. Nickel, so finely divided as to be pyrophoric and known as ' Raney nickel ', is very active as a catalyst, for it contains adsorbed hydrogen. It must be stored under alcohol. The discovery of this *hydrogenation* process in 1910 was the basis of the modern margarine industry (p. 152).]

Compounds containing double bonds can generally be reduced in the laboratory under quite mild conditions, using noble metal catalysts, e.g. palladium deposited on charcoal.

Addition of hydrogen to an olefin can also be achieved by a recently discovered type of reaction; in this, a compound containing hydrogen attached to boron is used, and the ethylene or other olefin adds on, thus:

$$>B{-}H + C_2H_4 \xrightarrow[\text{ether}]{\text{in}} >B{\cdot}CH_2{\cdot}CH_3$$

If this is now heated with a carboxylic acid (Chapter 10), the hydrogen of the acid forms an *alkane* (ethane):

$$>B{\cdot}CH_2{\cdot}CH_3 + (H) \longrightarrow >B{-} + CH_3{\cdot}CH_3$$

Alternatively, if the boron compound is treated with hydrogen peroxide, an *alchol* is formed:

$$>B{\cdot}CH_2{\cdot}CH_3 \xrightarrow{H_2O_2} >B{-} + HOCH_2{\cdot}CH_3$$
$$\text{ethyl alcohol}$$

The boron–hydrogen compound used is diborane, B_2H_6, generated from sodium borohydride, $NaBH_4$; and one of the ethers (Chapter 7) can be used as solvent. This method of obtaining an alkane or alcohol from an olefin is of great synthetic importance, because it is of very wide application. The reaction is called *hydroboration*.

(b) *Addition of the halogen elements:*—Dihalogen derivatives of ethane are formed. These compounds are colourless, so in the reaction the halogen is decolorized. Fluorine reacts violently; iodine adds on if the alkene is bubbled through an alcoholic solution of iodine. Chlorine and bromine (vapour or liquid) react readily as below:

$$\begin{array}{c}H\\{}\end{array}\!\!\diagdown\,\,C=C\,\diagup\!\!\begin{array}{c}H\\{}\end{array}\quad + \text{ } Cl_2 \longrightarrow \text{ } ClCH_2\!\cdot\!CH_2Cl$$
$$\text{1,2-dichloroethane}$$

The product, 1, 2-dichloroethane* (the old name, ethylene dichloride, is undesirable) is a symmetrical addition product, because each carbon atom carries a chlorine atom. It is isomeric with the unsymmetrical compound 1,1-dichloroethane (old name: ethylidene dichloride) which can be prepared by other methods.

With bromine, ethylene forms oily 1,2-dibromoethane (hence the name *olefiant* = oil forming) (see, further, p. 21).

(c) *Addition of the hydrogen halides:*—Ethyl halides are formed.

$$HI + \quad \begin{array}{c}H\\{}\end{array}\!\!\diagdown\,\,C=C\,\diagup\!\!\begin{array}{c}H\\{}\end{array} \longrightarrow CH_3\!\cdot\!CH_2I$$
$$\text{ethyl iodide}$$

(d) *Addition of hypochlorous acid:*—Ethylene chlorohydrin is formed. This can be done by passing a mixture of ethylene and carbon dioxide through a suspension of bleaching powder in water:

$$H_2C=CH_2 \longrightarrow CH_2Cl\!\cdot\!CH_2OH$$
$$\text{ethylene chlorohydrin}$$

By shaking up ethylene with bromine *water*, mixed products are formed, including ethylene bromohydrin and 1,2-dibromoethane.

(e) *Addition of sulphuric acid:*—Ethyl hydrogen sulphate is formed. With the concentrated acid, the reaction is slow, but it proceeds quite rapidly with the fuming acid:

$$H_2C=CH_2 + HO\!\cdot\!SO_2\!\cdot\!OH \longrightarrow CH_3\!\cdot\!CH_2\!\cdot\!O\!\cdot\!SO_2\!\cdot\!OH$$
$$\text{ethyl hydrogen sulphate}$$

* Note that numbers are used to indicate the positions of substituent atoms or groups. The numbers are assigned to the carbon atoms in the chain so as to give the lowest numbers to substituents, i.e. $CH_3\!\cdot\!CH\!\cdot\!CH_2Cl$ is 1,2-dichloropro-
$$\qquad\qquad\qquad\qquad\qquad\qquad\quad |$$
$$\qquad\qquad\qquad\qquad\qquad\qquad\quad Cl$$
pane. In Chapter 1 $(CH_3)_2CH\!\cdot\!CH_2\!\cdot\!CH_3$ (one of the isomeric pentanes) was named systematically as 2-methylbutane. The general rule in such cases is to select the longest chain in the molecule and to number this in such a way as to give any attached groups the lowest numbers, i.e.

(1) (2) (3)   (4)   (4) (3) (2) (1)

CH_3—CH—CH_2—CH_3 CH_3—CH—CH_2—CH_3

       CH_3 *not* CH_3

2-methylbutane   3-methylbutane

This reaction is of industrial importance for it is used to separate ethylene from other gases produced when petroleum is cracked (p. 43). Note that the reaction involves the addition of H and $O \cdot SO_2 \cdot OH$ fragments. The way in which this is thought to occur is discussed later (p. 23).

(*f*) *Addition of certain metal chlorides: e.g., copper(I) (cuprous) chloride:*—

$$C_2H_4 + CuCl \longrightarrow CuCl \cdot C_2H_4$$

This reaction occurs with many other transition metal salts, and other olefins react similarly. It is believed that the double bond of the olefin is attached 'sideways on' to the metal ion (M), thus:

These reactions are of industrial importance. When oxygen and ethylene are passed over palladium(II) chloride ($PdCl_2$) and copper(I) chloride, the ethylene adds on and is then oxidized to acetaldehyde, $CH_3 \cdot CHO$. Olefin–metal chloride reactions are also of great importance in polymerization processes (p. 25).

(*g*) *Oxidation by dilute alkaline potassium permanganate:*—The permanganate turns first green (due to reduction to the manganate) and then brown owing to precipitation of hydrated manganese dioxide. The reaction is a complicated one, but may be represented as:

$$H_2C{=}CH_2 + H_2O + [O] \longrightarrow CH_2OH \cdot CH_2OH$$
$$\text{from } KMnO_4 \qquad \text{ethylene glycol}$$

Baeyer gave this reaction as a test for a double bond, but the test is not specific for it is given by a triple bond compound also (e.g. acetylene) and also by any compound containing an easily oxidizable group. It is thus only conclusive for a double bond compound if these others are known to be absent.

Under more vigorous conditions, permanganate causes fission of the molecule at the double bond. In general, with homologues of ethylene, the products are aldehydes, ketones and carboxylic acids (see Chapters 8, 9, and 10):

A/547

R and R' represent alkyl groups, such as CH_3—, C_2H_5— etc.

Thus, by studying the structures of the oxidation products, it is possible to locate the position of the double bond in the original alkene.

(*h*) *Addition of ozone:*—Ethylene and its homologues react with ozonized oxygen to form unstable addition products known as *ozonides*, which are hydrolysed by water, forming aldehydes and ketones. Since hydrogen peroxide is also formed in the hydrolysis, and since aldehydes are sensitive to this reagent, steps must be taken to remove hydrogen peroxide if the aldehydes are the required products. One procedure is to add zinc dust to the hydrolysis medium.

Rubber contains double bonds and is attacked by ozone; for this reason, rubber connections should not be used in the preparation of the ozonide.

Note that ozonide formation differs from the addition reactions discussed previously in that the double bond is broken. *Ozonolysis* (i.e. fission by ozone) can be used in the same way as vigorous oxidation with potassium permanganate to study the position of a double bond in a carbon chain.

(4) **Polymerization.** In the presence of various catalysts, different polymers of ethylene can be obtained, the best known one being polythene (see p. 26).

Mechanisms of addition reactions

The addition reactions described above for ethylene are, in fact, general reactions for alkenes and for other compounds (of various classes) which contain a carbon-carbon double bond.

The reactions with halogens, hydrogen halides, hypochlorous acid, bromine water and sulphuric acid are all regarded as ionic reactions (Chapter 1, p. 12) since they are believed to involve shifts of pairs of electrons. Taking the addition of bromine as an example, it is known that the bromine molecule *does not* add as a whole, in a kind of broadside attack on the double bond. Evidence has accumulated that the reaction occurs in stages, the first step being:

The curved arrows in this diagram have the same significance as before: they represent movements of *electron pairs*.

In the alkene double bond, one pair of electrons is sufficient to bind the carbon atoms together. The other pair, although also involved in bonding, is available for combination with suitable reagents. These two electrons, in the present example, form a new bond between carbon and one of the bromine atoms, allowing the bromine molecule to undergo ionic fission (heterolysis, p. 12). The immediate products, as shown, are a bromide anion and an organic positive ion (called a *carbonium ion*, by analogy with ammonium etc.). The reaction is completed by combination of the carbonium ion with a bromide anion (not necessarily the anion which is formed in the first step):

In the second stage, the bromide anion (with an octet of valency electrons) provides *two* electrons for the second carbon-bromine bond. When these electrons are shared, the positive and negative charges are cancelled. The product, 1,2-dibromoethane, is represented here with two bromine atoms on opposite sides of the carbon-carbon bond to emphasize the two separate stages of the reaction. In fact, with certain more complicated alkenes, the structure of the addition product shows that the bromine atoms become attached to opposite sides of the original double bond. We infer that the same is true in ionic additions of bromine to simple alkenes like ethylene. In this particular case, addition at opposite sides, or on the same side, of the double bond would give the same product, 1,2-dibromoethane, because *the two —CH$_2$Br groups can rotate with respect to one another:*

(A)

(B)

Structures (A) and (B) are equivalent, i.e. they are both representations of the same molecule, 1,2-dibromoethane (the point is easily understood with the aid of models which allow the groups to rotate).

When ethylene is treated with bromine water, the reactions discussed above can still occur, but in the presence of large numbers of water molecules a different second step is possible and is favoured:

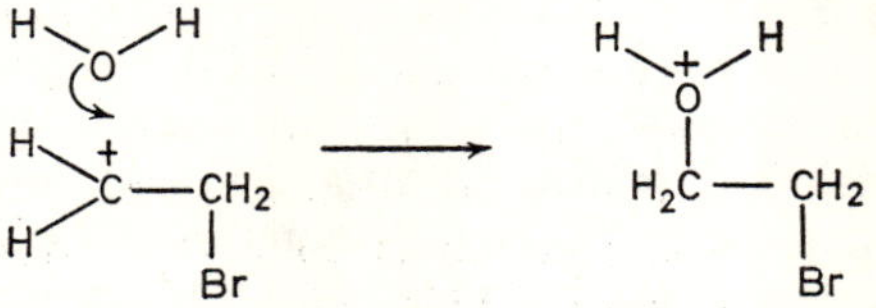

Again the curved arrow indicates that one of the electron pairs of the oxygen atom in the water molecule is shared with the positively charged carbon atom. A new carbon-oxygen bond is thus formed and the positive charge on the carbon is neutralized, but the product must contain a positive oxygen atom (the structure of the product is analogous to that of the hydrated proton or hydroxonium ion). The final step is a *proton transfer* (e.g. to another water molecule):

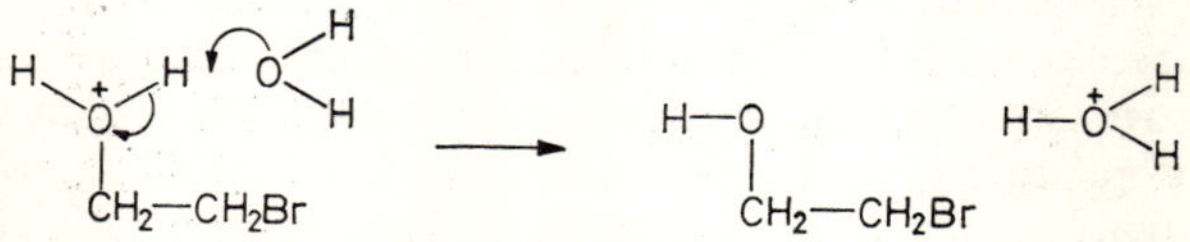

Inspection of the diagram will reveal that this also implies the re-distribution of certain electron pairs.

Proton transfer is also involved in the addition of hydrogen halides and of sulphuric acid to alkenes:

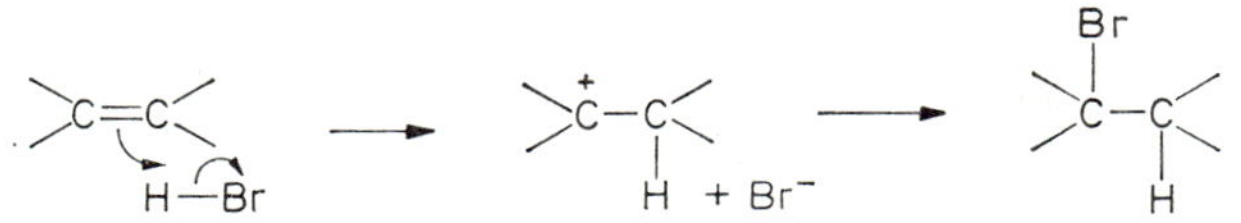

the addition being completed, as in the addition of bromine, by the attack of an anion (bromine anion in the diagram) on the positively charged carbon atom or carbonium ion.

With sulphuric acid the process may be represented:

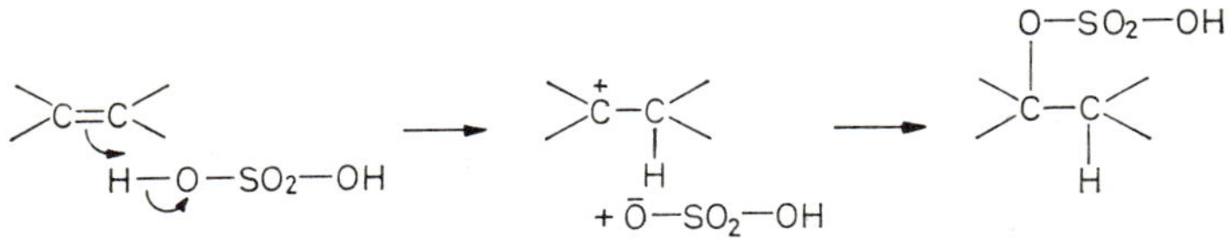

Addition of hydrogen to the alkene double bond, on the other hand, almost certainly involves *atoms* of hydrogen, which are supplied by the catalyst to the alkene *adsorbed* on its surface. Although the question cannot be pursued here, it is interesting to note that when hydrogen is added to a double bond, both hydrogen atoms are initially attached to the *same side* of the original double bond, in marked contrast to the addition of bromine. The process is represented diagrammatically thus:

Catalyst surface

Sources and uses of ethylene

Within relatively recent years, ethylene has become a very valuable raw material in chemical industry. Although both coal gas and natural gas contain small amounts of ethylene, the bulk of the ethylene of industry is produced by the cracking of petroleum (p. 43).

Ethylene is used for a wide variety of purposes, perhaps the most important of which is in the manufacture of polythene (see below). In small concentrations, it is used to ripen fruits, particularly citrus fruits, during shipment by acting as a stimulant to the enzymes which bring about the process. In larger amounts, it is used in the manufacture of ethylene glycol, commonly called *glycol* (strictly the name of the series of such compounds), a coolant for aircraft engines and an anti-freeze for motor cars, detergents and plastics (p. 24).

Hydrocarbons for use in motor spirit are manufactured from the gas by two methods: (*a*) by polymerization (p. 44), and (*b*) by reaction with an alkane, the process being called *alkylation* (p. 44).

A lubricant (a hydrocarbon of higher molecular weight than those in motor spirit) is also made by polymerizing ethylene under the influence of aluminium chloride as a catalyst.

Ethylene can be converted into ethyl alcohol. The gas is absorbed in concentrated sulphuric acid with which it forms a mixture of ethyl hydrogen sulphate, $C_2H_5O \cdot SO_2 \cdot OH$, and diethyl sulphate $(C_2H_5)_2SO_4$. Steam is then blown into the mixture when, by hydrolysis, ethyl alcohol is produced. At the high temperature, the alcohol distils over.

$$C_2H_5O \cdot SO_2 \cdot OH + H_2O \longrightarrow C_2H_5OH + H_2SO_4$$
ethyl alcohol

Processes of this kind, with ethylene and other alkenes, are used for the large-scale manufacture of alcohols.

HIGH POLYMERS AND PLASTICS

The ability of carbon atoms to form long chains has been mentioned already; the higher hydrocarbons contain such long chains and are waxy solids. It is possible to have such chains containing many hundreds of thousands of carbon atoms, and the substance is then considered to be a *high polymer*. *Cellulose* and *rubber* are examples of naturally-occurring high polymers. In recent times, methods of synthesizing high polymers from simple molecules (monomers) have become available, and many of these synthetic high polymers have important commercial uses, e.g. nylon, Bakelite, Perspex, Terylene; they are called *plastics* This term seems to have been used because of the plastic nature of many high polymers when heated, enabling them to be moulded easily. Some can be softened and hardened many times over by heat treatment and cooling; these are called *thermoplastics*. Others which soften on heating undergo some chemical change and become permanently hard; these are called *thermosetting* plastics, and are also referred to sometimes as synthetic *resins*; bakelite is an example of this class.

The properties of a plastic depend upon the nature of the carbon chains; these can be 'straight', giving rod-like molecules, i.e.:

| | | | | |
—C—C—C—C—C—
| | | | |

or branched, giving 'tree-like' molecules, i.e.:

$$
\begin{array}{c}
\mid \\
-\text{C}- \\
\mid \quad \mid \quad \mid \quad \mid \quad \mid \\
-\text{C}-\text{C}-\text{C}-\text{C}-\text{C}- \\
\mid \quad \mid \quad \mid \quad \mid \quad \mid \\
-\text{C}- \\
\mid \\
-\text{C}- \\
\mid
\end{array}
$$

There may also be cross-linking, joining the chains together in a three-dimensional network, i.e.:

$$
\begin{array}{c}
\mid \quad \mid \quad \mid \quad \mid \quad \mid \\
-\text{C}-\text{C}-\text{C}-\text{C}-\text{C}- \\
\mid \quad \mid \quad \mid \quad \mid \quad \mid \\
-\text{C}- \\
\mid \\
-\text{C}- \\
\mid \quad \mid \quad \mid \quad \mid \quad \mid \\
-\text{C}-\text{C}-\text{C}-\text{C}-\text{C}- \\
\mid \quad \mid \quad \mid \quad \mid \quad \mid
\end{array}
$$

When rubber is *vulcanized* by the addition of sulphur, the long carbon chains in the natural rubber are cross-linked by sulphur atoms, and the rubber is hardened. Cross-linked polymers are generally insoluble in organic solvents; straight or branched chain polymers dissolve or swell in some solvents. The thermosetting polymers are cross-linked.

To obtain a polymer, a *catalyst* or *initiator* is added to, or formed in the monomer from which the polymer is built up. There are two ways in which the monomers can join together to give long chains, first, by simple *addition*, and secondly by *condensation*. In the condensation process, the monomers react together with elimination of water or some other simple molecule. By the simple addition method, the polymer is the only product. Condensation reactions are considered later in the book; at the moment only addition polymerization will be considered, and for this there must be either a double bond, i.e. an alkene, or *conjugated double bonds* (p. 29) between the carbon atoms. The simplest example of an alkene is ethylene, which can be polymerized at ordinary temperatures and pressures by the addition of a catalyst [a mixture of titanium tetra-chloride and aluminium triethyl, $Al(C_2H_5)_3$]. Ethylene can also be polymerized under high pressure in the presence of a trace of oxygen as a catalyst. The product here is *polyethylene*, or *polythene*, a white solid showing great resistance to chemical attack.

$$CH_2{=}CH_2 \longrightarrow$$

$$
\begin{array}{ccccc}
\text{H} & \text{H} & \text{H} & \text{H} & \text{H} \\
| & | & | & | & | \\
\cdots{-}\text{C}{-}\text{C}{-}\text{C}{-}\text{C}{-}\text{C}{-}\cdots \\
| & | & | & | & | \\
\text{H} & \text{H} & \text{H} & \text{H} & \text{H}
\end{array}
$$

Polythene is a good insulator and can be bent without breaking.

However, ethylene is made much easier to polymerize if one of the hydrogen atoms is replaced by chlorine (to give *vinyl chloride*, p. 33) or by a phenyl group (p. 211) to give *styrene*, or by a cyanide group (CN) to give *acrylonitrile* (p. 33):

$$ClCH{=}CH_2 \longrightarrow$$
vinyl chloride

$$
\begin{array}{ccccc}
\text{Cl} & \text{H} & \text{Cl} & \text{H} & \text{Cl} \\
| & | & | & | & | \\
\cdots{-}\text{C}{-}\text{C}{-}\text{C}{-}\text{C}{-}\text{C}{-}\cdots \\
| & | & | & | & | \\
\text{H} & \text{H} & \text{H} & \text{H} & \text{H}
\end{array}
$$
polyvinyl chloride ('P.V.C')

$$C_6H_5{\cdot}CH{=}CH_2 \longrightarrow$$
styrene

$$
\begin{array}{cccc}
\text{C}_6\text{H}_5 & \text{H} & \text{C}_6\text{H}_5 & \text{H} \\
| & | & | & | \\
\cdots{-}\text{C}{-}\!\!-\!\!-\text{C}{-}\!\!-\!\!-\text{C}{-}\!\!-\!\!-\text{C}{-}\cdots \\
| & | & | & | \\
\text{H} & \text{H} & \text{H} & \text{H}
\end{array}
$$
polystyrene

$$NC{\cdot}HC{=}CH_2 \longrightarrow$$
acrylonitrile

$$
\begin{array}{ccccccc}
\text{H} & \text{H} & \text{H} & \text{H} & \text{H} & \text{H} & \text{H} \\
| & | & | & | & | & | & | \\
\cdots{-}\text{C}{-}\text{C}{-}\text{C}{-}\text{C}{-}\text{C}{-}\text{C}{-}\text{C}{-}\cdots \\
| & | & | & | & | & | & | \\
\text{CN} & \text{H} & \text{CN} & \text{H} & \text{CN} & \text{H} & \text{CN}
\end{array}
$$
polyacrylonitrile   ('Orlon')

Each of these polymers can have chain lengths of many hundreds of thousands of carbon atoms. *It has to be remembered that every carbon atom in such a chain has the tetrahedral configuration, and hence the chain is really a zig-zag*—sometimes called the ' backbone ' of the polymer. In a polymer, the particular group (say —CN in the ' Orlon ' polymer above) can be attached *randomly* along the chain, or (as in the diagram above) *regularly* (here, all the —CN groups are on the same side, arranged alternately). The arrangement of these groups affects the properties of the polymer—

a *regular* arrangement often gives a more useful polymer than a *random* arrangement. *Polymethylene* $(CH_2)_n$, can be made by eliminating nitrogen from the substance *diazomethane*:

$$nCH_2N_2 \longrightarrow nN_2 + (CH_2)_n$$
diazomethane

Another useful polymer is made from tetrafluorethylene, $CF_2{=}CF_2$, and is called polytetrafluorethylene, or ' P.T.F.E. ' or ' Teflon '.

Recently, propylene has been polymerized to give polypropylene which resembles polythene but possesses greater tensile strength and has a higher softening point.

By polymerizing two different monomers together, *co-polymers* can be obtained; and it is now possible to obtain these so that a definite pattern is repeated along the chain, e.g. 50 monomer A units, then 25 monomer B units, and so on. These are called *block co-polymers*.

The addition of hydrogen (as atoms) to ethylene has been considered (p. 23). If just *one* atom or group (X, say) could be attached, we should get the possibility

i.e. a ' free ' valency is left. The entity surrounded by the dotted line would then be a *free radical* (p. 10) and it could add on another molecule of ethylene, i.e.

$$H_2C{-}CH_2 + CH_2{=}CH_2 \longrightarrow H_2C{-}CH_2{-}CH_2{-}CH_2 \cdot$$
$$\quad | \qquad\qquad\qquad\qquad\qquad\qquad | \qquad$$
$$\quad X \qquad\qquad\qquad\qquad\qquad\qquad X \qquad$$

This new entity would still be a free radical, but now having a chain of four carbon atoms instead of two. Further addition of ethylene would then result in a longer and larger carbon chain—and we should obtain a polymer eventually. It is a process of this kind which actually *initiates* polymerization—the initiator X can be a free atom (e.g. sodium) or can itself be a free radical; for example, X might be the *methyl radical*, $CH_3 \cdot$, liberated by heat or light on substances from which methyl groups are easily detached.

Usually when a polymer is formed, the chains may grow to some more or less random length and are then terminated before all the monomer is used up. It has been found possible, however, to keep the chains growing in length until all the monomer has gone; when more monomer is added, the chains grow again. Such polymers are called 'living' polymers but they are not living, of course, in the usual sense of the word.

Artificial fibres

Examples of these are well-known; nylon and Terylene are two common ones. These are high polymers made by the condensation method and which are thermoplastic. As prepared, these polymers are not fibrous because the long carbon chains in the molecules are tangled up (a). When the melted polymer is extruded through a fine hole, and the resulting fibre solidifies, the carbon chains are straightened out or *oriented* (b); this orientation is further improved by stretching the raw fibre (c); a little disorder is allowed to remain to give the fibre flexibility.

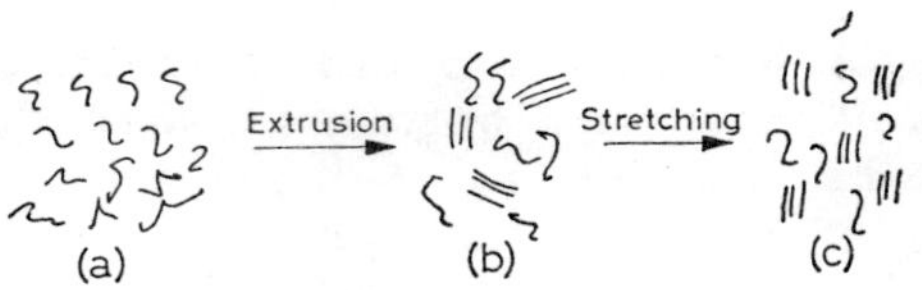

(a)   Extrusion →   (b)   Stretching →   (c)

CYCLOPARAFFINS

This is a series of hydrocarbons having a general formula the same as that of the alkenes, viz. C_nH_{2n}. The different members of the series are thus isomeric with the corresponding members of the alkene series. Some compounds of this type are formed during the refining of petroleum.

Although *cyclopropane*, the first member of the series, shows some similarities to the alkenes, in general the cycloparaffins or *cycloalkanes* behave as saturated compounds. Thus, cyclopentane and cyclohexane resemble the alkanes; their molecules clearly do not contain carbon-carbon double bonds and they must therefore be formulated as cyclic compounds $[(CH_2)_n]$.

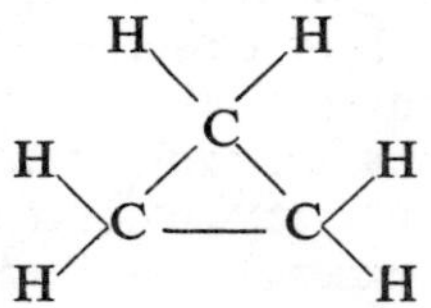

cyclopropane

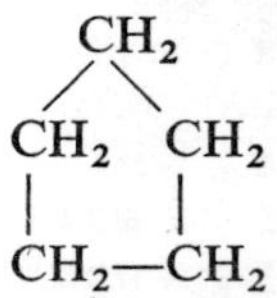

cyclopentane

Cyclopropane has been used as an anaesthetic.

When additive properties are shown (particularly with cyclopropane), the ring is broken to give a straight-chain product which is a paraffin derivative.

DI-OLEFINS OR DIENES

These are unsaturated hydrocarbons, the molecules of which contain two double bonds. The most important representatives are those in which the two double bonds are separated by a single bond. Two are of special interest:

$$\textit{butadiene} \quad CH_2{=}CH{-}CH{=}CH_2$$

and

$$\textit{isoprene} \quad CH_2{=}C{-}CH{=}CH_2$$
$$\underset{\displaystyle CH_3}{\vert}$$

Natural rubber has a formula of $(C_5H_8)_n$ where n is very high. It is a polymer of isoprene. When natural rubber is subjected to destructive distillation, isoprene is a product.

The linkage $C{=}C{-}C{=}C$ is known as a *conjugated system*. Such a linkage possesses characteristic properties, and polymerization is often easy to initiate owing to bond rearrangement. In 1910, isoprene was polymerized under the catalytic effect of sodium to give a rubber-like substance. This was the first synthetic rubber. Butadiene was later polymerized, again under the influence of sodium, to give a better product. The action may be represented as follows:

$$nCH_2{=}CH{-}CH{=}CH_2 \xrightarrow[\text{catalyst}]{Na} (-CH_2{-}CH{=}CH{-}CH_2{-})_n$$

This was the first of the ' Buna ' rubbers—the word being coined from ' butadiene ' and ' natrium '. Newer and better rubbers are obtained by linking up butadiene with other compounds and then polymerizing. This is an example of co-polymerization (p. 27).

When rubber is vulcanized to give greater strength and elasticity, sulphur adds on to the double bonds and causes the cross-linking of the molecular chains. This process was accidentally discovered by Goodyear in 1839.

Most of the butadiene used in the manufacture of synthetic rubber in this country is obtained directly from the cracking of petroleum and the dehydrogenation of butane under the influence of chromic oxide as a catalyst.

SUMMARY OF THE REACTIONS OF OLEFINS OR ALKENES

The general formula is C_nH_{2n}.

Reactions of ethylene, C_2H_4

C_2H_4

$\xrightarrow{\text{burns in air}} CO_2 + H_2O$

$\xrightarrow{\text{polymerizes}} (CH_2)_n$ ('polythene')

$\xrightarrow{H_2 \text{(catalyst)}} C_2H_6$

$\xrightarrow[\text{Cl}_2 \text{ (or Br}_2)]{\text{halogens}} ClCH_2 \cdot CH_2Cl$ (or $BrCH_2 \cdot CH_2Br$)
1,2-dichloroethane

$\xrightarrow{\text{HCl}} C_2H_5Cl$
ethyl chloride

$\xrightarrow{\text{HOCl}} HOCH_2 \cdot CH_2Cl$
ethylene chlorohydrin

$\xrightarrow{Br_2 \text{ water}} HOCH_2 \cdot CH_2Br$

$\xrightarrow{\text{KMnO}_4} HOCH_2 \cdot CH_2OH$ (fission occurs under
ethylene glycol   more vigorous
                               conditions)

$\xrightarrow{O_3}$ ozonide $\xrightarrow{H_2O} 2HC \overset{\displaystyle O}{\underset{\displaystyle H}{\big\langle}}$ $(+ H_2O_2)$
formaldehyde

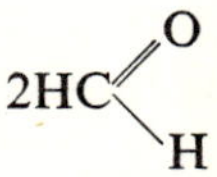

$\xrightarrow{\text{CuCl}} C_2H_4 \cdot CuCl$

The *addition reactions* shown above are general reactions for alkenes (and for many compounds which contain a carbon-carbon double bond in addition to other groups).

Oxidation (e.g. with permanganate or ozone) is used as a method of studying the structures of unsaturated compounds. The oxidation products (aldehydes, ketones, acids) result from fission at the double bond.

3

ACETYLENES OR ALKYNES

Functional Group: —C≡C—

ACETYLENE, of molecular formula C_2H_2, is the first member of this series of highly unsaturated hydrocarbons of general formula C_nH_{2n-2}. The structural formula of acetylene is H—C≡C—H with a *triple bond* between the two carbon atoms. The molecule is linear, and the triple bond is shorter than the olefinic double bond but in some respects is even more reactive. Acetylenes are of great industrial importance, and are widely used in organic synthesis.

Preparation of acetylene

The usual laboratory method is to drop water on to calcium ' carbide ' (strictly called calcium acetylide).

$$CaC_2 + 2H_2O \longrightarrow Ca(OH)_2 + C_2H_2$$

The apparatus shown in *Figure 3.1* can be used. As acetylene is only slightly soluble in water, it can be collected as shown.

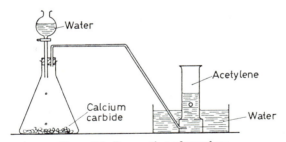

Figure 3.1. Preparation of acetylene.

The gas so obtained is impure due to the presence of phosphine (PH_3) and hydrogen sulphide, which are derived from calcium phosphide and calcium sulphide respectively. These substances are impurities in commercial calcium carbide. The rather objectionable smell of acetylene so prepared is due to the presence of these gaseous impurities; they may be removed by passage of the gas through a tube containing loosely-packed slaked lime.

Other methods of preparation are given later.

Industrially, acetylene can be produced from natural gas (p. 41).

Properties of acetylene

Acetylene is a colourless gas, slightly soluble in water and very reactive chemically.

(1) Acetylene is an endothermic compound*, and is very unstable. If the gas is compressed beyond two atmospheres, it explodes violently. The cylinders of acetylene for use in engineering shops or in laboratories contain solutions of the gas, under pressure of about twelve atmospheres, in acetone, the latter being absorbed in kieselguhr (a finely divided naturally occurring form of silica).

(2) Acetylene burns with a luminous flame and forms a dangerously explosive mixture with air or oxygen:

$$2C_2H_2 + 5O_2 \longrightarrow 4CO_2 + 2H_2O$$

The highly endothermic character of acetylene is in part responsible for the large amount of heat given out in this combustion.

(3) In presence of *acid* copper(I) (cuprous) chloride, acetylene polymerizes (dimerizes) to give *vinyl acetylene*, ' vinyl ' being the name given to the $CH_2=CH-$ group:

$$2HC \equiv CH \xrightarrow[\substack{\text{copper (I)} \\ \text{chloride}}]{\text{acid}} HC \equiv C-\underset{\underset{H}{|}}{C}=CH_2$$

Addition of hydrogen chloride to this vinylacetylene gives ' chloroprene ' which when polymerized gives a synthetic rubber.

Nearly 100 years ago, Berthelot discovered that acetylene polymerizes at red heat to give the more stable aromatic hydrocarbon benzene, C_6H_6:

$$3C_2H_2 \longrightarrow C_6H_6$$

This *trimerization* of acetylene can now be carried out at low temperatures by using polymerization catalysts of the type described on p. 25, and higher polymers can be prepared. By using substituted acetylenes (where one or both hydrogens are replaced by other groups) substituted benzene compounds can be made.

(4) Acetylene forms addition compounds (cf. alkenes; see also p. 35).

(*a*) Addition of *pairs of hydrogen atoms:*—In the presence of platinum black, palladium, or finely divided nickel, hydrogen adds to acetylene giving first ethylene and then ethane:

$$C_2H_2 \xrightarrow{H_2} C_2H_4 \xrightarrow{H_2} C_2H_6$$

* $2C+H_2 \longrightarrow C_2H_2$; $\Delta H = 54\cdot8$ kcal.

(*b*) Addition of *chlorine*:—Acetylene reacts explosively with chlorine liberating carbon:

$$C_2H_2 + Cl_2 \longrightarrow 2HCl + 2C$$

but under certain conditions both *chlorine* and *bromine* can be added to form ultimately a tetrahalogen derivative of ethane:

$$C_2H_2 + 2Cl_2 \longrightarrow CHCl_2 \cdot CHCl_2$$

This compound, *1,1,2,2-tetrachloroethane*, is an important solvent for oils and resins.

(*c*) Addition of *hydrogen chloride* and *hydrogen bromide*:— Addition is slow but that of hydrogen chloride is catalysed by mercury(I) chloride and activated charcoal:

$$C_2H_2 + HCl \xrightarrow{\text{Hg}_2\text{Cl}_2} H_2C{=}CHCl$$
vinyl chloride

(This compound, of great commercial importance, has been referred to on p. 26 and is again mentioned on p. 37.)

Further reaction gives the unsymmetrical dichlor derivative:

$$H_2C{=}CHCl + HCl \longrightarrow CH_3 \cdot CHCl_2$$
1,1-dichloroethane

Hydrogen cyanide adds on to acetylene in presence of acid copper(I) chloride to give vinyl cyanide, more commonly called *acrylonitrile* $CH_2{=}CH \cdot CN$. This polymerizes to give 'Orlon' (p. 26).

(*d*) Addition of *water*:—In the presence of sulphuric acid (30 per cent) and mercury(II) (mercuric) sulphate, acetylene adds water to form acetaldehyde; the reaction may be represented as shown below, but this is undoubtedly an over-simplification.

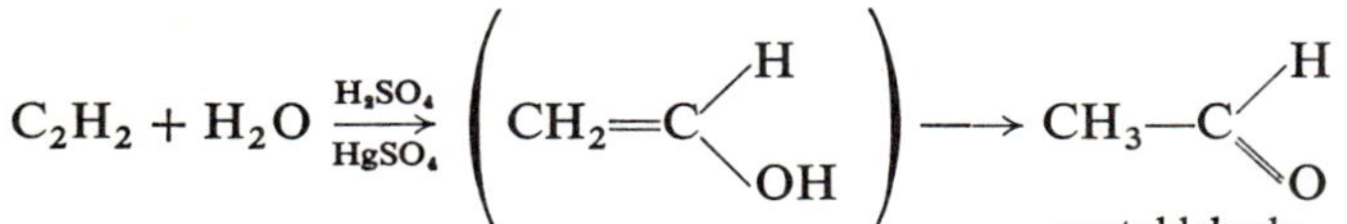

$$C_2H_2 + H_2O \xrightarrow[\text{HgSO}_4]{\text{H}_2\text{SO}_4} \left(CH_2{=}C{\Big\langle}{}^{H}_{OH} \right) \longrightarrow CH_3{-}C{\Big\langle}{}^{H}_{O}$$
acetaldehyde

The reaction provides an important commercial method of converting acetylene into acetaldehyde.

Mercury(II) sulphate also catalyses the addition of acetic acid ($CH_3 \cdot CO_2H$) to acetylene:

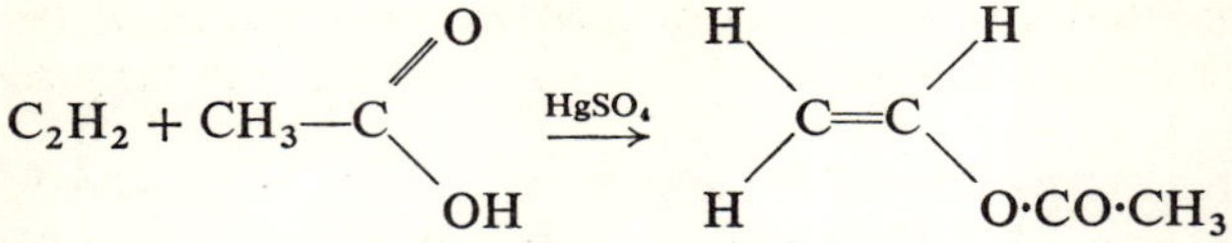

$$C_2H_2 + CH_3-C \overset{O}{\underset{OH}{\diagup\diagdown}} \xrightarrow{\text{HgSO}_4} \overset{H}{\underset{H}{\diagdown}}C=C\overset{H}{\underset{O\cdot CO\cdot CH_3}{\diagup}}$$

acetic acid   vinyl acetate

The product, *vinyl acetate*, like vinyl chloride can be converted to useful polymers.

(*e*) Oxidation by *potassium permanganate*:—Like alkenes, acetylene and its homologues are sensitive to oxidizing agents.

(5) The hydrogen atoms of acetylenes can be replaced by metals, i.e. acetylene shows some *acidic** behaviour. The metal derivatives of acetylene are of several types:

(*a*) The sodium derivative can be obtained by passing acetylene through a solution of the metal in liquid ammonia, or better, a solution of sodamide in liquid ammonia:

$$2C_2H_2 + 2Na \longrightarrow 2\,HC \equiv C^- Na^+ + H_2$$

$$C_2H_2 + NaNH_2 \longrightarrow HC \equiv C^- Na^+ + NH_3$$

In the latter case, the process actually occurring is a proton transfer:

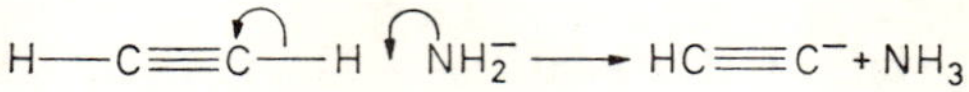

$$H-C \equiv C-H \quad NH_2^- \longrightarrow HC \equiv C^- + NH_3$$

As before, the curved arrows imply shifts of electron pairs.

Potassium and calcium derivatives of acetylene ($HC \equiv C^- K^+$, $^-C \equiv C^- Ca^{2+}$, calcium ' carbide ') resemble the sodium derivative in being salt-like compounds. They are immediately decomposed by water but are not explosive.

(*b*) Copper and silver derivatives of acetylene are easily prepared, by passing the gas through ammoniacal solutions of either copper(I) (cuprous) chloride or silver nitrate. The copper(I) derivative is red and the silver derivative, white; *both are precipitated from the aqueous solution*:

$$C_2H_2 \xrightarrow[\text{AgNO}_3]{\text{ammoniacal}} Ag_2C_2 \downarrow$$

silver acetylide

$$C_2H_2 \xrightarrow[\text{CuCl}]{\text{ammoniacal}} Cu_2C_2 \downarrow$$

copper(I) acetylide

* The use of the term ' acidic ' here must be noted carefully. It does *not* imply that acetylene dissociates in solution to form (solvated) protons and $(HC \equiv C)^-$ anions. Nevertheless, with *very* strong bases, e.g. sodamide, $Na^+NH_2^-$, acetylene does give up a proton and form the acetylide ion.

These compounds, in contrast to the alkali metal derivatives, are clearly *not* decomposed by water, but when dry they are dangerously explosive.

The ready formation of the copper and silver derivatives is the basis of simple laboratory tests for acetylene or any of its homologues which contain the grouping —C≡C—H.   Disubstituted acetylenes, RC≡CR, do not react, nor do alkenes. Thus ethylene and acetylene, which are similar in many ways, can easily be distinguished.

Mechanism of alkyne addition reactions

Some of the addition reactions of alkenes, discussed in detail in Chapter 2, have obvious counterparts in the reactions of acetylene and its homologues. Thus the addition of bromine to acetylene, which occurs in stages:

$$H\text{—}C\equiv C\text{—}H \longrightarrow BrCH=CHBr \longrightarrow Br_2CH\text{—}CHBr_2$$

1,2-dibromo-
ethylene

is similar to the addition of bromine to ethylene, and the *mechanisms* of the additions are also believed to be closely related. It is of considerable interest that the first product above, BrCH=CHBr, actually has the structure

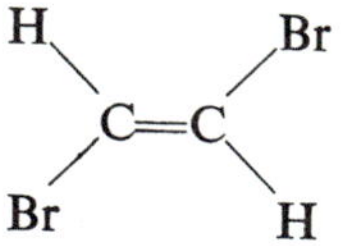

in which the bromine atoms are on *opposite sides* of the planar molecule. The addition can be pictured thus:

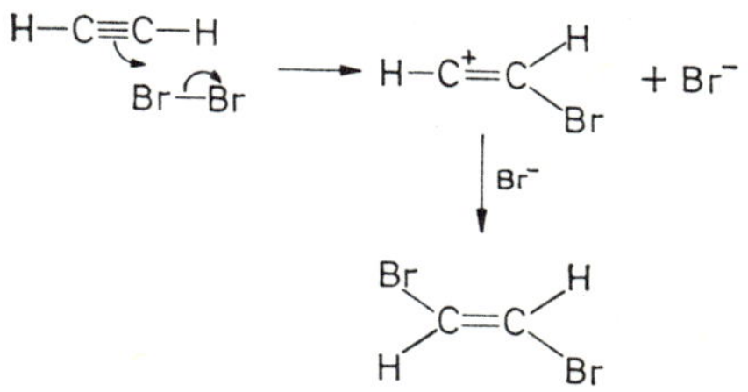

It should be noted that the presence of the double bond in this product imposes rigidity on the molecule. Under normal conditions, groups attached by a double bond cannot rotate with respect to one another (compare groups attached by a single bond, which can so rotate).   Two isomers of 1,2-dibromoethylene are therefore possible, distinguished by the names *cis* and *trans*:

cis-1,2-dibromoethylene   trans-1,2-dibromoethylene

The compound formed by the addition of bromine to acetylene is the *trans* isomer.

Many similarities in the reactions of ethylene and acetylene arise from the fact that in both molecules the carbon-carbon multiple bonds (double or triple) can act as sources of *electron pairs* which are used to form new bonds in ' ionic ' addition reactions.

Careful examination of the behaviour of alkenes and alkynes in catalytic reactions with hydrogen (catalytic hydrogenation) also reveals an interesting and important similarity. In Chapter 2, we indicated that alkenes add two hydrogen atoms on the same side of the double bond. The hydrogenation of a disubstituted acetylene, $R.C{\equiv}C.R'$, where in principle R and R' can be any groups we wish, leads first to an alkene; again we find that the two hydrogen atoms are added to the same side of the multiple bond:

cis isomer

The hydrogenation of an alkyne can usually be stopped at this half-way stage. Consequently the method is a very general one for the laboratory synthesis of alkenes. The preparation of the starting materials, $R{-}C{\equiv}C{-}R'$ or $R{-}C{\equiv}C{-}H$ (where R = alkyl, e.g. $CH_3{-}$ or $C_2H_5{-}$) will be described later. Note that with $R{-}C{\equiv}C{-}H$, only one product, $R{-}CH{=}CH_2$, is formed by partial reduction.

Complete hydrogenation of alkynes leads to alkanes, e.g.

$$R{-}C{\equiv}C{-}R' \longrightarrow R{\cdot}CH_2{\cdot}CH_2{\cdot}R'$$

and this is another useful, general procedure.

Uses

For many years acetylene was used as an illuminant and is still widely used for welding and cutting. The temperature of the oxyacetylene flame is well above 2000°C.

Within the last thirty years, acetylene has become a valuable raw material for the manufacture of a range of organic compounds of increasing importance. From acetylene are obtained *grease solvents* (e.g. Westron and Westrol), *vinyl plastics* and *synthetic rubbers**. Under the catalytic influence of a peroxide, vinyl chloride polymerizes to the thermoplastic polyvinyl chloride (p. 26):

$$ClCH{=}CH_2 \longrightarrow \quad \cdots - \overset{\displaystyle Cl}{\underset{\displaystyle H}{\overset{\displaystyle |}{\underset{\displaystyle |}{C}}}} - \overset{\displaystyle H}{\underset{\displaystyle H}{\overset{\displaystyle |}{\underset{\displaystyle |}{C}}}} - \overset{\displaystyle Cl}{\underset{\displaystyle H}{\overset{\displaystyle |}{\underset{\displaystyle |}{C}}}} - \overset{\displaystyle H}{\underset{\displaystyle H}{\overset{\displaystyle |}{\underset{\displaystyle |}{C}}}} - \overset{\displaystyle Cl}{\underset{\displaystyle H}{\overset{\displaystyle |}{\underset{\displaystyle |}{C}}}} - \cdots$$

polyvinyl chloride

Other vinyl plastics, which are co-polymers (p. 27) are also manufactured for a variety of uses, e.g. in sponge bags and raincoats.

Manufacture

(1) Coke and quicklime are heated in an electric furnace to a temperature of about 2000°C:

$$3C + CaO \xrightarrow{\text{2000°C}} CaC_2 + CO$$

Water is added to the calcium carbide obtained when acetylene is evolved:

$$CaC_2 + 2H_2O \longrightarrow C_2H_2 + Ca(OH)_2$$

The process can be carried on economically only in countries like Norway and Switzerland where electric power is readily available.

(2) Acetylene is obtained as a by-product in the cracking of the higher-boiling fractions in the refining of petroleum. Increasingly large amounts of acetylene are being obtained by this method in countries where the carbide process is uneconomic.

(3) Acetylene is obtained by the partial oxidation of methane:

$$4CH_4 + 3O_2 \longrightarrow 2C_2H_2 + 6H_2O$$

This process has been developed on a large scale in America where ample supplies of methane are available as natural gas.

SUMMARY OF THE CHEMISTRY OF ACETYLENES OR ALKYNES

The general formula is C_nH_{2n-2}

---

* During World War II, acetylene was used in Germany as the starting point in the preparation of artificial blood plasma.

(1) *Acetylene,* C_2H_2

$$CaC_2 + H_2O \longrightarrow C_2H_2 \rightarrow$$

- burns in air $\longrightarrow CO_2 + H_2O$
- polymerizes $\longrightarrow HC\equiv C-CH=CH_2,$ C_6H_6, etc.
- Na or NaNH$_2$ $\longrightarrow HC\equiv C^- Na^+$
- ammoniacal AgNO$_3$ $\longrightarrow Ag_2C_2$ (white)
- ammoniacal CuCl $\longrightarrow Cu_2C_2$ (red)

The following properties of acetylene are due to the triple bond in the molecule, giving it additive power, and making it sensitive to oxidation (e.g. by $KMnO_4$).

$$C_2H_2 \xrightarrow[\text{Ni, etc}]{H_2} C_2H_4 \longrightarrow C_2H_6$$

$$\xrightarrow{Cl_2} CHCl_2 \cdot CHCl_2 \text{ or explodes} \longrightarrow HCl + C$$

$$\xrightarrow{Br_2}$$

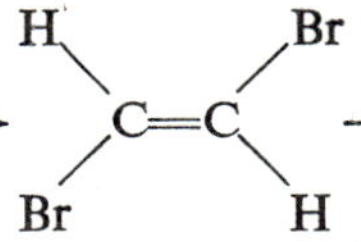

$$\begin{array}{c} H \diagdown \qquad Br \diagup \\ C = C \\ Br \diagup \qquad \diagdown H \end{array} \longrightarrow Br_2CH \cdot CHBr_2$$

trans-1,2-dibromo-ethylene

$$\xrightarrow{HCl} H_2C = CHCl \longrightarrow CH_3 \cdot CHCl_2$$

vinyl chloride   1,1-dichloroethane

$$\xrightarrow{HCN} H_2C = CH \cdot CN$$

acrylonitrile

$$\xrightarrow[H_2SO_4 + HgSO_4]{H_2O} CH_3CHO$$

acetaldehyde

(2) *Manufacture*

    (*a*) Coke and quicklime are heated in an electric furnace:

$$3C + CaO \xrightarrow{2000°C} CaC_2 + CO$$

Water is added to the product:

$$CaC_2 + 2H_2O \longrightarrow Ca(OH)_2 + C_2H_2$$

    (*b*) Product of petroleum cracking.

    (*c*) Partial oxidation of methane:

$$4CH_4 + 3O_2 \longrightarrow 2C_2H_2 + 6H_2O$$

(3) *Uses*

Cutting and welding; in the manufacture of many organic compounds, e.g. acetaldehyde, grease solvents, vinyl plastics, synthetic rubber.

EXAMINATION QUESTIONS

(1) Describe the preparation of acetylene free from hydrogen sulphide and phosphine.

Acetylene is an endothermic compound. What does this mean? What may happen if acetylene is highly compressed? How is acetylene stored in steel cylinders?

How is acetylene converted into (*a*) acetaldehyde, (*b*) cuprous acetylide, (*c*) vinyl bromide, $CH_2=CH\cdot Br$? (O. A.)

*(2) Starting from acetylene, how could the following substances be obtained: (*a*) benzene, (*b*) chloroform, (*c*) a commercial plastic material?

A hydrocarbon *Y* has molecular formula C_4H_6. A sample of *Y* (0·25 g) was shaken with hydrogen and palladium until uptake of hydrogen ceased; 207·4 ml (measured at 0°C and 760 mm mercury pressure) of hydrogen were absorbed. Write **three** structures for *Y* that fit these observations and state how you would differentiate between these structures. (H = 1, C = 12; the gram-molecular volume of a gas at S.T.P. = 22·41.) (O. and C.A.)

* (3) Outline the manufacture of acetylene, starting from natural sources. Give **two** properties of acetylene that distinguish it from ethylene.

By what reactions, or series of reactions, may the following be prepared from acetylene: (*a*) acetaldehyde, (*b*) trichlorethylene, (*c*) vinyl chloride, (*d*) vinyl acetate? Give the equations and reaction conditions.

Indicate the general chemical structure of polyvinyl compounds and mention **one** practical use for any one of them (J.M.B.A.)

(4) Suggest chemical tests which would enable you to distinguish between the four hydrocarbons

$$CH_3\cdot CH_2\cdot CH_2\cdot CH_2\cdot CH_2\cdot CH_3$$
$$CH_3\cdot CH_2\cdot CH_2\cdot CH_2\cdot CH=CH_2$$
$$CH_3\cdot CH_2CH_2\cdot CH_2\cdot C\equiv CH$$
$$CH_3\cdot CH_2\cdot CH_2\cdot C\equiv C\cdot CH_3$$

(5) Give examples to illustrate the statement that reactive carbon compounds are generally those in the formulae of which some degree of unsaturation is indicated. (A. S.)

* Part (*b*) in each of these questions requires a knowledge of reactions which are given in later chapters.

4

PETROLEUM

CRUDE OIL AND NATURAL GAS

IN various parts of the world, escape of inflammable gases and oils from the earth's surface has been taking place for centuries. The ' eternal fires of Baku ', caused by the burning of these substances, were visited by fire-worshippers as long ago as 600 B.C. The Americas, the Middle East and Russia are the chief oil-bearing parts of the world. Wherever oil is found, natural gas occurs above it. The compositions of both vary in different wells. Natural gas consists mainly of methane mixed with the higher paraffins, ethane, propane and butane. The oil, i.e. the crude petroleum which is a black treacle-like liquid with a greenish sheen, contains liquid and solid members of the paraffin series, cyclic hydrocarbons and derivatives of the aromatic hydrocarbon benzene.

It is now widely believed that petroleum is a product of the decomposition of the fatty constituents of vegetable and animal matter (probably of marine origin) under pressure and at high temperature. The necessary conditions for the production of petroleum in this way are shallow seas (rich in animal and vegetable life) and the burying of the dead organisms under a layer of mud or silt; here bacterial decay can occur, followed by other reactions, which take place when the pressure and temperature rise following the formation of a very deep layer of mud or silt.

Natural gas

Natural gas is now an important and world-wide source of energy; it is used in a variety of ways:

(1) It is used as a *fuel*. Natural gas may be piped from gas wells to cities hundreds of miles away and used directly as a fuel. It has a higher calorific value than coal gas. Liquefied natural gas is also transported in special tankers from the developing Sahara oilfields; it may be added to coal gas, or the higher hydrocarbons may be liquefied under relatively low pressure and stored in ' bottles ' or cylinders for use as portable fuel.

(2) The higher hydrocarbons—propane, the butanes and pentanes— are removed and condensed with olefins by the process of *alkylation* (p. 44) to make volatile liquid hydrocarbons which are very valuable for blending with petrol.

(3) It is used in synthetic chemistry. The methane portion may be passed with steam over heated nickel and converted into a mixture of carbon monoxide and hydrogen (p. 13):

$$CH_4 + H_2O \xrightarrow[\text{Ni}]{450°C} CO + 3H_2$$

The mixture is of value directly as a fuel or as the starting point in the manufacture of a variety of organic compounds (p. 195). Alternatively the reactions can be made to follow the path

$$CH_4 + 2H_2O \longrightarrow CO_2 + 4H_2$$

and the hydrogen used in synthesis of ammonia.

(4) It is used in the manufacture of *carbon black*. The gas is burned in an insufficient supply of air:

$$CH_4 + O_2 \longrightarrow C + 2H_2O$$
$$\text{(air)}$$

The carbon black so prepared is in a very finely divided condition suitable for making printers' ink, shoe polishes, gramophone records, and as a filler for the rubber of motor tyres to increase their strength and prevent wear. Large amounts of carbon black are also made by burning an atomized spray of aromatic hydrocarbons in a specially designed furnace.

By a partial oxidation similar to the above, attempts have been made in America to manufacture aldehydes, ketones, and acids.

(5) It is used to manufacture *acetylene*. The reaction

$$2CH_4 \rightleftharpoons C_2H_2 + 3H_2$$

goes from left to right at temperatures above 1300°, and the natural gas is heated (e.g. by an electric arc) to about 1500° and then quenched rapidly with sprays of water, and the acetylene removed by a suitable solvent—water under pressure is sometimes used. Alternatively the methane may be partially oxidized (p. 37).

Treatment of crude petroleum

There are two main processes in the treatment of crude petroleum. These are (1) Refining and (2) Cracking.

Refining

This is a straight *distillation process* designed, not to separate the viscous liquid into its numerous individual constituents but rather into *fractions* which boil within definite ranges of temperature.

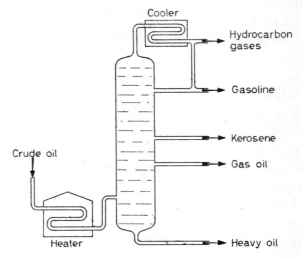

Figure 4.1. Refining of petroleum.

Figure 4.1 is a representation of a refining plant.

The crude oil is pumped through heated pipes and the vapours formed pass into a tall fractionating column in which the temperature decreases towards the top. Different fractions are withdrawn at different levels of the column and cooled. The column is divided into compartments to make separation easier and contains devices allowing the heavier, less volatile constituents to flow to the compartment below, and the lighter, more volatile ones to pass into the one above. Five fractions are usually collected. Bitumen (used for making roofing materials, binding road surfaces, lining sea walls, jetties and the like) is a residue.

The following are the distilled fractions, named in order as they are drawn off the column from the bottom upwards.

(1) *Heavy oil:*—This fraction runs out from the bottom of the column. It is subjected to further distillation under reduced pressure (to prevent decomposition) when various grades of *lubricating oils*, *fuel oil* (for firing furnaces) and *diesel oil* for ships' engines are obtained. *Vaseline* and *paraffin wax* (for the manufacture of candles, waterproofing paper and cardboard containers) are obtained by crystallization of oil and a solvent ('dewaxing').

(2) *Gas oil:*—From this fraction high speed diesel fuel is obtained.

(3) *Kerosene or Naphtha:*—Prior to World War I, this was the most valuable fraction obtained from petroleum, but with the development of the motor car it has become less valuable and some

of it is ' cracked ' or *reformed* to give the lower boiling hydrocarbons suitable for use in motor spirit. Treatment (solvent extraction) with liquid sulphur dioxide removes aromatic (p. 204) components from kerosene, leaving ' paraffin ' which is used as a fuel for heating and cooking, for tractor engines and for jet engines. (The aromatic components would cause ' smoky ' burning).

(4) *Gasoline:*—This most valuable fraction is the basic material from which petrol is made. It contains small amounts of sulphur which are first removed. (In the internal combustion engine sulphur compounds would burn to yield sulphur dioxide which would cause excessive cylinder wear.)

(5) *Hydrocarbon gases:*—These are the lower boiling members of the paraffin and olefin series. The former were dissolved in the liquid hydrocarbons of the petroleum and have been released, and the olefins are formed by some thermal decomposition which has occurred during the distillation.

Some portion of them is used as fuel for the distillation process. The remainder is mixed with the gases obtained from the cracking unit (below).

From some of the above fractions, many different solvents are obtained. These are used as paint thinners, solvents for lacquers, explosives, plastics, photographic films and in the manufacture of numerous other substances.

The separation into the various fractions by distillation is thus followed by other methods of separation, e.g. crystallization and solvent extraction, and adsorption (by activated charcoal or silica gel) may also be used.

Cracking

Cracking is the thermal breakdown of a hydrocarbon into two or more hydrocarbons each containing a smaller number of carbon atoms in their molecules. An example has already been given (p. 9):

$$C_3H_8 \longrightarrow CH_4 + C_2H_4$$

The decomposition products, with their smaller number of carbon atoms per molecule, are more volatile. Cracking has been used increasingly in industry during the last fifty years* to meet the growing demand for motor spirit. Solid hydrocarbons can be cracked to form liquids, and liquid hydrocarbons are similarly decomposed to more volatile ones.

In principle, the cracking unit resembles the refining unit. The material to be cracked is fed through a heating coil at about 500°C

* The first plant was set up in America in 1912.

and then under pressure into a tank to allow the decomposition to occur. Here different catalysts, e.g. natural or synthetic clays, are present to help the process. Sometimes, catalysts are used to permit low-temperature cracking, giving different products to the high-temperature process. In the presence of certain catalysts some of the gaseous products of the cracking process polymerize to form large molecules and aromatic compounds can be formed also; a long period in the cracking unit favours the production of aromatic hydrocarbons (Chapter 17). Cracking under strictly controlled conditions, so that conversion is limited, is called *reforming*.

The final products of the pyrolysis are led into a fractionating column similar to that used in the refining process. Here separation occurs. Higher boiling fractions are returned to the heating coil. The gasoline portion and the gaseous hydrocarbons are led off from the plant.

While the cracking process was originally designed to obtain gasoline for use in motor spirit, the gaseous hydrocarbons also obtained have proved to be extremely valuable as raw materials for the manufacture of other organic compounds in demand. These gaseous compounds and also those from the refining unit contain not only paraffin hydrocarbons but unsaturated olefins also. They may be treated in a variety of ways as follows:

(1) *Polymerization:*—By passing the gases at 200°C and under pressure over phosphoric acid as a catalyst, the unsaturated hydrocarbons polymerize into higher hydrocarbons. These were used for blending with petrol, but are now more commonly used for other synthetic processes.

(2) *Alkylation:*—Unsaturated hydrocarbons, by virtue of their additive power, can be made to combine with paraffins. A mixture of isobutane and isobutylene at 30°C in the presence of sulphuric acid, combines to form an iso-octane, valuable for blending in motor spirit. The process is known as *alkylation*. Other catalysts, e.g. hydrogen peroxide, boron trifluoride, and the aluminium halides permit reduction of the temperature and pressure of the process:

$$
\begin{array}{ccc}
\underset{\text{isobutane}}{\overset{\displaystyle CH_3}{\underset{\displaystyle CH_3}{CH_3{-}CH}}} + \underset{\text{isobutylene}}{\overset{\displaystyle CH_3}{\underset{\displaystyle CH_3}{CH_2{=}C}}} & \longrightarrow & \underset{\text{iso-octane}}{\overset{\displaystyle CH_3}{\underset{\displaystyle CH_3}{CH_3{-}C{-}CH_2{-}\overset{\displaystyle CH_3}{\underset{\displaystyle CH_3}{CH}}}}}
\end{array}
$$

Isobutane for this reaction can be obtained from butane by *isomerization*, a similar process to alkylation, in which the catalyst is a mixture of aluminium chloride and hydrogen chloride.

(3) *For the manufacture of alcohol and glycol:*—Ethylene is absorbed in concentrated sulphuric acid and the mixture then diluted and boiled, when the ethyl hydrogen sulphate is hydrolysed to alcohol which distils over (p. 74). This process is primarily an American one where ethylene is in plentiful supply. Other alcohols, valuable as solvents, are similarly obtained from other olefins, for by altering the concentration of the sulphuric acid, other olefins than ethylene can be absorbed successively.

Ethylene is also converted into ethylene glycol (p. 77).

The aromatic compounds produced by cracking are sometimes alkylated, e.g. ethylbenzene may be produced rather than benzene. These are subjected to a process called 'hydrodealkylation' whereby treatment with hydrogen removes the alkyl group, thus

$$C_6H_5 \cdot C_2H_5 \xrightarrow{H_2} C_6H_6 + C_2H_6$$

Over the last two decades, there has been an enormous increase in the quantity, and also in the range of chemicals produced from petroleum, and even the production of inorganic substances such as sulphur and ammonia from petroleum amounts to several million tons per year.

Knock inhibitors

As the piston moves down the cylinder of an internal combustion engine, a mixture of petrol and air is taken in. On the return stroke of the piston, the mixture is compressed, and at the point of maximum compression, a spark is passed to ignite the mixture. Now, if the petrol contains certain substances, the mixture may ignite by compression alone (as in a diesel engine) a fraction of a second before the spark passes. This premature explosion is the cause of 'knocking' or 'pinking' which results in loss of efficiency as well as excessive engine wear.

Straight-chain paraffins, for some reason, cause pinking to a very much greater extent than do branched chain compounds. n-Heptane has a poor anti-knock value as against iso-octane which has a very high one. These two hydrocarbons have been selected to make an arbitrary scale on which the value of a petrol can be measured. Heptane has been given a value of 0 and iso-octane a value of 100. A mixture of 20 per cent heptane and 80 per cent iso-octane is said to have an 'octane value (or number) of 80'. Attempts are made to raise the octane number of a fuel (particularly for aeroplane engines) by the addition of a variety of substances. Benzene, an aromatic hydrocarbon, has a high anti-knock value. Ethyl and methyl alcohols have high values also. Octane numbers of about 200 can be reached. The most efficient knock inhibitor is the compound lead tetraethyl of formula $Pb(C_2H_5)_4$*. Unfortunately, when combustion in the cylinder occurs, litharge is deposited but this is prevented by the addition of ethylene dibromide ($CH_2Br \cdot CH_2Br$) which converts the lead to the fairly volatile lead tetrabromide which escapes. Lead tetramethyl, $Pb(CH_3)_4$, which is more volatile than the tetraethyl compound, is now being used also.

---

* Lead tetraethyl is made by the action of a sodium–lead alloy on ethyl chloride (p. 52).

5

HALOGEN DERIVATIVES OF THE ALKANES

Functional Groups: —F, —Cl, —Br, —I

THE successive substitution of the hydrogen atoms of the methane molecule by chlorine atoms has been referred to previously (p. 10). This reaction is used on the industrial scale, particularly for the manufacture of methyl chloride, CH_3Cl, methylene chloride, CH_2Cl_2, and carbon tetrachloride CCl_4. Similar reactions occur with other alkanes but, in general, the reactions are not easily controlled and mixtures are formed; consequently halogenated alkanes are not made in the laboratory in this way.

Other methods of obtaining halogen derivatives of alkanes were mentioned in Chapters 2 and 3. These are:

(*a*) Addition of halogens (Br_2 and Cl_2) to alkenes, e.g.

$$H_2C\!\!=\!\!CH_2 + Br_2 \longrightarrow BrCH_2\!-\!CH_2Br$$

(*b*) Addition of hydrogen halides to alkenes, e.g.

$$H_2C\!\!=\!\!CH_2 + HBr \longrightarrow CH_3\!-\!CH_2Br$$

(*c*) Addition of halogens to acetylene, e.g.

$$HC\!\!\equiv\!\!CH + 2Cl_2 \longrightarrow Cl_2CH\!-\!CHCl_2$$

Reactions of these types are often used as laboratory methods of preparation.

Chlorine, bromine and iodine derivatives of alkanes are commonly encountered. Many of them are valuable solvents and are used in industry for this purpose; the chlorine derivatives are the most important in this respect because they are relatively cheap. In the laboratory, alkyl halides (mono-halogenated alkanes) are very frequently used in synthetic work; the bromides and iodides are usually preferred because they are more reactive than the chlorides.

Within relatively recent years, many fluorine compounds have been prepared and are being increasingly used in industry for a variety of purposes. By contrast with the great reactivity of fluorine, these compounds are very inert and their value as refrigerants and solvents due, in part, to this inactivity.

Alkyl halides

These compounds have the general formula of $C_nH_{2n+1}X$, where X is a halogen atom.

Methyl chloride, CH_3Cl, is a gas; ethyl chloride is a volatile liquid of b.p. 13°C. The boiling point of a bromide is higher than that of the corresponding chloride and that of the iodide higher still. Thus methyl bromide boils at 5°C and methyl iodide at 42°C (p. 293). The alkyl halides are thus more volatile than the corresponding alcohols (methyl alcohol, CH_3OH, b.p. 64°, ethyl alcohol, C_2H_5OH, b.p. 78°C). This is due to the fact that the molecules of alcohols are associated (p. 294).

It will be clear from the above that organic halides are very different from the ionic metal halides commonly studied in inorganic chemistry. Ionic halides are often soluble in water, whereas alkyl halides are all virtually insoluble; alkyl halides are soluble in organic solvents, such as hydrocarbons, but ionic halides are insoluble.

Properties of alkyl halides

Alkyl halides undergo many reactions but, fortunately, it is possible to classify these into a relatively small number of reaction types:

(1) *Displacement reactions:*—These are very characteristic and important reactions. In fact, an example has already been given of a displacement reaction, although not under this name; the reaction of ethyl chloride with alkalis (effectively, with the hydroxide ion) to give ethyl alcohol was discussed in Chapter 1 and was represented:

$$HO^- \quad \overset{H}{\underset{CH_3}{\overset{|}{C}}}\!\!-\!\!Cl \quad \longrightarrow \quad HO\!-\!\overset{H}{\underset{CH_3}{\overset{|}{C}}} \quad + \; Cl^-$$

Alkanes are quite stable towards alkalis because their atoms are (essentially) electrically neutral. Alkyl halides react with alkalis because in these molecules the carbon atom to which the halogen atom is attached carries a small positive charge:

$$CH_3\!-\!\overset{\overset{\displaystyle H}{|}}{\underset{\underset{\displaystyle H}{|}}{C}}\!\overset{\delta+\;\;\;\delta-}{-Cl} \quad \text{or} \quad CH_3\!-\!\overset{\overset{\displaystyle H}{|}}{\underset{\underset{\displaystyle H}{|}}{C}}\!\rightarrow\!-Cl$$

and also because the halogen atom can form the stable halide anion. Thus, in general, alkyl halides are 'hydrolysed' by alkalis; the reaction is a *displacement* of X^- by OH^-:

$$R\!-\!X + OH^- \longrightarrow R\!-\!OH + X^-$$

The alkali (sodium or potassium hydroxide) is usually dissolved in water and the solution heated with the alkyl halide. ' Moist silver oxide ' (probably the hydroxide) is a milder reagent which gives a similar result:

$$CH_3Br + {}^{\prime}AgOH^{\prime} \longrightarrow CH_3OH + AgBr$$

Another, and in practice more important, displacement reaction is that between an alkyl halide and potassium cyanide, often carried out by heating the reactants together in a mixture of ethyl alcohol and water:

$$R-X + CN^- \longrightarrow R-CN + X^-$$

The product, $R-C\equiv N$, is an alkyl cyanide (Chapter 13).

Similar reactions carried out in liquid ammonia, are:

$$R-X + C\equiv\bar{C}H \longrightarrow RC\equiv CH + X^-$$

$$R-X + \bar{C}\equiv CR' \longrightarrow RC\equiv CR' + X^-$$

The sodium derivatives of the acetylenes are used.

Many other reactions are known; we will give two further examples. Ethyl alcohol reacts with metallic sodium to give the ionic compound *sodium ethoxide*, $Na^+OC_2H_5{}^-$, analogous in its structure to sodium hydroxide. When an alkyl halide is heated with sodium ethoxide, the halogen atom is displaced:

$$R-X + \bar{O}C_2H_5 \longrightarrow R-O-C_2H_5 + X^-$$

The product is an *ether*, the general name for the series of compounds in which two alkyl groups are linked to an oxygen atom; the best known member of the series is *diethyl ether*, $C_2H_5-O-C_2H_5$, which is often referred to simply as ' ether '. Diethyl ether could be prepared, for example, by reaction between ethyl iodide and sodium ethoxide. This particular type of displacement reaction is known as the *Williamson ether synthesis* (p. 85).

Alkyl halides react with ammonia (an alcoholic solution can be used) to give a mixture of products which we shall not discuss in detail at this point. In the first stage of the reaction, the nitrogen atom of the ammonia provides an electron pair to form a bond to a carbon atom, and as in the other displacement reactions considered above, the halogen atom separates from the same carbon atom as a halide ion:

It will be noticed that in this reaction the attacking reagent is an uncharged molecule (NH_3) whereas in the other reactions discussed the effective reagent was an anion. Since in the reaction the nitrogen atom shares two of its electrons with a carbon atom, the nitrogen atom must become positively charged. The system as a whole must remain electrically neutral (since the starting materials were neutral); the positive charge which appears on the nitrogen atom is balanced by the negative charge on the halide ion. The product is an *alkyl substituted ammonium salt*, which could also be produced by reaction between an alkyl substituted ammonia, $R—NH_2$ (an *amine*, see Chapter 14) and the hydrogen halide:

$$R—NH_2 + HX \xrightarrow[\text{slow}]{\text{very fast}} R—NH_3{}^+X^-$$
$$R—X + NH_3 \xrightarrow{\hspace{2cm}}$$

(2) *Elimination reactions:*—With hot concentrated solutions of caustic alkalis (especially with ethyl alcohol as the solvent rather than water) alkyl halides can undergo elimination reactions. Here, the alkali removes the elements of the hydrogen halide (HX) from the molecule, instead of replacing the halogen atom by the hydroxyl group, as previously discussed, e.g.

$$CH_3 \cdot \underset{\underset{I}{|}}{CH} \cdot CH_3 \xrightarrow[\text{alkali}]{\text{conc. alcoholic}} CH_3 \cdot CH = CH_2 + \underset{\substack{\text{(removed by} \\ \text{alkali)}}}{HI}$$

This type of reaction, which is the reversal of one of the reactions of alkenes (Chapter 2), obviously cannot occur with methyl halides and is not important with ethyl halides, but becomes more important with halides of higher molecular weight. It provides a general method of synthesis of alkenes. The processes involved in the elimination reactions of alkyl halides are represented below:

As before, the curved arrows represent shifts of electron pairs, and the reaction is regarded as an ionic reaction.

(3) *Reactions with metals:*—Alkyl halides, especially bromides and iodides, react readily with certain metals. The most important reaction of this kind occurs with magnesium, giving a product which

is known as a Grignard compound or *Grignard reagent*, after the French chemist who discovered this class of compounds:

$$H_3C{-}I + Mg \xrightarrow[\text{ether}]{\text{in}} `H_3C{-}Mg{-}I'$$

or generally

$$R{-}X + Mg \xrightarrow[\text{ether}]{\text{in}} `R{-}Mg{-}X'$$
Grignard reagent

The formulae are shown as ' $H_3C{-}Mg{-}I$ ', ' $R{-}Mg{-}X$ ' because the Grignard reagents are known to have more complex structures; the structure RMgX is an approximation but one adequate to explain the reactions of these very important substances.

In practice, Grignard reagents are prepared by gradually adding an alkyl halide to magnesium turnings suspended (by stirring) in ether. The reaction often requires heat initially, but once it has commenced it usually continues without further heating, since the reaction is exothermic. The apparatus and materials must all be dry, because Grignard reagents react immediately with water:

$$R{-}Mg{-}X + H_2O \longrightarrow RH + Mg(OH)X$$

Formation of a Grignard reagent and subsequent decomposition with water provides a method of ' reducing ' an alkyl halide RX to an alkane RH.

Further examples of the use of Grignard reagents will be given in later chapters, but one other simple application can be illustrated here:

$$RMgX + CO_2 \longrightarrow R{\cdot}C{\overset{O}{\underset{OMgX}{\Big\langle}}} \xrightarrow[\text{acid}]{H_2O} R{\cdot}C{\overset{O}{\underset{OH}{\Big\langle}}}$$
a carboxylic acid

Carbon dioxide, either as the gas or in the form of ' dry ice ', reacts readily with Grignard reagents as shown, providing a useful synthesis of carboxylic acids (Chapter 10).

Alkyl halides can be converted to alkanes by other metallic reducing agents, e.g. with sodium–mercury amalgam and with the zinc–copper couple. Catalytic reduction is also possible:

$$RX + H_2 \longrightarrow R{\cdot}H + HX$$

Reaction of an alkyl halide in ether solution with metallic sodium yields hydrocarbons:

$$2RX + 2Na \longrightarrow R{-}R + 2NaX$$

This reaction, known as the *Wurtz synthesis*, is chiefly of historical importance. In practice, the reaction is less simple than would appear from the equation given above, and other products can be formed.

N.B. Although alkyl halides are not ionic like many metal halides, they readily react so as to give halide anions. Thus, they differ from metallic halides in not giving immediate precipitates of silver halide with aqueous silver nitrate solution; but precipitates are formed when the alkyl halide is heated with alcoholic silver nitrate. This test serves to distinguish alkyl halides from some other organic halogen compounds (e.g. halogen derivatives of benzene, Chapter 19) which are much less reactive, and do not produce silver halide even on prolonged heating with silver nitrate solution.

Uses of the alkyl halides

The use of alkyl halides as solvents has been referred to.

Methyl chloride, although a gas at room temperature and pressure, is readily compressible to a liquid and is thus used in refrigeration. It has been used as a local anaesthetic, being sprayed on to the skin.

Ethyl chloride possesses similar anaesthetic properties. It is also used in the manufacture of lead tetraethyl (p. 46).

$$4C_2H_5Cl + Na_4Pb \longrightarrow 4NaCl + Pb(C_2H_5)_4$$
$$\text{lead tetraethyl}$$

(Na_4Pb is a simple representation of a sodium/lead alloy.)

Dihalogen derivatives

The best known compounds of this class are those from ethane of formula $C_2H_4X_2$ where X is the halogen chlorine or bromine. Two isomers of each derivative are known, one symmetrical in structure and the other unsymmetrical.

$CH_2Cl \cdot CH_2Cl$	$CH_3 \cdot CHCl_2$
1,2-dichloroethane	1,1-dichloroethane
$CH_2Br \cdot CH_2Br$	$CH_3 \cdot CHBr_2$
1,2-dibromoethane	1,1-dibromoethane

The symmetrical derivative is obtained by the addition of the halogen to ethylene. When ethylene is passed into liquid bromine (covered with a layer of water), decolorization occurs and the oily liquid left is 1,2-dibromoethane (the compound used in petrol containing lead tetraethyl, p. 46):

$$CH_2{=}CH_2 + Br_2 \longrightarrow CH_2Br \cdot CH_2Br$$
$$\text{1,2-dibromoethane}$$

When, on the other hand, hydrogen bromide adds on to acetylene the unsymmetrical compound is formed:

$$CH{\equiv}CH + 2HBr \longrightarrow CH_3{\cdot}CHBr_2$$
1,1-dibromoethane

In general, these dihalogen compounds resemble the alkyl halides discussed above.

Trihalogen derivatives

The commonest members of this class are *chloroform*, $CHCl_3$, and *iodoform*, CHI_3, although the corresponding fluorine and bromine compounds are known also. Laboratory methods for the preparation of these *haloform* compounds will be discussed later.

Chloroform, $CHCl_3$

This colourless volatile liquid (b.p. 61°C) was first prepared by Liebig in 1831. It was first used as an anaesthetic by Simpson in 1847.

Chloroform is a heavy liquid (sp. gr. 1·525) with a sweet sickly smell. It is a good solvent. Iodine dissolves in it to form a purple solution. The liquid is non-flammable but the vapour can be ignited in air to burn with a characteristic greenish flame.

The following are the chief chemical properties:

(1) In air and in the presence of light, chloroform undergoes oxidation slowly to the very poisonous gas carbonyl chloride (' phosgene '), $COCl_2$, which dissolves in the chloroform. Chlorine is also formed:

$$CHCl_3 \xrightarrow[\text{sunlight}]{O_2 \text{ (air)}} COCl_2 + H_2O + Cl_2$$

To prevent this reaction, chloroform is usually stored in dark glass bottles. This is particularly necessary if the chloroform is to be used as an anaesthetic, for both phosgene and chlorine are dangerously poisonous.

(2) Chloroform is hydrolysed on boiling with aqueous alkalis to a salt of formic acid*.

$$CHCl_3 + 4NaOH \longrightarrow HCO\overset{-}{O}Na^+ + 3NaCl + 2H_2O$$
sodium formate

(3) With a primary amine and alcoholic potash, chloroform gives a *carbylamine*, readily recognizable by its objectionable smell.

* It is because chloroform and also the corresponding bromine and iodine derivatives give formic acid on hydrolysis that Dumas gave these compounds the respective names of chloro*form*, bromo*form* and iodo*form*.

This is *Hofmann's carbylamine reaction*, described fully on p. 158. The reaction provides a simple but unpleasant test for chloroform.

(4) Chloroform reduces *Fehling's solution* (p. 94)

Iodoform, CHI_3

Iodoform is a pale yellow solid with a pleasant characteristic smell. It is almost insoluble in water, but readily soluble in alcohol. It has been used as an antiseptic.

The compound is obtained if ethyl alcohol is heated with a few crystals of iodine in the presence of alkali. This is the *iodoform test* described on p. 101 for the recognition of ethyl alcohol and some related compounds.

Tetrahalogen derivatives

Carbon tetrachloride, CCl_4

Carbon tetrachloride is the best known member of this group of compounds. It is a liquid (b.p. 77°C). Because it is non-flammable, it is used extensively as a solvent in ' dry cleaning ', degreasing and in the rubber industry. It is also used in some portable fire extinguishers but it may prove dangerous, for if partially oxidized the highly poisonous gas phosgene is a product.

Large quantities of it are used to manufacture dichlorodifluoromethane, CCl_2F_2, used as a refrigerant under the name of ' Freon '.

Carbon tetrachloride is one of the few compounds containing chlorine and one other element only, which cannot be directly synthesized*. It is manufactured by the action of chlorine on carbon disulphide. More recently it has been obtained on the large scale as the final product of the action of chlorine on methane.

ALKYL DERIVATIVES OF SULPHURIC ACID

The alkyl halides discussed in this chapter can be regarded as derivatives of hydrogen halides. Similarly, sulphuric acid can be regarded as the parent of two series of compounds known as *alkyl sulphates* and *alkyl hydrogen sulphates:*

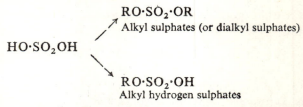

$$RO \cdot SO_2 \cdot OR$$
Alkyl sulphates (or dialkyl sulphates)

$$HO \cdot SO_2 OH$$

$$RO \cdot SO_2 \cdot OH$$
Alkyl hydrogen sulphates

* In Humphry Davy's classical experiment at the beginning of the last century, when he heated carbon to a high temperature in an atmosphere of chlorine in order to prove that chlorine was an element, no carbon tetrachloride was formed.

In many respects, these compounds resemble alkyl halides, but since they are derivatives of alcohols and are, in fact, usually made from alcohols, discussion of their reactions is deferred until the next chapter.

Ethyl hydrogen sulphate was mentioned previously (pp. 18, 23) as the product of the addition of sulphuric acid to ethylene:

$$H_2C{=}CH_2 + HO{\cdot}SO_2{\cdot}OH \longrightarrow CH_3{\cdot}CH_2O{\cdot}SO_2{\cdot}OH$$
ethyl hydrogen sulphate

SUMMARY OF THE REACTIONS OF ALKYL HALIDES

The general formula is $C_nH_{2n+1}X$ where X = halogen.

Reactions can be classified as:

(1) *Displacement reactions*

(*a*) Hydrolysis by aqueous alkali or moist silver oxide:

$$R{-}X + OH^- \longrightarrow R{-}OH + X^-$$
(an alcohol)

(*b*) Reaction with metal cyanides:

$$R{-}X + CN^- \longrightarrow R{-}CN + X^-$$
(alkyl cyanide)

(*c*) Reaction with metal alkoxides:

$$R{-}X + {}^-OC_2H_5 \longrightarrow R{-}O{-}C_2H_5 + X^-$$
(ethoxide anion) (an ether)

(*d*) Reaction with ammonia:

$$R{-}X + NH_3 \longrightarrow R{-}NH_3^+ X^-$$
(an alkyl ammonium salt)

Only the first stage is shown; subsequent reactions are discussed in Chapter 14.

(*e*) Reaction with sodium acetylide:

$$RX + \bar{C}{\equiv}CH \longrightarrow RC{\equiv}CH + X^-$$

(2) *Elimination reactions*

Reaction with alcoholic alkali (method of alkene synthesis):

$$CH_3{\cdot}CH{\cdot}CH_3 + OH^- \longrightarrow CH_3{\cdot}CH{=}CH_2 + H_2O + X^-$$
$$|$$
$$X$$

(3) *Reactions with metals*

(a) Formation of Grignard reagents:

$$R\text{---}X + Mg \xrightarrow[\text{ether}]{\text{dry}} RMgX$$

Grignard reagents are decomposed by water:

$$RMgX + H_2O \longrightarrow RH + Mg(OH)X$$

and react with carbon dioxide:

$$RMgX + CO_2 \longrightarrow RCO\cdot OMgX \xrightarrow[\text{(acid)}]{H_2O} RCO\cdot OH$$
$$\text{(a carboxylic acid)}$$

(b) Reduction by metals (e.g. Na/Hg or Cu/Zn couple) and by catalytic methods:

$$RX + H_2 \longrightarrow RH + HX$$

(c) Wurtz synthesis (of symmetrical alkanes):

$$2RX + 2Na \longrightarrow R\text{---}R + 2NaX$$

N.B. Alkyl halides, being covalent compounds, do not react immediately with cold aqueous silver nitrate, in contrast to ionic metal halides.

SUMMARY OF THE REACTIONS OF CHLOROFORM, $CHCl_3$

(1) Slowly oxidized in air:

$$CHCl_3 \xrightarrow{\text{air}} COCl_2 + H_2O + Cl_2$$

(2) Hydrolysed by alkali:

$$CHCl_3 + 4NaOH \longrightarrow HCOONa + 3NaCl + 2H_2O$$
$$\text{sodium formate}$$

(3) Gives the Hofmann carbylamine reaction (p. 158).

(4) Reduces Fehling's solution (p. 94).

EXAMINATION QUESTIONS

(1) How do alkyl halides react with metals and with metallic compounds? (L.S.).

(2) An alkyl iodide X contains 26·09% C, 4·89% H and 69·02% I. On treatment with alcoholic potash an unsaturated hydrocarbon Y is obtained which contains 85·71% C. On oxidation Y gives acetone, carbon dioxide and water. Addition of hydriodic acid to Y gives Z which is isomeric with X. Suggest structures for X, Y and Z.

$$[C = 12\cdot0, \qquad H = 1\cdot00, \qquad I = 127\cdot0] \qquad \text{(L.A.)}$$

(3) n-Propyl iodide undergoes displacement reactions with metallic nitrites $(M^+NO^-_2)$ to form two compounds, both having the formula $C_3H_7O_2N$. Show how these reactions could be explained in electronic terms, and suggest structures for the two products.

6

ALIPHATIC ALCOHOLS

Functional Group —OH

THE *alcohols* are compounds containing the monovalent hydroxyl group —OH attached to a hydrocarbon radical. In some respects they resemble water, where the functional group —OH is attached to hydrogen. The commonest alcohol is ethyl alcohol of formula C_2H_5OH—which is often called simply ' alcohol '. Since it contains one hydroxyl group in its molecule, it is said to be a *monohydric* alcohol.

There may be more than one hydroxyl group in a molecule, e.g. ethylene glycol of structural formula $CH_2OH \cdot CH_2OH$ has two such groups and is said to be a *dihydric* alcohol. Glycerol of formula $CH_2OH \cdot CHOH \cdot CH_2OH$ is a *trihydric* alcohol. Many other *polyhydric* alcohols are known.

MONOHYDRIC ALCOHOLS

The monohydric alcohols derived from the paraffin or alkane series of saturated hydrocarbons have the general formula of $C_nH_{2n+1}OH$. The first member of the series, where $n = 1$, is *methyl alcohol*, CH_3OH. This nomenclature is based on the name of the alkyl group in the molecule, but others are in use. Kolbe suggested the name of *carbinol* for this first member, CH_3OH. Other alcohols were named as derivatives of this. Thus, ethyl alcohol, $CH_3 \cdot CH_2OH$ is called *methyl carbinol*. A third system, now generally accepted, consists in replacing the terminal *-e* of the parent paraffin by *-ol*. Thus CH_3OH is *methanol*, C_2H_5OH is *ethanol*, etc.

Methyl alcohol	(methanol)	CH_3OH	(b.p. 64°C)
Ethyl alcohol	(ethanol)	C_2H_5OH	(b.p. 78°C)
Propyl alcohol	(propanol)	C_3H_7OH 2 isomers	
Butyl alcohol	(butanol)	C_4H_9OH 4 isomers	

The structural formulae of the two propyl alcohols are:

$$CH_3 \cdot CH_2 \cdot CH_2OH \quad \text{and} \quad \begin{matrix} CH_3 \\ \diagdown \\ \diagup \\ CH_3 \end{matrix} CH\text{---}OH$$

n-propyl alcohol   isopropyl alcohol

and the four isomers of butyl alcohol are:

(1) $CH_3 \cdot CH_2 \cdot CH_2 \cdot CH_2OH$,   (2) $CH_3 \cdot CH_2 \cdot CH(OH) \cdot CH_3$,
n-butyl alcohol   secondary butyl alcohol

(3) 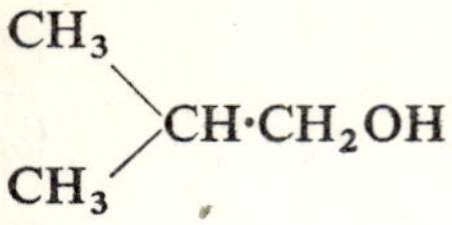 CH₃
$$\begin{array}{c} CH_3 \\ \\ CH_3 \end{array}\Big\rangle CH\cdot CH_2OH$$ and   (4) 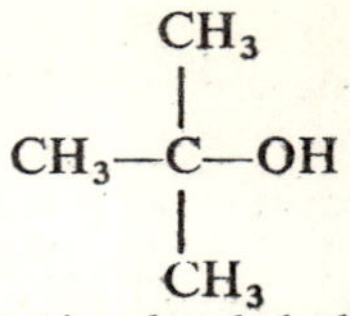 $$\begin{array}{c} CH_3 \\ | \\ CH_3-C-OH \\ | \\ CH_3 \end{array}$$

     isobutyl alcohol   tertiary butyl alcohol

In isomers (1) and (3), the hydroxyl group is attached to a carbon atom to which are also linked two hydrogen atoms and an alkyl group, i.e. in both isomers, the group

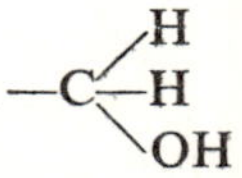

$$-C\begin{array}{c} H \\ -H \\ OH \end{array}$$ is present.

Alcohols containing the monovalent $-CH_2OH$ group are called *primary* alcohols.

In isomer (2), the characteristic group is the divalent $\begin{array}{c} H \\ \diagdown C \\ \diagup \diagdown \\ OH \end{array}$

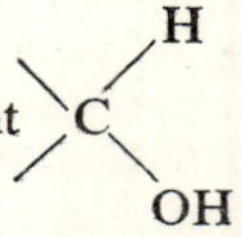

group. An alcohol containing this group is said to be a *secondary* alcohol.

Isomer (4) possesses the trivalent group $\geqslant$C–OH, which is characteristic of a *tertiary* alcohol.

Since, in a primary alcohol, the hydroxyl group is linked to a carbon atom to which two hydrogen atoms are also attached, the $-CH_2OH$ group is monovalent, and the carbon atom is at the end of the chain. In secondary and tertiary alcohols, the hydroxyl group is of necessity attached to a carbon atom somewhere within the chain.

A further reference to this classification of monohydric alcohols is made later (p. 64) when means of distinguishing between the three classes are given.

Preparation of Alcohols

Simple alcohols are rarely prepared in the laboratory—the important ones are manufactured on a large scale and are commercially available. Consequently alcohols are commonly used as starting materials in organic syntheses.

Some methods of preparing alcohols have already been discussed and are summarized here:

(*i*) Addition of sulphuric acid to an alkene and subsequent decomposition of the addition product with steam (pp. 18, 24):

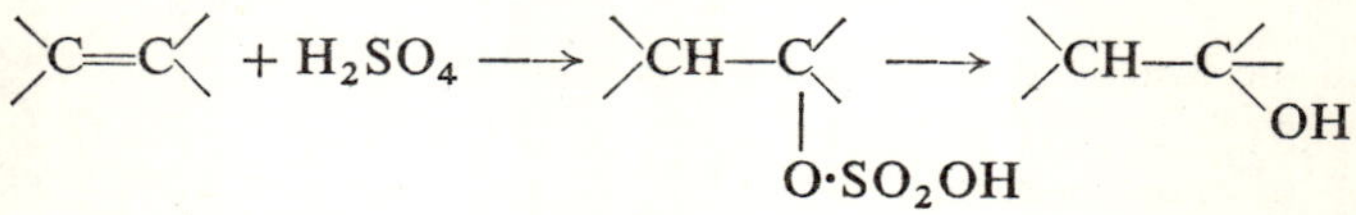

$$\begin{array}{c} \rangle C{=}C\langle + H_2SO_4 \longrightarrow \rangle CH{-}C\langle \longrightarrow \rangle CH{-}C\langle_{OH} \\ | \\ O{\cdot}SO_2OH \end{array}$$

The overall result is the addition of water to the double bond.

(*ii*) Addition of diborane to an alkene (hydroboration), followed by oxidition of the addition product with hydrogen peroxide (p. 17):

$$\text{>C=C<} + \tfrac{1}{2}\text{B}_2\text{H}_6 \longrightarrow \underset{\underset{\text{BH}_2}{|}}{\text{>CH—C<}} \longrightarrow \text{a trialkylborane } R_3B$$

$$\downarrow \text{H}_2\text{O}_2$$

$$\text{>CH—C<}_{\text{OH}}$$

(*iii*) Hydrolysis of an alkyl halide with aqueous alkali or with moist silver oxide (p. 49):

$$R—X + OH^- \longrightarrow R—OH + X^-$$

Other methods of preparing alcohols will be described in relevant places in the text and collected in Chapter 16 (which deals generally with preparative methods).

General characteristics of alcohols

Physical properties

As in all homologous series, the boiling point rises as *n* increases. Methyl and ethyl alcohols are colourless volatile liquids of b.p. 64° and 78°C respectively*. Higher members are oily liquids and those where *n* is greater than 11 are solids. Straight chain alcohols boil at higher temperatures than do their branched chain isomers.

Lower members (up to *n* = 3) mix readily with water, but as the percentage oxygen content decreases (i.e. as *n* increases), ability to mix decreases. When methyl or ethyl alcohol is mixed with water, heat is evolved (p. 294). There is also a considerable contraction in volume.

Ethyl alcohol forms a constant boiling mixture with water, making separation of the anhydrous alcohol from its aqueous mixture by simple fractional distillation impossible. The constant boiling mixture obtained at 760 mm has a composition of 95·6 per cent by weight of alcohol and 4·4 per cent of water. The mixture is known as *rectified spirit*. *Absolute alcohol* is anhydrous alcohol (p. 75).

* The relatively high boiling points of alcohols compared with other organic compounds of equal molecular weight are due to association between the molecules, because of hydrogen bonding (p. 294)—similar to the association occurring in liquid water and responsible for its high boiling point. The miscibility of alcohol with water arises from hydrogen bonds between alcohol and water molecules.

Chemical properties

(1) Because alcohols contain hydroxyl groups they show some resemblance to water. Thus they form addition compounds with certain salts, notably calcium chloride and magnesium chloride. With calcium chloride, methyl and ethyl alcohols form respectively $CaCl_2 \cdot 4CH_3OH$ and $CaCl_2 \cdot 3C_2H_5OH$ (cf. $CaCl_2 \cdot 6H_2O$); hence anhydrous calcium chloride cannot be used to dry either of these alcohols. Alcohols react (less violently than water does) with sodium and potassium to form ionic compounds, sodium and potassium *alkoxides* (or alcoholates):

$$2K + 2CH_3OH \longrightarrow 2CH_3O^-K^+ + H_2$$
potassium
methoxide

$$2Na + 2CH_3 \cdot CH_2OH \longrightarrow 2CH_3 \cdot CH_2O^-Na^+ + H_2$$
sodium
ethoxide

With less electropositive metals (e.g. magnesium, aluminium) reaction can be induced in the presence of mercury(II) chloride. Aluminium alkoxides are useful reagents (pp. 95, 188).

Alcohols react with phosphorus halides, thionyl chloride and similar compounds which are attacked by water. Some examples are:

$$3C_2H_5OH + PBr_3 \longrightarrow 3C_2H_5Br + P(OH)_3$$
ethyl
bromide

$$C_2H_5OH + PCl_5 \longrightarrow C_2H_5Cl + HCl + POCl_3$$
ethyl
chloride

$$C_2H_5OH + SOCl_2 \longrightarrow C_2H_5Cl + SO_2 + HCl$$
thionyl   ethyl
chloride   chloride

Reactions of these types are used for the preparation of alkyl halides. The reaction with phosphorus tribromide can be brought about conveniently by adding bromine to a mixture of the alcohol and red phosphorus (p. 68).

(2) Alcohols react with hydrogen halides to form alkyl halides:

$$ROH + HX \rightleftharpoons RX + H_2O$$

The reaction is quite different from the neutralization of a metal hydroxide, e.g.

$$NaOH + HX \longrightarrow NaX + H_2O$$

although the equations as written are similar. In the inorganic reaction both starting materials exist as ions in aqueous solution and the essential change occurring is

$$OH^- + H^+ \text{(solvated)} \longrightarrow H_2O$$

a very rapid and exothermic reaction. In the organic reaction, the alcohol is a covalent substance containing no hydroxide ions. Formation of the alkyl halide requires the breaking of covalent bonds (between carbon and oxygen) and new covalent bonds (between carbon and halogen) appear. Consequently the reaction is a slow one and it is not necessarily complete.

Nevertheless, the reaction of an alcohol with a hydrogen halide does involve ionic intermediates and it can be regarded as a *displacement reaction*:

$$CH_3 \cdot CH_2\,OH + HX \longrightarrow CH_3 \cdot CH_2\,\overset{+}{O}H_2 + X^-$$
protonated alcohol

$$CH_3 \cdot CH_2 \overset{\overset{+}{\frown}}{-} \overset{+}{O}H_2 \quad \longrightarrow CH_3 \!-\!\!\!-\!\! CH_2 + H_2O$$
$$\underset{X^-}{} \qquad\qquad\qquad \underset{X}{|}$$

This should be compared with the displacement reactions discussed in the previous chapter. In this example, the halide anion displaces a neutral water molecule from the protonated alcohol. The effect of attaching a proton to the oxygen of the alcohol is clearly to weaken the link between carbon and oxygen, i.e.

$$CH_3 \cdot CH_2\!-\!O\!-\!H \xrightarrow{\;\;acid\;\;} CH_3 \cdot CH_2 \overset{+}{\rightarrow}\!-\!O\!\!\begin{smallmatrix} \diagup H \\ \diagdown H \end{smallmatrix}$$

making the protonated alcohol more like an alkyl halide, $CH_3CH_2 \rightarrow X$ in respect of its reactivity towards anions.

Alkyl halides are commonly prepared from alcohols in the laboratory by reaction with hydrogen halides. The ease of the reaction depends both on the nature of HX (the order of reactivity is $HI > HBr > HCl$) and on the structure of the alcohol. Thus tertiary-butyl alcohol reacts readily with concentrated aqueous hydrochloric acid at room temperature,

$$(CH_3)_3OH + HCl \longrightarrow (CH_3)_3CCl + H_2O$$

but reaction of primary alcohols requires heating with anhydrous hydrogen chloride, in the presence of zinc chloride*. Concentrated

* The function of the zinc chloride is complicated; one effect of its presence is probably to assist in the fission of the carbon-oxygen link.

hydrobromic acid (mixed with a little sulphuric acid) is a satisfactory reagent for the preparation of bromides but a convenient modification is to treat the alcohol with potassium bromide and sulphuric acid (p. 67).

(3) Reaction of alcohols with concentrated sulphuric acid alone, under mild conditions, leads to mono- and di-alkyl derivatives of the acid. With ethyl alcohol, for example, the products are ethyl hydrogen sulphate, $C_2H_5O \cdot SO_2 \cdot OH$, and diethyl sulphate, $C_2H_5O \cdot SO_2 \cdot OC_2H_5$ (obtained by heating the mixture under reduced pressure so that the product is removed by distillation):

$$C_2H_5OH + H_2SO_4 \rightleftharpoons C_2H_5O \cdot SO_2 \cdot OH + H_2O$$

$$C_2H_5OH + HO \cdot SO_2 \cdot OC_2H_5 \rightleftharpoons C_2H_5O \cdot SO_2 \cdot OC_2H_5 + H_2O$$

Like the reactions with hydrogen halides, these are displacement reactions. Some practical details for the preparation of ethyl hydrogen sulphate as its potassium salt are given later (p. 69).

(4) Under more vigorous conditions, ethyl alcohol is *dehydrated* by concentrated sulphuric acid:

$$C_2H_5\,OH \xrightarrow[\text{H}_2\text{SO}_4 \text{ at } 170°C]{\text{excess conc.}} H_2O + H_2C{=}CH_2 \quad \text{(p. 63, 65)}$$
$$\text{ethylene}$$

$$2C_2H_5OH \xrightarrow[\text{at } 140°C]{\text{conc. H}_2\text{SO}_4} H_2O + C_2H_5OC_2H_5 \quad \text{(p. 83)}$$
$$\text{diethyl ether}$$

The latter reaction can be made a ' continuous ' process by gradual addition of alcohol to the reaction mixture. Hence it is known as *the continuous etherification* process, but in practice the sulphuric acid is gradually diluted by the water formed. The reactions occurring are probably:

(Proton transfer)
$$C_2H_5OH + H_2SO_4 \rightleftharpoons C_2H_5\overset{+}{O}H_2 + \overset{-}{O} \cdot SO_2 \cdot OH$$

(Displacement)
$$C_2H_5\overset{+}{O}H_2 + \overset{-}{O} \cdot SO_2 \cdot OH \rightleftharpoons C_2H_5{-}O \cdot SO_2 \cdot OH + H_2O$$

(Displacement)
$$C_2H_5O \cdot SO_2 \cdot OH + C_2H_5OH \rightleftharpoons C_2H_5\overset{+}{O}H + \overset{-}{O} \cdot SO_2 \cdot OH$$
$$\underset{C_2H_5}{|}$$

(Proton transfer)

$$C_2H_5\overset{+}{O}\underset{C_2H_5}{\overset{H}{\big\langle}} + \overset{-}{O}\cdot SO_2\cdot OH \longrightarrow C_2H_5OC_2H_5 + H_2SO_4$$

N.B. The *Williamson ether synthesis* (p. 85), involving reaction between an alkoxide, RO^-Na^+, and an alkyl halide, is a more general method for the preparation of ethers.

Dehydration of alcohols to alkenes is a very important general procedure. In the example given above, formation of ethylene from ethanol, vigorous conditions are used. The ease of dehydration of an alcohol depends markedly on its structure; tertiary alcohols, e.g. tertiary butanol $(CH_3)_3COH$, dehydrate under mild conditions (moderate temperature, diluted acid); secondary alcohols, e.g. isopropyl alcohol, $(CH_3)_2CH\cdot OH$, are intermediate in reactivity; primary alcohols generally require concentrated acids and relatively high temperatures.

An alternative method of dehydration is to pass the vapour of the alcohol over strongly heated alumina.

(5) Alcohols react with strong oxidizing agents, such as ' chromic acid ' (acidified sodium or potassium dichromate) or chromium(VI) oxide CrO_3, to give products which depend on the structure of the alcohol:

$$-CH_2-OH \longrightarrow -C\overset{H}{\underset{O}{\big\langle}} \longrightarrow -C\overset{OH}{\underset{O}{\big\langle}}$$

primary alcohol   aldehyde   carboxylic acid

$$\big\rangle CH-OH \longrightarrow \big\rangle C=O$$

secondary alcohol   ketone

A primary alcohol gives first an *aldehyde*; this can often be isolated by distilling it from the hot reaction mixture, but if the aldehyde is left in contact with the oxidizing agent, further oxidation occurs, giving a *carboxylic acid*, e.g.

$$C_2H_5OH \longrightarrow CH_3CHO \longrightarrow CH_3COOH$$
ethyl alcohol   acetaldehyde   acetic acid

Secondary alcohols give *ketones* on oxidation and these products are relatively stable towards further oxidation, e.g.

$$
\begin{array}{c}
CH_3 \\
 \diagdown \\
 CHOH \longrightarrow \\
 \diagup \\
CH_3
\end{array}
\qquad
\begin{array}{c}
CH_3 \\
 \diagdown \\
 C{=}O \\
 \diagup \\
CH_3
\end{array}
$$

isopropyl   acetone
alcohol

Aldehydes and ketones are examples of *carbonyl compounds*— their molecules contain the group $\diagup C{=}O$ and in consequence they show many characteristic reactions (see Chapters 8 and 9).

Tertiary alcohols, containing the group $\diagup C{-}OH$, clearly cannot be oxidized in the same way as primary and secondary alcohols. In fact, they are rather more stable towards oxidizing agents; but, under vigorous conditions, the molecule is broken up (i.e. carbon-carbon bonds are split) and various products can result, including ketones, carboxylic acids, and even carbon dioxide.

Since simple tests are available for aldehydes and for ketones, *study of the behaviour of an alcohol on oxidation is one way of distinguishing between primary, secondary, and tertiary alcohols.* Oxidation of primary and secondary alcohols is also *a very important general preparative method for aldehydes and ketones.*

(6) Similar products (i.e. aldehydes and ketones) are obtained when alcohols are *dehydrogenated*; this can be brought about by passing the vapour of the alcohol over certain metals (e.g. copper) heated to about 300°C, e.g.

$$
CH_3OH \longrightarrow HC{\diagup}^{O}_{\diagdown H} + H_2
$$

methyl   formaldehyde
alcohol

$$
\begin{array}{c}
CH_3 \\
\diagdown \\
CH{\cdot}OH \longrightarrow \\
\diagup \\
CH_3
\end{array}
\qquad
\begin{array}{c}
CH_3 \\
\diagdown \\
C{=}O + H_2 \\
\diagup \\
CH_3
\end{array}
$$

acetone
(a ketone)

Under these conditions, *tertiary* alcohols are dehydrated, not dehydrogenated:

$$CH_3-\underset{\underset{CH_3}{|}}{\overset{\overset{CH_3}{|}}{C}}-OH \longrightarrow \underset{CH_3}{\overset{CH_3}{>}}C=CH_2$$

isobutylene

(7) Further reactions of alcohols, e.g. with halogens (haloform reactions) and with carboxylic acids (ester formation) are discussed in later chapters.

PREPARATIONS ILLUSTRATING SOME REACTIONS OF ALCOHOLS
Preparation of ethylene by dehydration of ethyl alcohol
(1) With concentrated sulphuric acid. This is a convenient laboratory method.

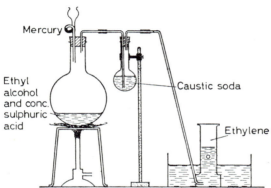

Figure 6.1. Preparation of ethylene.

About 20 ml of ethyl alcohol* are added carefully, in small amounts, to about 60 ml of concentrated sulphuric acid, contained in a 2-litre flask. (These quantities are in the approximate molecular proportions of 1 : 3 respectively.) Much heat is evolved during the mixing and to avoid the temperature rising too much, the flask may be dipped at intervals in a trough of cold water. During the mixing, ethyl hydrogen sulphate is formed:

$$C_2H_5OH + H_2SO_4 \longrightarrow C_2H_5O{\cdot}SO_2\ OH + H_2O$$
ethyl hydrogen sulphate

A little sand is added to the mixture. The flask is fitted with a safety funnel and connected up as shown in *Figure 6.1*.

The mixture is heated over a sand bath and when the temperature

* ' Rectified spirit ' which is ethyl alcohol containing just over 4 per cent of water, is suitable.

reaches about 170°C, ethylene comes off in a steady stream, the unstable ethyl hydrogen sulphate decomposing:

$$C_2H_5 \cdot HSO_4 \longrightarrow C_2H_4 + H_2SO_4$$

As is usual in organic chemistry, side reactions, in addition to the main one, take place. Almost always when an organic compound is heated with concentrated sulphuric acid, some carbon is liberated, giving a dark brown colour to the mixture. 'Charring', as this process is called, is seen to occur in the course of the above preparation, and since sulphuric acid is an oxidizing agent, carbon dioxide and sulphur dioxide are formed:

$$C + 2H_2SO_4 \longrightarrow CO_2 + 2SO_2 + 2H_2O$$

These two acidic anhydrides are absorbed in the wash bottle containing a 10 per cent aqueous solution of caustic soda.

Other impurities coming off are ether and unchanged alcohol. Most of these condense as liquids in either the wash bottle or the pneumatic trough.

(2) With ' syrupy ' phosphoric acid (i.e. a concentrated solution of orthophosphoric acid, H_3PO_4, in water) which is another convenient dehydrating agent. The solution of syrupy phosphoric acid is heated to about 200°C and ethyl alcohol is dropped into it slowly from a tap funnel:

$$C_2H_5OH \xrightarrow{H_3PO_4} C_2H_4 + H_2O$$

No carbon or sulphur dioxides are formed here but a certain amount of ether and alcohol distil over.

(3) With alumina or thoria, in a hard glass tube heated by a furnace to about 350–400°C:

$$C_2H_5OH \xrightarrow[400°C]{Al_2O_3} C_2H_4 + H_2O$$

The apparatus is shown in *Figure 6.2*.

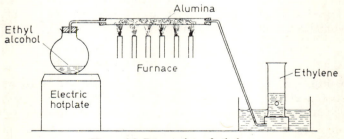

Figure 6.2. Preparation of ethylene.

Preparation of ethyl bromide

The following are two suitable laboratory methods of preparing ethyl bromide:

(1) Fifty ml of ethyl alcohol (rectified spirit) are added in small amounts to an equal volume of concentrated sulphuric acid, cooling between additions. The mixture is then poured into a 1-litre distillation flask containing 100 g of potassium bromide. The distillation flask is connected to a Liebig condenser fitted with an adapter dipping under water in the conical flask as shown in *Figure 6.3.*

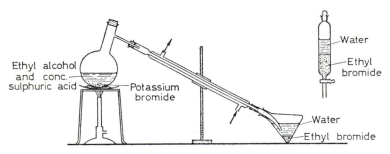

Figure 6.3. Preparation of ethyl bromide (1).

The distillation flask is heated over a sand bath when ethyl bromide distils over and collects as a colourless liquid under water in the conical flask.

The reactions occurring are:

$$KBr + H_2SO_4 \longrightarrow HBr + KHSO_4$$

$$HB + C_2H_5OH \longrightarrow C_2H_5Br + H_2O$$

The above quantities used in the preparation are approximately those calculated from these equations. Care is taken during the preparation to keep the temperature as low as possible to prevent oxidation of the hydrogen bromide by the concentrated sulphuric acid to liberate bromine. The oxidation is evident by the appearance of brown fumes:

$$2HBr + H_2SO_4 \longrightarrow 2H_2O + Br_2 + SO_2$$

The ethyl bromide so obtained may contain as impurities acid fumes, ethyl alcohol, ether and water. It is first separated from the water. The contents of the conical flask are transferred to a separating funnel, collecting the lower layer of denser ethyl bromide. The ethyl bromide is then shaken up with aqueous sodium carbonate to remove any acid gases present, the carbonate being added until there is no further effervescence. The ethyl bromide layer is again separated. A few pieces of calcium chloride are now added, and the mixture allowed to stand for a few days in a stoppered vessel. The calcium chloride removes any water and also ethyl alcohol which may be present. The remaining ethyl bromide is distilled over a water bath, the thermometer bulb being opposite the side tube of the distillation flask. The boiling point of ethyl bromide is about 38°C and the fraction distilling over between the temperatures of 35° and 43°C can be collected. Because of the low boiling point, it is advisable to surround the receiver by ice.

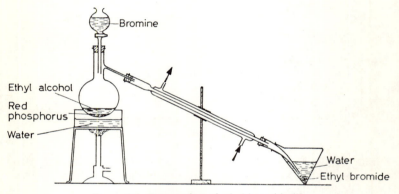

Figure 6.4. Preparation of ethyl bromide (2).

(2) A distillation flask containing approximately 70 ml of ethyl alcohol and 10 g of red phosphorus is fitted with a tap funnel and connected to a condenser and adapter as shown in *Figure 6.4.*

Twenty ml of bromine* are run in from the tap funnel drop by drop while the flask is immersed in a trough of cold water. After the addition of the bromine, the mixture is allowed to stand for some hours and then distilled over a water bath. Oily ethyl bromide collects in the receiver. It is purified as in method (1).

The phosphorus and bromine combine and the product reacts with the alcohol. Red phosphorus is used instead of the white allotrope which reacts rather violently. Assuming the bromine and phosphorus combine to form the tribromide, the equations may be written as below. The quantities used in the preparation are approximately those calculated from the equations:

$$2P + 3Br_2 \longrightarrow 2PBr_3$$

* Great care is necessary when bromine is used. The vapour attacks the throat and lungs, and the liquid can cause severe burns on the skin.

followed by:

$$PBr_3 + 3C_2H_5OH \longrightarrow 3C_2H_5Br + P(OH)_3$$

orthophos-
phorous acid

Preparation of alkyl iodides, $C_nH_{2n+1}I$

Ethyl iodide, C_2H_5I, of b.p. 72°C, is prepared by a method comparable to (2) above. Iodine is added in small quantities to a mixture of ethyl alcohol and red phosphorus. The mixture is heated under a reflux condenser for some time and the ethyl iodide formed is then distilled from the mixture.

Preparation of potassium ethyl sulphate

A mixture of about 80 ml ethyl alcohol and 20 ml concentrated sulphuric acid in a 500 ml flask is heated over a water bath and under a reflux condenser for about two hours. The following equilibrium is set up:

$$C_2H_5OH + H_2SO_4 \rightleftharpoons C_2H_5 \cdot O \cdot SO_2 \cdot OH + H_2O$$

When cool the equilibrium mixture is poured into a large beaker containing about half a litre of water. Chalk is now added until there is no further effervescence. Two reactions occur:

$$H_2SO_4 + CaCO_3 \longrightarrow H_2O + CO_2 + CaSO_4\downarrow$$

and $2C_2H_5 \cdot O \cdot SO_2 \cdot OH + CaCO_3 \longrightarrow Ca(C_2H_5SO_4)_2 + CO_2 + H_2O$

calcium ethyl sulphate

The sulphuric acid is thus removed from the equilibrium mixture as the (almost) insoluble calcium sulphate, and the ethyl hydrogen sulphate is converted into the soluble calcium ethyl sulphate. The precipitated calcium sulphate is removed by filtering, using a Buchner funnel and flask connected to a pump to suck the filtrate through and thus speed up the process (*Figure 6.5*).

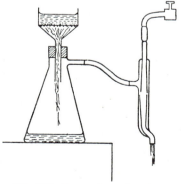

Figure 6.5. Filtration under reduced pressure.

The filtrate contains calcium ethyl sulphate and unchanged alcohol.

Potassium carbonate is added to it until no further precipitate of calcium carbonate is formed, but care is taken to avoid addition of excess potassium carbonate for, being soluble, it is difficult to remove:

$$Ca(C_2H_5SO_4)_2 + K_2CO_3 \longrightarrow CaCO_3\downarrow + 2KC_2H_5SO_4$$

To test completion of this reaction, about 1 ml of the mixture may be filtered into a test-tube and a little potassium carbonate added. If the reaction is complete, there is no precipitate.

The insoluble calcium carbonate is now filtered off by the Buchner apparatus using a filter paper cut to fit the Buchner funnel exactly. The filtrate contains potassium ethyl sulphate, unchanged alcohol, a little calcium sulphate and possibly some potassium carbonate. If it is at all brown (due to charring in the original reaction) it may be boiled with a little decolorizing charcoal and filtered. (This is a standard method of removing traces of coloured impurities in preparative organic chemistry.)

The filtrate is now evaporated to crystallizing point in a large evaporating basin over a water bath. After standing, crystals of potassium ethyl sulphate separate. These may still contain a little calcium sulphate and potassium carbonate as impurity. These impurities may be removed by a further crystallization from alcohol in which the potassium ethyl sulphate (but not the calcium sulphate or the potassium carbonate) is soluble.

Tests for potassium ethyl sulphate

(1) To an aqueous solution of the compound, barium chloride solution is added. There is no precipitate since the compound does not give rise to the sulphate ion.

(2) A little of the solid is heated in an ignition tube. Ethylene, recognizable by its burning at the mouth of the tube with a luminous flame, is given off.

Uses of alcohols

The lower alcohols in particular are used extensively in industry for a variety of purposes. They, and also the esters derived from them, are extremely valuable solvents in the manufacture of lacquers and varnishes. They are the raw materials for the manufacture of important organic compounds, e.g. methyl alcohol is oxidized to formaldehyde, the starting point in the making of some plastics; ethyl alcohol is the raw material in the manufacture of one of the synthetic rubbers. Both methyl and ethyl alcohols are constituents of some anti-freeze mixtures. Ethyl alcohol is of use as a fuel, as a substitute for petrol and in rocket propulsion.

Manufacture of methyl alcohol

Most of the methyl alcohol used in industry is now manufactured from water gas, which is obtained by passing steam over white hot coke:

$$C + H_2O \longrightarrow \underbrace{CO + H_2}_{\text{water gas}}$$

Water gas is mixed with half its volume of hydrogen* and the mixture passed, under a pressure of between two and three hundred atmospheres and at a temperature between 300° and 400°C, over zinc oxide mixed with chromium(III) oxide† to 'promote' its catalytic action:

$$\underbrace{CO + H_2 + H_2}_{\text{water gas}} \xrightarrow[\substack{\text{Cr}_2\text{O}_3 \text{ promoter} \\ 300°\text{--}400°\text{C} \\ 200\text{--}300 \text{ atm}}]{\text{ZnO catalyst}} CH_3OH$$

This method, which is discussed in more detail on pp. 195–196, displaced the older method of distillation of wood which is the way in which Boyle first prepared methyl alcohol in 1661. Dry hardwood, such as birch, beech and oak, was heated in ovens at temperatures between 200° and 400°C when destructive distillation occurred. Gases, chiefly hydrocarbons, were given off. A liquid which separated into two layers—one aqueous and the other tarry, distilled over and charcoal was left. The aqueous liquid contained up to 6 per cent methyl alcohol, up to 10 per cent of acetic acid and a little acetone. Lime was added to 'fix' the acetic acid as calcium acetate, and the liquid was then fractionally distilled. By further fractionation, the methyl alcohol and acetone content could be increased to about 60 and 10 per cent respectively. Without further separation, this mixture was sold as a solvent.

Pure methyl alcohol could be obtained from it by adding calcium chloride to form $CaCl_2\cdot4CH_3OH$. The methyl alcohol is obtainable from this complex by heating.

Manufacture of Ethyl Alcohol

Alcohols containing a number of carbon atoms between two and five can be obtained by the fermentation of carbohydrates (i.e. compounds of carbon, hydrogen and oxygen only, where the hydrogen and oxygen are in the same proportions as in water),

---

* The hydrogen is obtained by various methods, e.g. the Bosch process, electrolysis of brine, steam over red hot iron.

† It is interesting to note that if the catalyst used is a metal which forms a carbonyl, e.g. nickel, methane is the product:

$$CO + 3H_2 \longrightarrow CH_4 + H_2O$$

different ferments being used. The carbohydrates occurring naturally in large amounts and used in the manufacture of alcohols are *cellulose*, which is the chief constituent of wood, straw and various grasses (the hairy covering of the cotton seed is almost pure cellulose), *starch*, $(C_6H_{10}O_5)_n$*, which is present in maize, rice, cereals, potatoes, *sucrose*, $C_{12}H_{22}O_{11}$, which is in sugar cane and sugar beet and *fructose*, $C_6H_{12}O_6$, which is in many fruits, e.g. the grape.

Although known from the earliest times, the process of fermentation (from the Latin *fervere* = to boil) was first explained by Pasteur (1822–95), and shown to be a decomposition of the carbohydrate into simpler compounds (finally into alcohols and carbon dioxide) under the influence of catalysts, known as *enzymes*. The word *enzyme* comes from the Greek and means *in yeast*; the earliest enzymes known were those produced by the yeast plant. It is now known that enzymes play an essential part in the life processes of all plants and animals. Enzymes are very unstable organic compounds all of which contain nitrogen and are related chemically to proteins. They are secretion products of living cells; Buchner showed that they were not living as Pasteur at first thought. No enzyme has as yet been prepared artificially. They are specific in their action as are other catalysts, but in contrast with inorganic catalysts, can only exert their influence under certain conditions, e.g. certain degrees of acidity and alkalinity and within definite temperature ranges. Enzyme actions are in general, exothermic.

Ethyl alcohol is manufactured in enormous amounts for use as a beverage and in industry.

Alcoholic beverages

Beer:—Beer is made from barley. The grain is steeped in water, then spread out in layers to dry and kept for a few days at a temperature suitable for germination. During germination, the enzyme *diastase* is formed in the grain. When it is found that a certain amount of diastase has been produced, the temperature is raised to kill the plant and the grain is dried. This dry partly-germinated grain is called *malt*.

The malt is now put into water which is maintained at a temperature between 60° and 65°C. The diastase acts upon the starch of the grain hydrolysing it to the soluble carbohydrate maltose (which is isomeric with sucrose)

$$(C_6H_{10}O_5)_n + \frac{n}{2} H_2O \longrightarrow \frac{n}{2} C_{12}H_{22}O_{11}$$
$$\text{starch} \qquad\qquad\qquad \text{maltose}$$

* *n* is of high value and varies in starches from different sources. Starch and cellulose are examples of naturally-occurring high polymers (p. 24). Carbohydrate-structures are discussed briefly in Chapter 25.

The liquid, called *wort*, now containing maltose, is separated from the insoluble part of the grain and boiled in copper pans with hops*. This gives a bitter taste to the beer and also acts as a preservative.

The liquid is now cooled quickly and transferred to vats maintained at a temperature of about 18°C. Yeast is then added. The yeast plant produces (among others) the two enzymes *maltase* and *zymase*. The maltase hydrolyses the maltose to glucose:

$$C_{12}H_{22}O_{11} + H_2O \xrightarrow{\text{maltase}} 2C_6H_{12}O_6$$
$$\text{maltose} \qquad\qquad\qquad \text{glucose}$$

The zymase converts the glucose into ethyl alcohol and carbon dioxide. Certain side reactions occur also, but the alcoholic fermentation may be represented as follows:

$$C_6H_{12}O_6 \xrightarrow{\text{zymase}} 2C_2H_5OH + 2CO_2$$
$$\text{glucose} \qquad\quad \text{ethyl alcohol}$$

At the temperature of the fermentation, a considerable amount of the carbon dioxide remains in solution. In some breweries, the carbon dioxide evolved is compressed into cylinders for use in the mineral water industry or to make ' dry ice ' for refrigeration. The yeast grows and is another by-product of the industry, being sold for bread making, the manufacture of cattle food, Marmite, etc.

Wines:—Wines are made from grapes which are pressed to extract the juice which contains the isomeric carbohydrates *glucose* and *fructose*. The fruit juice is put into open vats where alcoholic fermentation takes place. There is no need to add yeast, for the ' bloom ' on the grape contains spores of the yeast plant.

Spirits:—By contrast with beer (of alcoholic content about 4 per cent) and wines (of varying alcohol content about 14–20 per cent), spirits contain up to 40 per cent alcohol. The high alcohol content of some wines, e.g. port, is reached by the direct addition of alcohol, and in spirits by distillation.

Whisky and gin are made by the alcoholic fermentation of the starch in the rye and barley grain, by a process similar to the manufacture of beer. The actual fermentation is allowed to continue for a longer period to get more alcohol. The maximum limit of alcohol concentration by yeast fermentation is 15 per cent. The liquid is then distilled to concentrate it. Rum is made by the fermentation of molasses, the carbohydrate residue after the purification of the sucrose extract from the sugar cane.

In most countries, a tax is levied on alcoholic beverages, the amount dependent on the percentage alcohol present. At one time, the test to determine the alcohol content consisted in pouring the

* Hops are the dried flowers of the hop plant.

alcoholic liquor over gunpowder and then applying a light. The alcohol burned away first leaving the water. If the powder was left dry enough to ignite, it was ' proof ' that the spirit did not contain too much water. Such spirit was said to be *proof*. If too damp it was ' underproof '. Now the excise officer (who is responsible for collecting the alcohol tax) determines the amount of alcohol present by measuring the specific gravity.

Industrial spirit

A very large proportion of alcohol for industrial use is made by the fermentation of carbohydrates, which vary from one country to another. In Europe, starch from potatoes is the common starting point; in America, cane sugar molasses and starch from maize are used extensively; in Asia, starch is obtained from rice; in Sweden, alcohol is manufactured from the cellulose in wood.

Potatoes are crushed and stirred with water. The mixture is then forced through sieves to remove the cellulose fibres. The starch is allowed to settle from the suspension passing through the sieves and is then hydrolysed to maltose by the enzyme diastase as above. The diastase is obtained from malt or in some countries from a mould. The alcoholic fermentation of the maltose is brought about by yeast.

The cellulose (of wood, straw, esparto grass) is hydrolysed to glucose by digestion with dilute sulphuric acid and steam under a pressure of six to seven atmospheres. The glucose is then fermented by yeast.

In America, increasing amounts of industrial spirit are being made by chemical methods. At the beginning of last century, Michael Faraday gave a solution of ethylene in concentrated sulphuric acid to the English apothecary Hennell. Hennell diluted it with water and then boiled it, when the ethyl hydrogen sulphate originally present was hydrolysed to ethyl alcohol, which distilled over. This (in 1828) was the first time alcohol had been prepared by a method other than fermentation. Now, over one hundred years later, the method is being used on the commercial scale. Ethylene, from the petroleum cracking plant, is absorbed in concentrated sulphuric acid at 100°C:

$$C_2H_4 + H_2SO_4 \longrightarrow C_2H_5 \cdot O \cdot SO_2 \cdot OH$$

Water is added and the solution heated when alcohol distils over

$$C_2H_5 \cdot O \cdot SO_2 OH + H_2O \longrightarrow C_2H_5OH + H_2SO_4$$

The sulphuric acid left is concentrated by heating.

Rectified spirit

This is the constant boiling mixture of ethyl alcohol and water of composition 95·6 per cent alcohol (by weight) and 4·4 per cent water, which is obtained by fractionation of an aqueous solution of alcohol.   Under permit, it can be obtained duty free for the manufacture of certain pharmaceutical products and also for purposes of instruction in science.

Absolute alcohol may be prepared from it by refluxing for some time with quicklime. The quicklime* should be previously heated to decompose any slaked lime which may be present. The mixture is then distilled. The distillate contains 0·2 per cent water which may be removed by the addition of metallic calcium and subsequent distillation. Care should be taken in this last operation, since ethyl alcohol is strongly hygroscopic.

Methylated spirit:—Most of the duty-free alcohol for use in industry is in the form of methylated spirit which is rendered unfit to drink by the addition of a little methyl alcohol to rectified spirit—a process called *denaturizing*. Methyl alcohol is very poisonous and in small amounts may cause insanity and blindness.

Two kinds of methylated spirit are obtainable, *mineralized spirit* and *surgical spirit*.

Mineralized spirit consists of 90 per cent rectified spirit, 9 per cent methyl alcohol, a little of some substance, e.g. pyridine, to give it an objectionable taste, and a purple dye. It turns milky when mixed with water. It is very difficult, and also illegal, to attempt to obtain pure ethyl alcohol from it.

Surgical spirit consists of 95 per cent rectified spirit and 5 per cent methyl alcohol. It is colourless.

Manufacture of other Alcohols

A brief reference has already been made to the conversion of ethylene to ethyl alcohol and the conversion to glycols is described on p. 77. Higher olefins can be used to synthesize higher alcohols; in one process, carbon monoxide, water vapour and the olefin are passed over a catalyst containing a metal carbonyl (i.e. a metal to which carbon monoxide molecules are bound); thus, propene gives butyl alcohol, or butanol:

$$CH_3 \cdot CH{=}CH_2 \xrightarrow[H_2O]{CO} CH_3 \cdot CH_2 \cdot CH_2 \cdot OH$$

This is known as the ' oxo ' reaction.

* Barium oxide and anhydrous copper(II) sulphate may also be used.

Structural formula of ethyl alcohol

The molecular formula of C_2H_6O is obtained by one or other of the methods described in Chapter 28.

If the action of sodium on ethyl alcohol is investigated quantitatively, it will be found that however much sodium is added, only one-sixth of the hydrogen is displaced. The conclusion from this is that one atom of hydrogen is grouped within the molecule in a manner different from the grouping of the remaining five. This we can write $(C_2H_5O)H$.

When phosphorus pentachloride acts on absolute alcohol it can be shown by analysis of the product (of formula C_2H_5Cl) that one atom of hydrogen and the one atom of oxygen of the alcohol molecule have been replaced by the chlorine atom.

Since ethyl chloride, C_2H_5Cl, can be converted back to ethyl alcohol by the action of alkalis, it is reasonable to assume that ethyl alcohol contains the —OH group. The only possible structure for ethyl alcohol which gives the constituent atoms their normal valencies is thus $CH_3 \cdot CH_2 \cdot OH$, and clearly the hydrogen atom which is displaced by sodium is the one attached to the oxygen atom.

Modern methods of structural analysis, e.g. the method known as Nuclear Magnetic Resonance Spectroscopy, have confirmed this structural formula, and have shown that there are three ' different ' ways in which hydrogen is linked in the molecule, i.e. hydrogens in the CH_3—group, hydrogens attached in the —CH_2—group and one hydrogen in the—OH group.

An entirely different arrangement is possible for a molecule of formula C_2H_6O, viz.:

$$
\begin{array}{ccc}
 & H & & H \\
 & | & & | \\
H- & C & -O- & C & -H \\
 & | & & | \\
 & H & & H \\
\end{array}
$$

This structure represents the compound known as dimethyl ether. It should be clear that such a structure could not account for the observed reactions of ethyl alcohol; in the ether structure, all the hydrogen atoms are equivalent (carbon has tetrahedrally disposed valencies) and the formation of ethyl chloride from the ether is inconceivable.

DIHYDRIC ALCOHOLS

The commonest is *ethylene glycol*, $CH_2OH \cdot CH_2OH$, which is used as a coolant for aircraft engines and for preventing ice formation on the wings of aircraft, and in anti-freeze mixtures. It is a sweet-tasting, colourless, hygroscopic liquid of b.p. 197°C and is soluble in water.

Since the molecule contains two primary alcohol groups, each of which is oxidizable to an aldehyde group and then a carboxyl group (p. 63), a number of intermediate oxidation products* can be obtained before the final oxidation to oxalic acid is reached.

$$\begin{array}{ccc} CH_2OH & \xrightarrow{\text{oxidation}} & COOH \\ | & & | \\ CH_2OH & & COOH \end{array}$$
$$\text{oxalic acid}$$

Ethylene glycol reacts with sodium and potassium, the halides of phosphorus, thionyl chloride, and forms esters as do the monohydric alcohols.

It may be prepared as follows:

(1) By the oxidation of ethylene by potassium permanganate in alkaline (sodium carbonate) solution (p. 19):

$$\begin{array}{ccc} CH_2 & & CH_2OH \\ \| & \longrightarrow & | \\ CH_2 & & CH_2OH \end{array}$$

(2) By the hydrolysis of ethylene dibromide by aqueous alkali:

$$CH_2Br \cdot CH_2Br \xrightarrow{\text{NaOH (aq.)}} CH_2OH \cdot CH_2OH$$

It is manufactured from ethylene by passing the gas along with carbon dioxide through a suspension of bleaching powder in water. The carbon dioxide acts upon the bleaching powder setting free hypochlorous acid which adds on to ethylene forming ethylene chlorohydrin (p. 18):

$$\begin{array}{ccc} CH_2 & OH \\ \| & | \longrightarrow CH_2OH \cdot CH_2Cl \\ CH_2 & Cl \quad \text{ethylene chlorohydrin} \end{array}$$

* Intermediate oxidation products are:

$$\begin{array}{cccc} CHO & CHO & COOH & COOH \\ | & | & | & | \\ CH_2OH & CHO & CH_2OH & CHO \\ \text{glycollic} & \text{glyoxal} & \text{glycollic} & \text{glyoxylic} \\ \text{aldehyde} & & \text{acid} & \text{acid} \end{array}$$

By means of different oxidizing agents, it is possible to prepare an individual product. For example, glyoxal can be obtained by oxidizing ethylene glycol by dilute nitric acid containing some nitrous acid, or by means of air in the presence of heated copper or silver as a catalyst.

This is then hydrolysed to glycol by boiling with milk of lime:

$$CH_2OH \cdot CH_2Cl \xrightarrow{H_2O} CH_2OH \cdot CH_2OH$$

If the chlorohydrin is boiled with aqueous caustic potash, *ethylene oxide* distils over:

$$CH_2OH \cdot CH_2Cl - HCl \xrightarrow{KOH} H_2C\underset{O}{\diagup\!\!\diagdown}CH_2$$

ethylene oxide

Note that this reaction is, in fact, an 'internal' displacement reaction:

Both ethylene oxide and ethylene glycol are valuable in industry. Ethylene glycol is used to make 'Terylene' (p. 252).

Ethylene oxide is also made now by direct oxidation of ethylene, using a silver catalyst; since the ethylene oxide is easily hydrolysed by aqueous acids to ethylene glycol, this constitutes another route to the dihydric alcohol:

$$CH_2{=}CH_2 \xrightarrow{oxidation} CH_2\underset{O}{\diagup\!\!\diagdown}CH_2 \xrightarrow[acid]{aq.} \underset{OH\ \ \ OH}{CH_2-CH_2}$$

Ethylene oxide is used to make acrylonitrile (p. 26) and in certain classes of detergents (p. 154). Ethylene oxide polymers are now becoming important.

POLYHYDRIC ALCOHOLS

A common polyhydric alcohol of commercial importance is the trihydric alcohol, *glycerol*, $CH_2OH \cdot CHOH \cdot CH_2OH$. The oils and fats occurring in nature are the glycerol esters* of aliphatic

* The term ester is the general name given to compounds formed from acids and alcohols, e.g.

$$R \cdot OH + HO \cdot NO_2 \rightleftharpoons R \cdot O \cdot NO_2 + H_2O$$
alcohol   nitric acid   alkyl nitrate

$$C_2H_5 \cdot OH + \underset{O}{\overset{HO}{C}}{-}CH_3 \rightleftharpoons C_2H_5 \cdot O \cdot C \cdot CH_3 + H_2O$$
acetic acid   ethyl acetate

carboxylic acids containing a high number of carbon atoms in their molecules. Glycerol is obtained as a by-product in the manufacture of soap (p. 152), in which natural fats and oils are hydrolysed.

It possesses those properties of alcohols connected with the hydroxyl group. Since in its molecule there are two primary and one secondary alcohol group, it gives a variety of oxidation products, which may contain aldehyde, ketone and carboxylic acid groups.

Glycerol (also commonly called *glycerine*) is used as an antifreeze, as a food preservative, in various medicinal preparations and in the manufacture of the explosive ' nitroglycerin '.

' Nitroglycerin ' is wrongly named, for it is the glycerol ester of nitric acid and should be called *glyceryl trinitrate.* It is made by the action of a mixture of concentrated nitric and sulphuric acids in the cold on glycerol:

$$\begin{array}{ccccc}
CH_2{\cdot}OH & HONO_2 & & CH_2{\cdot}ONO_2 & \\
| & & & | & \\
CH{\cdot}OH & +\ \ HONO_2 & \longrightarrow & CH{\cdot}ONO_2 & +\ 3H_2O \\
| & & & | & \\
CH_2{\cdot}OH & HONO_2 & & CH_2{\cdot}ONO_2 &
\end{array}$$

$$\text{glyceryl trinitrate or ' nitroglycerin '}$$

Dynamite consists of some absorbent material, usually kieselguhr, impregnated with nitroglycerin. Kieselguhr is a finely divided form of silica, consisting of the skeletal remains of tiny sea animals.

SUMMARY OF THE CHEMISTRY OF ALIPHATIC ALCOHOLS

(1) *Definition*

An *alcohol* is a compound containing one or more hydroxy groups attached to hydrocarbon radicals. The number of hydroxy groups in the molecule serves to classify them into:

(a) *Monohydric alcohols*, e.g. C_2H_5OH, ethyl alcohol

(b) *Dihydric alcohols*, e.g. $CH_2OH{\cdot}CH_2OH$, ethylene glycol

(c) *Trihydric alcohols*, e.g. $CH_2OH{\cdot}CHOH{\cdot}CH_2OH$, glycerol, and

(d) *Polyhydric alcohols*, with a greater number of hydroxyl groups.

(2) *Monohydric alcohols, of general formula* $C_nH_{2n+1}OH$

These are classified on the nature of the group carrying the hydroxyl group:

(a) *Primary alcohols:*—These contain the group —CH_2OH.   On oxidation, they give first aldehydes and then carboxylic acids:

primary alcohol   aldehyde   carboxylic acid

(b) *Secondary alcohols:*—These contain the group >CHOH. On oxidation, ketones are formed:—

secondary alcohol   ketone

(c) *Tertiary alcohols:*—These contain the group —$>$COH. They are not affected by mild oxidizing agents; vigorous oxidation breaks up the molecule, giving products containing fewer carbon atoms per molecule.

On *dehydrogenation*, primary alcohols give aldehydes; secondary alcohols give ketones.

(3) General properties of alcohols

(a) Due to hydroxyl group and therefore common to all types of alcohol:

(i) Evolves hydrogen with sodium or potassium:

$$2C_nH_{2n+1}OH + 2Na \longrightarrow 2C_nH_{2n+1}ONa + H_2$$
an alkoxide

(ii) Forms halide with thionyl chloride or the phosphorus halides:

$$C_nH_{2n+1}OH + SOCl_2 \longrightarrow C_nH_{2n+1}Cl + SO_2 \uparrow + HCl \uparrow$$

$$3C_nH_{2n+1}OH + PBr_3 \longrightarrow 3C_nH_{2n+1}Br + P(OH)_3$$

(iii) Forms esters with acids, e.g.:

$$CH_3COOH + C_2H_5OH \rightleftharpoons CH_3COOC_2H_5 + H_2O$$
ethyl acetate

(b) Due to hydroxyl group and group to which attached:

(*i*) On dehydration, gives ether or alkene, e.g.:

$$2C_2H_5OH - H_2O \longrightarrow (C_2H_5)_2O$$
$$\text{diethyl ether}$$

$$C_2H_5OH - H_2O \longrightarrow C_2H_4$$
$$\text{ethylene}$$

(*ii*) Is oxidizable (above):

Primary alcohol $\longrightarrow$ aldehyde $\longrightarrow$ carboxylic acid

Secondary alcohol $\longrightarrow$ ketone

(*iii*) Can be dehydrogenated:
Primary alcohols give an *aldehyde:*

$$-CH_2OH \xrightarrow{-2H} -C\!\!\begin{array}{c} H \\ \diagdown \\ O \end{array}$$

Secondary alcohols give a *ketone:*

$$\text{>CHOH} \xrightarrow{-2H} \text{>C=O}$$

(4) *Manufacture of methyl alcohol,* CH_3OH

From coke: $C + H_2O \longrightarrow \underbrace{CO + H_2}_{\text{water gas}}$

Then: $(CO + H_2) + H_2 \xrightarrow[Cr_2O_3]{ZnO} CH_3OH$

(5) *Manufacture of ethyl alcohol,* C_2H_5OH
   (*a*) Fermentation of starch, sucrose, fructose or glucose.
   (*b*) From ethylene, by absorption in concentrated sulphuric acid:

$$C_2H_4 + H_2SO_4 \longrightarrow C_2H_5 \cdot O \cdot SO_2 \cdot OH$$
$$C_2H_5 \cdot O \cdot SO_2 \cdot OH + H_2O \longrightarrow C_2H_5OH + H_2SO_4$$

EXAMINATION QUESTIONS

(1) A colourless liquid *A* contains 52·18 per cent carbon, 13·06 per cent hydrogen and 34·76 per cent oxygen. It burns readily and reacts with a mixture of phosphorus and iodine to give an organic compound *B*. It also reacts with acetyl chloride, with sodium and with a mixture of iodine and potassium hydroxide.

Identify *A* and *B* and explain the reactions mentioned above. Describe **three** important reactions of *B* which are used in syntheses. (S. A.)

(2) Name the isomers of the alcohol which has the molecular formula $C_4H_{10}O$ and give the structural formula of each isomer. How do these isomers behave (*i*) on oxidation, (*ii*) with phosphorus pentachloride?

By what physical property may one of the isomers be distinguished from the others? (L. A.)

(3) 75 ml of oxygen were mixed with 12·5 ml of a gaseous hydrocarbon X. After exploding and cooling to room temperature, 50 ml of gas were left and when this was shaken with potassium hydroxide solution, 12·5 ml of oxygen remained. Calculate the formula of X. (All volumes were measured at the same temperature and pressure).

Explain the evidence on which the structural formula of ethylene is based. Briefly describe one preparation of ethylene and explain how it could be converted into any two other hydrocarbons. (S. A.)

(4) Describe with full practical details, the usual laboratory preparation of ethyl bromide from ethyl alcohol. State clearly how you would purify the product, giving reasons for each step.

Draw a careful sketch of the apparatus which you would use for the final distillation of the purified product, showing all details necessary to efficiency. (J. M. B. A.)

7

ALIPHATIC ETHERS

Functional Group $>$O

THE aliphatic ethers have the general formula $(C_nH_{2n+1})_2O$. The first member of the series, dimethyl ether, is a gas of formula $(CH_3)_2O$. The best known member is the diethyl derivative $(C_2H_5)_2O$, commonly known as 'ether'. The two alkyl groups in the general formula are not necessarily the same, for 'mixed' ethers, e.g. methyl ethyl ether, $CH_3 \cdot O \cdot C_2H_5$, can be prepared.

Isomerism within the series is common. It may arise not only because of isomerism within the alkyl group as in higher members of series previously studied, but also because of the occurrence of these unsymmetrical compounds. There are, for example, two ethers of formula $CH_3 \cdot O \cdot C_3H_7$, viz.:

$$CH_3 \cdot O \cdot CH_2 \cdot CH_2 \cdot CH_3 \text{ and } CH_3 \cdot O \cdot CH(CH_3)_2$$
methyl n-propyl ether   methyl isopropyl ether

which are isomeric with diethyl ether, $C_2H_5 \cdot O \cdot C_2H_5$. Ethers are also isomeric with the monohydric alcohols, e.g. dimethyl ether, $CH_3 \cdot O \cdot CH_3$, is isomeric with ethyl alcohol, C_2H_5OH. (Note that dimethyl ether, a gas, is much more volatile than ethyl alcohol, b.p. 78°C; alcohols, like water, are *associated liquids*; ethers, lacking the hydroxyl group, are more volatile than the alcohols with which they are isomeric. In addition, ethers are not miscible with water in contrast to the simple alcohols, e.g. methyl, ethyl and propyl alcohols, which are miscible.)

Ether was probably first prepared in 1544 by the German physician Cordus who obtained it by distilling spirits of wine with concentrated sulphuric acid—a method which is similar to the usual laboratory preparation of today.

Preparation of ether

(1) In the laboratory ether is prepared by the dehydration of alcohol with concentrated sulphuric acid (see Chapter 6):

$$2C_2H_5OH \xrightarrow{H_2SO_4} (C_2H_5)_2O$$

About 50 ml of rectified spirit are added in small amounts, shaking at intervals, to an equal volume of concentrated sulphuric acid. The mixture is then transferred to a 500 ml flask containing a little clean silver sand. The flask is fitted with a tap funnel containing rectified spirit, a thermometer dipping into the liquid, and is connected to a condenser as shown (*Figure 7.1*). The temperature is raised to 140°C and maintained there during the preparation. At this temperature, ether starts distilling over. Rectified spirit is then run in from the tap funnel at approximately the same rate as the ether distils over.

Theoretically, the preparation may be continued indefinitely and it is, in fact, called the *continuous etherification process*. A stage

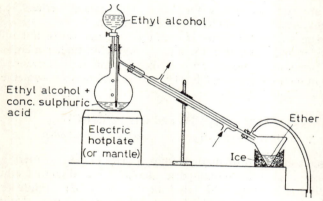

Figure 7.1. Preparation of ether.

is reached, however, when the sulphuric acid is diluted to such an extent that it becomes ineffective.

Ether is a dangerous liquid for it is very volatile (b.p. 34·6°C) and its vapour is very inflammable. Precautions to prevent contact between the ether vapour and any flame *must* be taken. A rapid stream of water must flow through the condenser to ensure good condensation. The receiver, corked to the condenser, is surrounded by ice or ice-water. A rubber tube is connected to the side tube of the receiver and the open end put down the sink into which a stream of water is flowing to carry away from the bench any uncondensed ether vapour.

In addition to ether, the distillate contains ethyl alcohol, water, carbon dioxide and sulphur dioxide. It is transferred to a separating funnel and shaken up with an aqueous solution of caustic soda saturated with common salt. The caustic soda absorbs the acid gases. Ether, although immiscible with water, does dissolve in water to some extent, but it is much less soluble in brine. The brine thus prevents loss of ether.

The lower aqueous layer is run off, and the ether is later transferred to a flask. A little fused calcium chloride is added and the mixture allowed to stand for a few days. The calcium chloride not only removes any water but also combines with the ethyl alcohol which may be present (p. 60).

The ether residue is finally purified by redistillation, collecting the fraction coming over within a few degrees round the boiling point. The distillation must be carried out over a water bath (ether should *never* be distilled over a naked flame) and the precautions mentioned above against fire must be taken.

(2) By Williamson's synthesis (see description of displacement reactions in Chapter 5).

Sodium ethoxide is first prepared by adding sodium to absolute alcohol:

$$2C_2H_5OH + 2Na \longrightarrow 2C_2H_5\overset{-\;+}{O}Na + H_2$$

Some considerable heat is developed during the reaction, and pieces of sodium should be added to the alcohol in a flask fitted with a reflux condenser.

Ethyl iodide is run in and the mixture heated over a water bath for some time:

$$C_2H_5\overset{-\;+}{O}Na + IC_2H_5 \longrightarrow C_2H_5{\cdot}O{\cdot}C_2H_5 + NaI$$

Being insoluble in ether and alcohol, the sodium iodide is precipitated.

The ether is recovered from the mixture by distillation and purification as in method (1).

(3) By the action of *dry* silver oxide on an alkyl iodide (cf. action of moist silver oxide p. 49):

$$2C_2H_5I + Ag_2O \longrightarrow 2AgI + (C_2H_5)_2O$$

Method (2) is the most general method of preparation since it can be used also for the preparation of an unsymmetrical ether. In this case an iodide of an alkyl group other than that from which the alkoxide was made must be used, thus:

$$CH_3\overset{-\;+}{O}Na + IC_2H_5 \longrightarrow CH_3{\cdot}O{\cdot}C_2H_5 + NaI$$

Properties of ether

Ether is a colourless, volatile, dangerously inflammable liquid of b.p. 34·6°C. Its anaesthetic properties, both local and general, were first discovered by Faraday in 1818 and first used in surgery in 1842 by the American Dr. Long. It was also used again in America in 1846 as an anaesthetic in dentistry by Morton who had no knowledge of its previous use.

Ether, and ethers in general, are relatively unreactive compounds. When ether burns, carbon dioxide and water are formed:

$$(C_2H_5)_2O + 6O_2 \longrightarrow 4CO_2 + 5H_2O$$

It does not react with metallic sodium nor, in the cold, with phosphorus pentachloride. This indicates that no hydroxyl group is present in the molecule. Although ether does not form addition compounds with calcium chloride as do water and alcohols, it can form addition compounds, called *etherates*, with some substances; a co-ordinate link is formed (p. xxii). Boron trifluoride, BF_3, forms the stable etherate, $(C_2H_5)_2O \cdot BF_3$.

Uses of ether

Ether is a very good solvent for fats and resins. *It is frequently used in preparative chemistry to extract a product from its reaction mixture,* and is also used as a solvent in certain important reactions (e.g. the preparation of Grignard reagents, see p. 51).

Structural formula of diethyl ether

The molecular formula of $C_4H_{10}O$ is found by one or other of the methods described in Chapter 28.

By comparison with the action of sodium on compounds, containing a hydroxyl group, it is concluded that since sodium has no action on ether, there is no such group in its molecule. Confirmation of this is given by the behaviour of ether and phosphorus pentachloride (see above).

Since no hydroxyl group is present and since the valencies of oxygen and hydrogen are 2 and 1 respectively, the oxygen atom can only be situated between the two carbon atoms, thus:

$$
\begin{array}{ccc}
| & & | \\
-C & -O- & C- \\
| & & |
\end{array}
$$

The remainder of the molecule, two atoms of carbon and ten of hydrogen, may be attached to this group in three ways:

$$(1) \quad H-\underset{\underset{H}{|}}{\overset{\overset{H}{|}}{C}}-\underset{\underset{H}{|}}{\overset{\overset{H}{|}}{C}}-O-\underset{\underset{H}{|}}{\overset{\overset{H}{|}}{C}}-\underset{\underset{H}{|}}{\overset{\overset{H}{|}}{C}}-H$$

$$(2) \quad H-\underset{\underset{H}{|}}{\overset{\overset{H}{|}}{C}}-O-\underset{\underset{H}{|}}{\overset{\overset{H}{|}}{C}}-\underset{\underset{H}{|}}{\overset{\overset{H}{|}}{C}}-\underset{\underset{H}{|}}{\overset{\overset{H}{|}}{C}}-H$$

$$(3) \quad H-\underset{\underset{H}{|}}{\overset{\overset{H}{|}}{C}}-O-\underset{\underset{\overset{|}{C}}{|}}{\overset{\overset{H}{|}}{C}}-\underset{\underset{H}{|}}{\overset{\overset{H}{|}}{C}}-H$$

Williamson's synthesis of ether (1850) determines which of the above formulae is a correct representation of the ether molecule. It can be assumed that since sodium reacts with ethyl alcohol at room temperature, no intramolecular rearrangement of atoms takes place and the formula of the sodium ethoxide may be written as:

$$H-\underset{\underset{H}{|}}{\overset{\overset{H}{|}}{C}}-\underset{\underset{H}{|}}{\overset{\overset{H}{|}}{C}}-\overset{-}{O}\ \overset{+}{Na} \quad \text{(analogous to } \overset{+}{N}a\overset{-}{O}H)$$

The formula of ethyl iodide can only be:

$$H-\underset{\underset{H}{|}}{\overset{\overset{H}{|}}{C}}-\underset{\underset{H}{|}}{\overset{\overset{H}{|}}{C}}-I$$

Since, in the interaction of these, sodium iodide is formed, the monovalent ethyl group from the ethyl iodide must become attached to the oxygen atom of the alkoxide, to give a molecule which can be written:

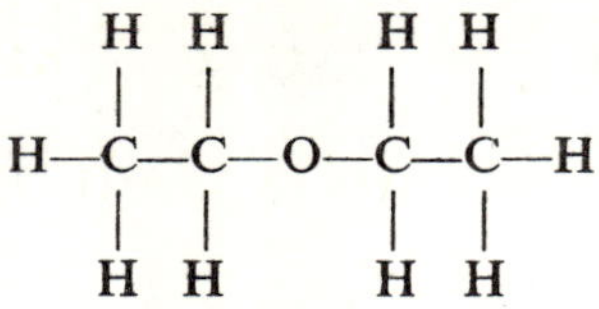

$$H-\underset{\underset{H}{|}}{\overset{\overset{H}{|}}{C}}-\underset{\underset{H}{|}}{\overset{\overset{H}{|}}{C}}-O-\underset{\underset{H}{|}}{\overset{\overset{H}{|}}{C}}-\underset{\underset{H}{|}}{\overset{\overset{H}{|}}{C}}-H$$

Williamson's synthesis did in fact establish the constitutional formula of ether for the first time.

SUMMARY OF THE CHEMISTRY OF ALIPHATIC ETHERS

The general formula is $(C_nH_{2n+1})_2O$.

The first member is $(CH_3)_2O$, dimethyl ether.

The representative member is $(C_2H_5)_2O$, diethyl ether or ' ether '.

Preparation and properties

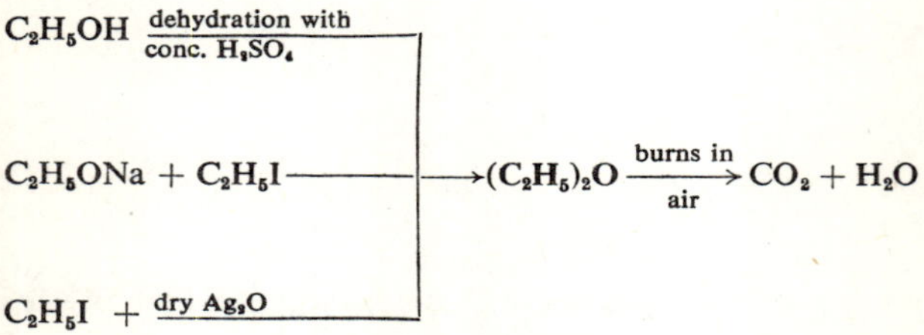

C_2H_5OH $\xrightarrow[\text{conc. } H_2SO_4]{\text{dehydration with}}$

$C_2H_5ONa + C_2H_5I \longrightarrow \quad \longrightarrow (C_2H_5)_2O \xrightarrow[\text{air}]{\text{burns in}} CO_2 + H_2O$

$C_2H_5I + \xrightarrow{\text{dry } Ag_2O}$

EXAMINATION QUESTIONS

(1) Outline **two** *general* methods for the preparation of ethers.

Name **three** properties of diethyl ether which make it suitable for use as an extraction solvent. Give reasons.

An organic compound gives on analysis the following figures:

$C = 60.0\%$; $H = 13.3\%$; $O = 26.7\%$; vapour density $= 30$.

Deduce the various possible structural formulae for this compound and suggest (in outline) observations and experiments by which you could determine which formula was correct. (J. M. B. A.)

(2) Outline a suitable method for the preparation of diethyl ether directly from ethyl alcohol. (Practical details are not required). What are the likely impurities in the crude product and how do they arise?

Describe ONE general synthetic method which is applicable to the preparation of ethers containing two *different* alkyl groups.

20 c.c. of a gaseous ether was mixed with 70 c.c. of oxygen and exploded in a eudiometer. After cooling to room temperature the residual gases occupied 50 c.c. of which 40 c.c. was absorbed by potassium hydroxide solution and the remainder completely absorbed by alkaline pyrogallol solution. All volumes were measured under identical conditions. Deduce the structural formula of the ether. (W. A.)

8

ALIPHATIC ALDEHYDES

Functional Group $-C\begin{smallmatrix}\diagup H \\ \diagdown O\end{smallmatrix}$

THE aliphatic aldehydes are compounds of the general formula $C_nH_{2n+1}CHO$ with the characteristic group having the structure of $-C\begin{smallmatrix}\diagup H \\ \diagdown O\end{smallmatrix}$.

They are closely related to *ketones* of general formula

$C_nH_{2n+1}\diagdown$
$\qquad\qquad CO$ and characteristic group $\diagdown C{=}O$.
$C_mH_{2m+1}\diagup$

Both series of compounds contain the group $\diagdown C = O$, known as the *carbonyl* group. Aldehydes and ketones would therefore be expected to possess certain properties in common, viz. the properties of the carbonyl group, and this is found to be so. It would be expected, too, that the important differences between aldehydes and ketones would be due to the former having a hydrogen atom attached to the carbonyl group, possibly influencing the group in some way. Again this is found to be the case.

Both classes of compounds are very reactive, aldehydes being the more active of the two.

The aldehyde series is the first one to be studied, in which an organic compound is obtained when $n = 0$ is substituted in the general formula, thus: $H-C\begin{smallmatrix}\diagup H \\ \diagdown O\end{smallmatrix}$. This is *formaldehyde*. If in both the alcohol and the ether series, $n = 0$ is substituted, the formula obtained is that of water. Formaldehyde, in which *two* hydrogen atoms are attached to the carbonyl group, does show some differences to other aldehydes.

In studying this series a detailed account is given of the second member, acetaldehyde, and from this, the properties of members where $n > 1$ can be deduced. A brief study is then made of formaldehyde, in particular referring to its peculiarities.

The commonly used nomenclature of the series is based on that of the carboxylic acid containing the same number of carbon atoms. Thus acetic acid, $CH_3 \cdot COOH$, is the carboxylic acid containing the same number of carbon atoms as acetaldehyde, $CH_3 \cdot CHO$, and formic acid, $HCOOH$, is similarly compared with formaldehyde, $HCHO$. The name of the aldehyde is obtained by substituting the word *-aldehyde* for *-ic* in the corresponding *acid*. This is general for other members, e.g.:

Formaldehyde	HCHO	gas (b.p. $- 21°C$)
Acetaldehyde	$CH_3 \cdot CHO$	} volatile liquids
Propionaldehyde	$C_2H_5 \cdot CHO$	
Butyraldehyde	$C_3H_7 \cdot CHO$	(isomerism starts here)

Acetaldehyde, $CH_3 \cdot CHO$

Although the existence of acetaldehyde was recognized by Scheele in 1774 as an oxidation product of ethyl alcohol, Liebig was the first to prepare and name it in 1835.

Preparation of acetaldehyde

Acetaldehyde may be obtained by the following reactions:
(1) By the oxidation of ethyl alcohol.
   (*a*) By air.
A mixture of air and the vapour of ethyl alcohol is passed over platinum, copper or silver, heated to temperatures varying between 300° and 600°C. The metals act as catalysts:

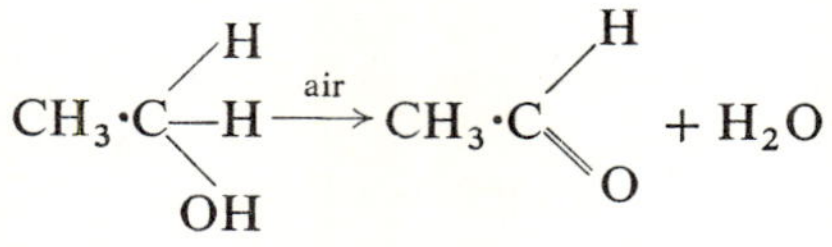

$$CH_3 \cdot C\underset{OH}{\overset{H}{-}}H \xrightarrow{\text{air}} CH_3 \cdot C\overset{H}{\underset{O}{\diagdown}} + H_2O$$

(*b*) By a mixture of sodium dichromate* and sulphuric acid.
Since the volatile acetaldehyde distils off (see details on p. 91), its oxidation to acetic acid is prevented to some extent.

(2) By the dehydrogenation of ethyl alcohol.
The vapour of ethyl alcohol is passed over copper heated to about 300°C or over zinc oxide heated to a slightly higher temperature:

$$CH_3 \cdot CH_2OH - 2H \xrightarrow[300°C]{Cu} CH_3 \cdot CHO$$

* This is preferred to potassium dichromate since it is more soluble in water.

There is a distinct advantage here over method (1*b*) for further dehydrogenation is not possible. In the first method, there is a reduced yield owing to some further oxidation to acetic acid.

(3) By the catalytic hydration of acetylene.

Acetylene is first absorbed in concentrated sulphuric acid containing a little mercury(II) sulphate which acts as a catalyst. The mixture is then diluted with water and distilled, when acetaldehyde comes over. The reaction may be represented as the addition of the elements of water to the triple bond:

$$C_2H_2 \xrightarrow[\text{HgSO}_4]{\text{H}_2\text{SO}_4} \left(CH_2{=}C\diagup{\overset{\text{H}}{}}_{\diagdown\text{OH}} \right) \longrightarrow CH_3{\cdot}C\diagup{\overset{\text{H}}{}}_{\diagdown\text{O}}$$

(The reaction probably involves an intermediate with a carbon-mercury bond, which is subsequently hydrolysed.)

Laboratory preparation (*Method* 1*b*):—The usual laboratory method is the oxidation of ethyl alcohol by sodium dichromate.

An aqueous solution of 60 g of sodium dichromate is brought up to boiling point in a 1-litre flask provided with a tap funnel and connected to a Liebig condenser as shown in *Figure 8.1*. The

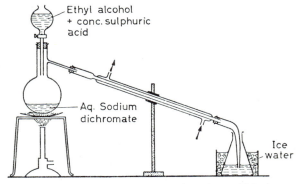

Figure 8.1. Laboratory preparation of acetaldehyde.

bunsen is now turned out and a mixture of 37 ml concentrated sulphuric acid and 60 ml rectified spirit (previously made by adding the alcohol to the acid slowly, with constant shaking) dropped in from the tap funnel. When the first vigorous action has moderated, the mixture is warmed. Heating is continued until about 60 ml of distillate have collected in the receiver, cooled by ice water.

Acetaldehyde, as we shall see later (p. 94), is a reducing agent, being readily oxidized to acetic acid. An attempt to prevent this

oxidation in the above preparation is made by using excess of alcohol
and thereby providing plenty of oxidizable material for the oxidizing
mixture. Further, since the first oxidation product of the alcohol,
viz. the acetaldehyde, is very volatile (b.p. 21°C), this tends to distil
over from the mixture before further oxidation occurs.

The distillate contains acetaldehyde, unchanged alcohol and
water. It is purified as follows:

The distillate is heated on a water bath, in a flask connected to
a reflux condenser through which water at about 35–40°C is flowing
from the upper pneumatic trough. To the slightly cooler water

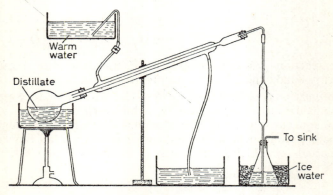

Figure 8.2. Purification of acetaldehyde.

collecting in the lower trough, a little boiling water is added and
the water then transferred to the upper trough as needed.

Any alcohol and water distilling from the mixture are condensed
and flow back into the flask. The acetaldehyde passes through the
condenser into the receiving flask containing ether, kept cool by
ice. For safety reasons, the outlet tube of this receiver is led into
the sink (cf. preparation of ether, p. 84).

The pipette is now disconnected from the condenser and connected
to an apparatus to generate dry ammonia. (The gas may be obtained
by warming ' 0·880 ammonia ' and drying it by quicklime.) The
ammonia adds on to the acetaldehyde (p. 96) forming colourless
crystals of acetaldehyde ammonia, which separate out.

$$CH_3 \cdot C \overset{H}{\underset{O}{\diagup}} \atop \overset{+}{H_2N-H} \longrightarrow CH_3 \cdot C \overset{H}{\underset{NH_2}{\diagdown}} OH$$

acetaldehyde ammonia

The crystals are filtered off and distilled with dilute sulphuric acid from a water bath (the temperature of which should be kept below 35°C) when the free acetaldehyde comes off, being collected in a receiver cooled in ice.

The yield in this preparation is well below the theoretical one, primarily because of loss of acetaldehyde during a long preparation and purification, on account of its high volatility.

Manufacture of acetaldehyde

(1) By the catalytic hydration of acetylene (p. 91):

$$C_2H_2 + H_2O \xrightarrow[H_2SO_4]{HgSO_4} CH_3 \cdot C \overset{H}{\underset{O}{\Big\langle}}$$

(2) By the oxidation of ethyl alcohol by air, using silver as a catalyst:

$$C_2H_5OH \xrightarrow[300-600°C]{\substack{air, \\ Ag}} CH_3 \cdot C \overset{H}{\underset{O}{\Big\langle}} + H_2O$$

(3) By the dehydrogenation of alcohol by copper:

$$CH_3 \cdot CH_2OH - 2H \xrightarrow[300°C]{Cu} CH_3 \cdot CHO$$

Properties of acetaldehyde

Acetaldehyde is a colourless volatile liquid (b.p. 21°C) possessing a characteristic pungent smell. It burns in air to form carbon dioxide and water. It mixes readily with water, alcohol, and ether.

The relatively long list of properties given below testifies to the great chemical activity of acetaldehyde. Most of these properties are general for aldehydes since they are properties of the $-C\overset{H}{\underset{O}{\Big\langle}}$ group (formaldehyde is studied separately). Some properties can be attributed to the single hydrogen atom attached to the carbonyl group, e.g. the reducing character. Others are due to the presence of the carbonyl group containing the doubly-bound oxygen atom. Since ketones also contain a carbonyl group, they possess many properties in common with aldehydes (Chapter 9).

(1) Acetaldehyde is a reducing agent, being itself oxidized to acetic acid:

$$CH_3 \cdot C{\overset{O}{\underset{H}{\lessgtr}}} + [O] \longrightarrow CH_3 \cdot C{\overset{O}{\underset{OH}{\lessgtr}}}$$

(*a*) It reduces *ammoniacal silver oxide*. A little dilute caustic soda is added to aqueous silver nitrate when a brown precipitate of silver oxide is formed. Ammonia is now added carefully until this precipitate is just dissolved. Acetaldehyde is then added. Silver is deposited either as a black precipitate or as a mirror on the walls of the containing vessel. The reaction may be represented as:

$$CH_3 \cdot C{\overset{O}{\underset{H}{\lessgtr}}} + Ag_2O \longrightarrow CH_3 \cdot COOH + 2Ag\downarrow$$

The acetic acid forms acetates with the alkalis present.

(In ammoniacal solution, the silver is present as a complex ammino ion, and the above equation is more correctly written thus:

$$CH_3 \cdot CHO + 2[Ag(NH_3)_2]^+ + OH^- \longrightarrow CH_3COO^- + 2NH_4^+ \\ + 2Ag\downarrow + 2NH_3)$$

(*b*) It reduces *Fehling's solution*, which consists of copper(II) sulphate and Rochelle salt (sodium potassium tartrate), to which caustic soda is added in an amount sufficient to make a deep blue solution. On adding acetaldehyde and warming, there is an orange-red precipitate of (hydrated) copper(I) oxide, the copper being reduced from the cupric to the cuprous state. The reaction may be represented as:

$$CH_3 \cdot C{\overset{O}{\underset{H}{\lessgtr}}} + 2CuO \longrightarrow CH_3 \cdot C{\overset{O}{\underset{OH}{\lessgtr}}} + Cu_2O\downarrow \\ \text{orange-red}$$

$$\downarrow NaOH$$

$$CH_3 \cdot CO\overset{+}{O}\overset{-}{N}a$$

(The copper(II) forms a tartrato–complex with the Rochelle salt, and it is this complex which is reduced.)

(*c*) It reduces aqueous *acidified dichromate*, giving green chromium(III) sulphate.

(*d*) It reduces aqueous *potassium permanganate*, acidified with dilute sulphuric acid, decolorizing it and forming manganese(II) ions.

$$2MnO_4^- + 5CH_3 \cdot CHO + 6H^+ \longrightarrow 2Mn^{2+} + 5CH_3 \cdot COOH$$

(purple)   (colourless) $+ 3H_2O$

(2) Acetaldehyde is reduced to ethyl alcohol:

$$CH_3 \cdot C{\overset{\displaystyle O}{\underset{\displaystyle H}{\big<}}} \longrightarrow CH_3 \cdot C{\overset{\displaystyle OH}{\underset{\displaystyle H}{\big<}}}H$$

The reduction can be brought about by catalytic hydrogenation (gaseous hydrogen in the presence of finely-divided metal catalysts, p. 16). In this respect the double bond between carbon and oxygen in the carbonyl group resembles the carbon-carbon double bond in ethylene; both are unsaturated and both add hydrogen. However, it is important to note that, *in general*, the reagents which undergo addition reactions with alkenes do not give addition products with carbonyl compounds (aldehydes and ketones); conversely, with the notable exception of hydrogen, those substances which form addition products with carbonyl compounds do not normally add to ethylene and similar compounds.

As an illustration of this principle, the carbonyl group can be reduced by a number of methods which are not applicable to the reduction of alkenes. Particularly important is reduction by complex metal hydrides like lithium aluminium hydride, $LiAlH_4$, and sodium borohydride, $NaBH_4$:

$$CH_3 \cdot C{\overset{\displaystyle O}{\underset{\displaystyle H}{\big<}}} + LiAlH_4 \longrightarrow CH_3 \cdot C{\overset{\displaystyle OAlH_3Li}{\underset{\displaystyle H}{\big<}}}H \xrightarrow{H_2O} CH_3 \cdot CH_2OH$$

These reducing agents are convenient in use and of general application, giving high yields of alcohols. Lithium aluminium hydride is decomposed by water; since it is soluble in ether it is commonly used in this solvent. Sodium borohydride is rather less reactive and can be used, for example, in solution in aqueous methanol; in the general case no serious problems arise in separating the alcohol produced by reduction from the methanol present as solvent.

Aluminium isopropoxide, $[(CH_3)_2CH \cdot O]_3Al$, in isopropyl alcohol is another useful, highly specific reagent for the reduction of aldehydes and ketones. This reagent, and the complex metal hydrides, are thought to act as hydride (H^-) transfer agents, but the mechanisms are beyond the scope of this book.

(3) Addition reactions other than reductions:

Aldehydes form typical *addition products* with hydrogen cyanide, sodium hydrogen sulphite (sodium bisulphite) and ammonia:

$$CH_3 \cdot C \overset{O}{\underset{H}{\big<}} + HCN \longrightarrow CH_3 \cdot C \overset{OH}{\underset{H}{\big<}} CN \quad \text{acetaldehyde cyanhydrin}$$

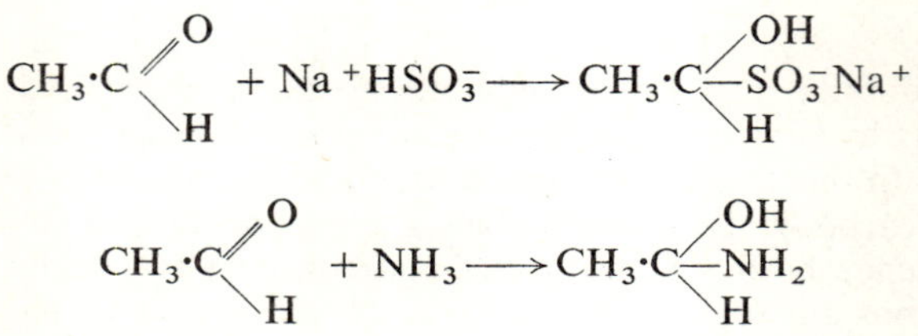

$$CH_3 \cdot C \overset{O}{\underset{H}{\big<}} + Na^+ HSO_3^- \longrightarrow CH_3 \cdot C \overset{OH}{\underset{H}{\big<}} SO_3^- Na^+$$

$$CH_3 \cdot C \overset{O}{\underset{H}{\big<}} + NH_3 \longrightarrow CH_3 \cdot C \overset{OH}{\underset{H}{\big<}} NH_2$$

The bisulphite addition product, for example, is prepared by mixing acetaldehyde with an aqueous solution of sodium bisulphite, which should be freshly prepared (by passing sulphur dioxide through a saturated solution of sodium carbonate). The addition product separates as a white crystalline solid. Since the aldehyde can easily be regenerated by heating the addition product with dilute sulphuric acid, formation of these products is sometimes used as a means of separating aldehydes from other compounds (which do not react with sodium bisulphite). Similarly, formation of aldehyde-ammonia compounds can be used in the isolation (i.e. purification) of aldehydes (p. 92).

The addition products with hydrogen cyanide have a different use; they are important as intermediates in the synthesis of compounds called *hydroxy-acids*:

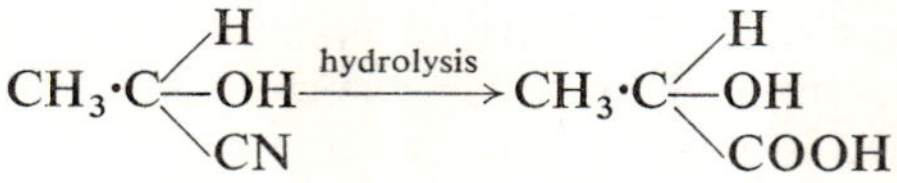

$$CH_3 \cdot C \overset{H}{\underset{CN}{\big<}} OH \xrightarrow{\text{hydrolysis}} CH_3 \cdot C \overset{H}{\underset{COOH}{\big<}} OH$$

The addition of hydrogen cyanide is catalysed by cyanide ions and the process is a stepwise one:

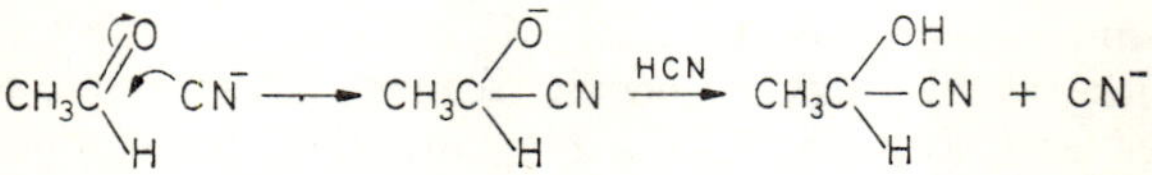

$$CH_3C \overset{O}{\underset{H}{\big<}} CN^- \longrightarrow CH_3C \overset{\bar{O}}{\underset{H}{\big<}} CN \xrightarrow{HCN} CH_3C \overset{OH}{\underset{H}{\big<}} CN + CN^-$$

In the second stage, the negatively charged oxygen atom abstracts a proton, either from hydrogen cyanide or from the solvent. Similarly, formation of the bisulphite addition compound involves attack on the carbon of the carbonyl group by the negatively-charged bisulphite ion:

$$CH_3C \overset{O}{\underset{H}{}} \quad HSO_3^- \longrightarrow CH_3C \overset{O^-}{\underset{H\ (Na^+)}{-SO_3H}} \longrightarrow CH_3C \overset{OH}{\underset{H\ (Na^+)}{-SO_3^-}}$$
$$(Na^+)$$

The intermediate product contains the strongly acidic sulphonic acid group—SO_3H and proton transfer occurs so that again the negatively-charged oxygen atom becomes a hydroxyl group.

The addition of ammonia can be formulated in a similar way:

$$CH_3C \overset{O}{\underset{H}{}} \quad :NH_3 \longrightarrow CH_3C \overset{O^-}{\underset{H}{-\overset{+}{N}H_3}} \longrightarrow CH_3C \overset{OH}{\underset{H}{-NH_2}}$$

but in this case the initial attack on the carbon atom of the carbonyl group is by an uncharged molecule, ammonia, not by an anion. However, the ammonia molecule, the bisulphite anion and the cyanide anion have one feature in common, *the possession of at least one pair of unshared electrons*, e.g. :NH_3, :C̄≡N; and in each addition reaction these unshared pairs are used to form a bond to carbon. Reagents like these, having an electron pair which can be shared with another nucleus, forming a new bond, are referred to as *nucleophilic reagents* (nucleophilic = nucleus-loving) or *nucleophiles.**

It has been pointed out above that alkenes and carbonyl compounds do not generally undergo the same addition reactions. In Chapter 2, 'ionic' addition reactions of alkenes were represented as stepwise processes in which the first step was the sharing of an electron pair from the ethylenic double bond with the reagent, e.g.

$$\underset{H}{\overset{H}{}}C = C\underset{H}{\overset{H}{}} \longrightarrow \underset{H}{\overset{H}{}}C^+ - CH_2 \\ Br—Br \qquad\qquad Br \quad Br^-$$

Here, the carbon-carbon double bond is acting as the source of an electron pair and the reagent (bromine in this case) is accepting a

---

* A substance with a molecule or ion having an unshared pair of electrons which can be donated to another molecule or ion is sometimes called a *Lewis base*, and the molecule or ion which accepts the electron pair is then a *Lewis acid*. These definitions, by G. N. Lewis, are used to classify reagents in inorganic chemistry.

share in those electrons. Thus, the ethylene double bond can be regarded as *nucleophilic* in character and it becomes less surprising that alkenes *do not* react with the nucleophilic reagents which attack carbonyl groups (in other words, nucleophiles do not react with nucleophiles).

Reagents like bromine in the above example, which accept a share in an electron pair, are called *electrophilic reagents* or *electrophiles*; the derivation of the name should be clear. Note that, in the second step of the ethylene–bromine addition reaction,

$$\begin{array}{c}
\overset{H}{\underset{H}{\text{C}}} \overset{+}{-} \text{CH}_2 \quad \overset{\curvearrowleft \text{Br}^-}{\underset{\text{Br}}{\big|}} \longrightarrow \quad \text{H}_2\text{C} \overset{\text{Br}}{\underset{\text{Br}}{\big|}} \text{CH}_2
\end{array}$$

the *bromide anion* is, according to our definition, acting as a nucleophile (providing an electron pair to form a bond); the positively-charged carbonium ion (with only a sextet of valency electrons) accepts a share in the electron pair and is clearly acting as an electrophile. 'Ionic' organic reactions are, in fact, always reactions between an electrophile and a nucleophile.

In the ethylene double bond, the electrons are shared equally between the two carbon atoms, but the system is easily *polarizable*; in the presence of a suitably reactive electrophile the double bond can provide an electron pair for bond formation. The reactions between the aldehyde carbonyl group and nucleophilic reagents (e.g. cyanide ion and ammonia) imply that the carbonyl group must have an electrophilic character, or more correctly, that the carbon atom of the carbonyl group must be an electrophilic centre. The carbon-oxygen double bond of the carbonyl group is clearly *unsymmetrical* and is already *polarized* so that the oxygen atom has the greater share in the bonding electrons. We can attempt to represent this situation by $\overset{\delta+}{\underset{}{}}\text{C}\overset{\delta-}{=}\text{O}$ or we can regard the true state of a carbonyl group as being intermediate between $\text{C}=\text{O}$ (with equal sharing) and $\overset{+}{\text{C}}-\overset{-}{\text{O}}$. A useful convention is:

$$\text{C}=\text{O} \longleftrightarrow \overset{+}{\text{C}}-\overset{-}{\text{O}}$$

where the double-headed arrow is taken to mean an 'in-between' state (fuller discussion of this symbolism is deferred to Chapter 26).

(4) Reaction with Grignard reagents:

Reactions similar to those discussed under (3) occur between aldehydes (or ketones) and Grignard reagents (p. 51):

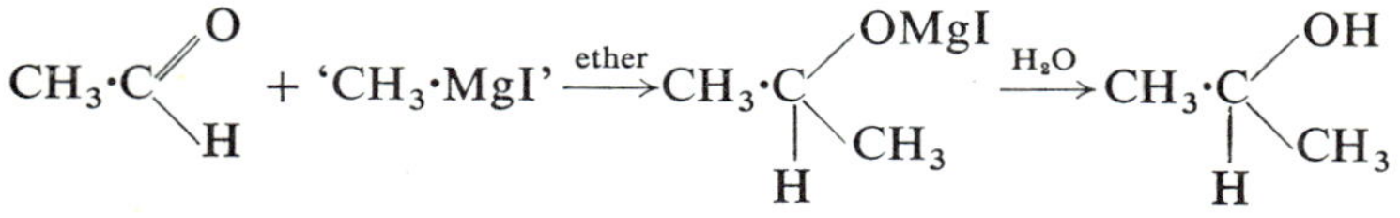

$$CH_3 \cdot C \overset{O}{\underset{H}{\diagdown}} + \text{`}CH_3 \cdot MgI\text{'} \xrightarrow{\text{ether}} CH_3 \cdot C \underset{\underset{H}{|}}{\overset{OMgI}{\diagup}} CH_3 \xrightarrow{H_2O} CH_3 \cdot C \underset{\underset{H}{|}}{\overset{OH}{\diagup}} CH_3$$

The addition products decompose on treatment with water to give alcohols. Note that by using formaldehyde in place of acetaldehyde, a primary alcohol is produced:

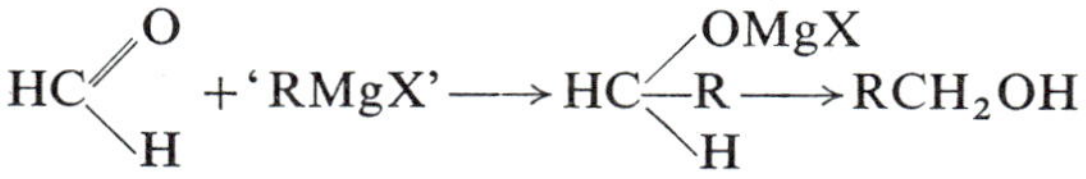

$$HC \overset{O}{\underset{H}{\diagdown}} + \text{`}RMgX\text{'} \longrightarrow HC \underset{\underset{H}{|}}{\overset{OMgX}{\diagup}} R \longrightarrow RCH_2OH$$

These reactions provide very widely applicable methods of synthesizing alcohols (see Chapter 16).

(5) Addition–Elimination reactions:

Acetaldehyde is often said to undergo ' condensation ' reactions or to give ' condensation products ', condensation here meaning combination of two molecules with elimination of water. These reactions are, in fact, addition reactions of essentially the same type as those discussed above, except that the addition product subsequently loses water, giving the final product.

With *hydroxylamine*, NH_2OH, acetaldehyde forms an *aldoxime*:

$$CH_3 \cdot \overset{\overset{\displaystyle H}{|}}{C}{=}O + H_2NOH \longrightarrow CH_3 \cdot \overset{\overset{\displaystyle H}{|}}{C}{=}NOH + H_2O$$
$$\text{acetaldehyde oxime}$$

With *hydrazine*, $NH_2 \cdot NH_2$, acetaldehyde forms a *hydrazone*:

$$CH_3 \overset{\overset{\displaystyle H}{|}}{C}{=}O + H_2N \cdot NH_2 \longrightarrow CH_3 \cdot \overset{\overset{\displaystyle H}{|}}{C}{=}N \cdot NH_2 + H_2O$$
$$\text{acetaldehyde hydrazone}$$

Similar products are formed with hydrazine derivatives, e.g. *phenyl-hydrazine*, $C_6H_5 \cdot NH \cdot NH_2$ (the phenyl group, C_6H_5, is derived from benzene, C_6H_6, p. 202), *2,4-dinitrophenylhydrazine* (Brady's reagent) and *semicarbazide*, $NH_2 \cdot NH \cdot CO \cdot NH_2$, e.g.

$$CH_3 \cdot C \overset{H}{\underset{O}{<}} + NH_2 \cdot NH \cdot C_6H_5 \longrightarrow CH_3 \cdot CH = N \cdot NH \cdot C_6H_5$$
acetaldehyde phenylhydrazone

$$CH_3 \cdot C \overset{H}{\underset{O}{<}} + NH_2 \cdot NH \cdot CO \cdot NH_2 \longrightarrow CH_3 \cdot CH = N \cdot NH \cdot CO \cdot NH_2$$
acetaldehyde semicarbazone

With hydroxylamine as an example, the reaction can be formulated as:

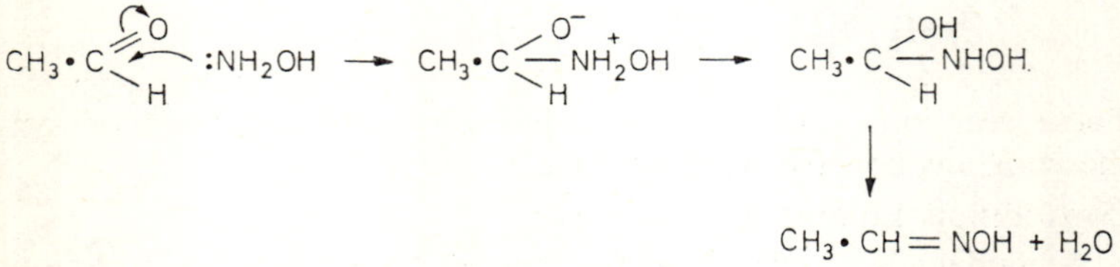

$$CH_3 \cdot C \overset{O}{\underset{H}{<}} :NH_2OH \rightarrow CH_3 \cdot C \overset{O^-}{\underset{H}{<}} \overset{+}{N}H_2OH \rightarrow CH_3 \cdot C \overset{OH}{\underset{H}{<}} NHOH$$

$$\downarrow$$

$$CH_3 \cdot CH = NOH + H_2O$$
oxime

Hydroxylamine, hydrazine and the substituted hydrazines act as *nucleophiles*, as discussed above.

The products formed (oximes, hydrazones, etc.) are of value as *crystalline derivatives* of aldehydes (and ketones). The simpler carbonyl compounds are liquids and it is useful to be able to identify them by the preparation of crystalline compounds which have definite melting points (the melting point depends on the nature of the parent carbonyl compound).

(6) Halogenation of acetaldehyde:

(*a*) The *hydrogen* atoms of the methyl group may be replaced by *chlorine*, by passing chlorine into acetaldehyde in the presence of calcium carbonate. The compound formed is trichloracetaldehyde, commonly called *chloral*.

$$CH_3 \cdot \overset{H}{\underset{}{C}} = O \overset{Cl_2}{\longrightarrow} CCl_3 \cdot \overset{H}{\underset{}{C}} = O$$
trichloracetaldehyde
or 'chloral'

Similarly, bromine gives *bromal* $CBr_3 \cdot CHO$, and iodine *iodal* $CI_3 \cdot CHO$. All these are hydrolysed by alkali, giving chloroform, bromoform or iodoform:

$$CX_3 \cdot CHO \xrightarrow{\text{NaOH}} CHX_3 + HCOONa$$

(b) Similar reactions are involved in the conversion of ethyl alcohol to chloroform or iodoform, respectively by the action of bleaching powder or of iodine in an alkaline medium. With bleaching powder, for example, the alcohol is first oxidized to acetaldehyde:

$$CH_3 \cdot CH_2OH \longrightarrow CH_3 \cdot C\overset{\displaystyle H}{\underset{\displaystyle O}{<}}$$

ethyl
alcohol   acetaldehyde

Chlorination then occurs:

$$CH_3 \cdot C\overset{\displaystyle H}{\underset{\displaystyle O}{<}} \longrightarrow CCl_3 \cdot C\overset{\displaystyle H}{\underset{\displaystyle O}{<}}$$

trichloracetaldehyde
or
' chloral '

This breaks down on hydrolysis to chloroform and formic acid (which is converted to calcium formate).

These reactions form the basis for a laboratory preparation of chloroform. The starting material, alcohol, can conveniently be replaced by acetone (no oxidation step is then involved). With acetone the reactions are:

$$CH_3 \cdot CO \cdot CH_3 \longrightarrow CH_3 \cdot CO \cdot CCl_3$$

acetone   trichloracetone

$$CH_3 \cdot CO \cdot CCl_3 \longrightarrow CH_3 \cdot C\overset{\displaystyle O}{\underset{\displaystyle OH}{<}} + CHCl_3$$

acetic acid
└────→calcium acetate

The reaction with iodine (under alkaline conditions) is used as a test reaction ('iodoform test') for the presence of the groups CH_3CO- or $CH_3 \cdot CH(OH)-$. Iodoform is a yellow crystalline solid which is easily recognized (by smell and melting point). Note that ethyl alcohol and isopropyl alcohol, $CH_3 \cdot CH(OH) \cdot CH_3$, give positive tests but methyl alcohol and n-propyl alcohol, $CH_3 \cdot CH_2 \cdot CH_2 \cdot OH$, fail to give iodoform.

(7) Other reactions of acetaldehyde:

(*a*) Acetaldehyde readily undergoes polymerization. If a few drops of concentrated sulphuric acid are added to acetaldehyde the liquid boils up suddenly. On adding water, a little pale yellow oil is seen to separate. This is a *trimer* of acetaldehyde called *paraldehyde*:

$$3CH_3 \cdot CHO \xrightarrow{\overset{\text{conc.}}{H_2SO_4}} (CH_3 \cdot CHO)_3$$
paraldehyde

The structural formula of this product is

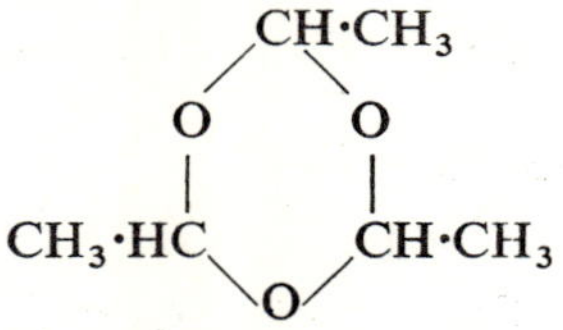

On standing in the presence of dilute aqueous alkalis or some salts, e.g. sodium carbonate or zinc chloride, two molecules of acetaldehyde unite to form the compound *aldol* (first prepared in 1872 by Wurtz):

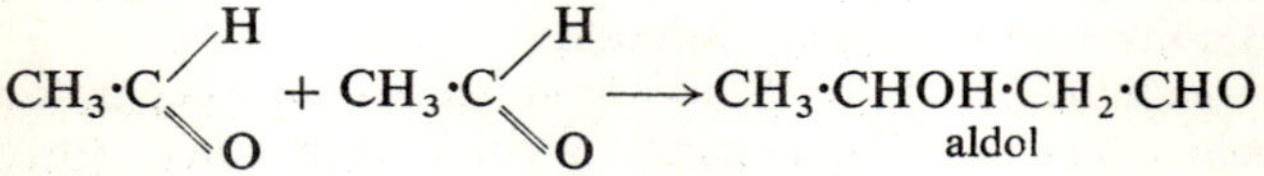

The reaction is usually referred to as the *aldol reaction*.* Since aldol still contains an aldehyde group, it may react in like manner with another molecule of acetaldehyde and continue to do so. With stronger alkalis (NaOH) acetaldehyde is converted to a resinous product of high molecular weight.

If acetaldehyde is treated with small amounts of acids or sulphur dioxide at temperatures below 0°C, a solid *tetramer* called *metaldehyde*, is obtained:

$$4CH_3 \cdot CHO \xrightarrow[\substack{\text{or} \\ SO_2}]{\text{e.g. HCl}} (CH_3 \cdot CHO)_4$$
metaldehyde

This is manufactured by the above method for sale as a solid fuel and also (mixed with bran) as a toxic garden bait for slugs and snails.

* The mechanism is known, but is not discussed here.

(b) With ethyl alcohol in the presence of an anhydrous acid (e.g. dry HCl) acetaldehyde condenses to give a compound known as *acetal*:

$$CH_3 \cdot \overset{\overset{\displaystyle H}{|}}{C}=O + \overset{HOC_2H_5}{\underset{HOC_2H_5}{}} \xrightarrow{\text{acid}} CH_3 \cdot C\underset{OC_2H_5}{\overset{H}{\underset{OC_2H_5}{<}}} + H_2O$$

acetal

(c) The *oxygen* atom of the carbonyl group is replaced by *two chlorine atoms* on treatment with phosphorus pentachloride:

$$CH_3 \cdot \overset{\overset{\displaystyle H}{|}}{C}=O + PCl_5 \longrightarrow CH_3 \cdot CHCl_2 + POCl_3$$

1,1-dichloroethane

(d) Acetaldehyde gives *Schiff*'s test.

An aqueous solution of the dye magenta or rosaniline is first decolorized by reducing it by sulphur dioxide. On adding a little acetaldehyde to this decolorized solution, a purple colour (not due to the original magenta) is obtained.

This is not a specific test for acetaldehyde, for it is given by most aldehydes.

Formaldehyde, H·CHO

Although formaldehyde exhibits many of the properties shown by aldehydes in general, it possesses certain properties not representative of the class. It differs, of course, in constitution from other aldehydes in that the functional group, —CHO, is not attached to a carbon atom.

Formaldehyde is a gas (b.p. $-21°C$) and hence the only aldehyde gaseous at room temperature. It is best known in its 40 per cent aqueous solution as *formalin*, in which it probably exists as a hydrate:

$$H-C\overset{\diagup O}{\underset{\diagdown H}{}} + H_2O \rightleftharpoons H-C\underset{\diagdown H}{\overset{\diagup OH}{-OH}}$$

Formaldehyde is a more effective reducing agent than other aldehydes. It reduces Fehling's solution*, ammoniacal silver oxide and solutions of platinum salts to the metal in each case, and dichromates and permanganates to the respective metal cation. It will also

* Whereas acetaldehyde reduces Fehling's solution to copper(I) oxide on warming, formaldehyde brings about this reaction in the cold. On warming and in presence of excess formaldehyde, metallic copper is deposited.

reduce an aqueous solution of mercury(II) chloride giving, first, the white insoluble mercury(I) chloride and then grey metallic mercury:

$$HgCl_2 \xrightarrow{\text{HCHO}} Hg_2Cl_2\downarrow \xrightarrow{\text{HCHO}} 2Hg\downarrow$$
$$\qquad\qquad\quad \text{white} \qquad\qquad \text{grey-black}$$

In all these reactions, the formaldehyde is oxidized to carbonic acid which decomposes to give carbon dioxide and water:

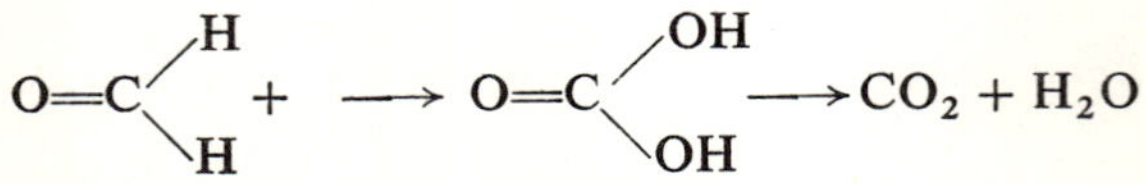

$$O=C\begin{smallmatrix}H\\H\end{smallmatrix} + \longrightarrow O=C\begin{smallmatrix}OH\\OH\end{smallmatrix} \longrightarrow CO_2 + H_2O$$

formaldehyde   carbonic acid

Like other aldehydes, formaldehyde adds on sodium bisulphite and hydrogen cyanide, and condenses with hydrazine and hydroxylamine, giving characteristic derivatives.

In the following reactions, formaldehyde behaves differently from other aldehydes:

(1) Formaldehyde undergoes polymerization as follows:

(a) Formaldehyde alone polymerizes much more readily than other aldehydes; both as a liquid and as a gas, polymerization may be quite rapid, even at low temperatures.

(b) If an aqueous solution of formaldehyde, i.e. formalin, is evaporated to dryness, a white solid called *paraformaldehyde* of high and variable molecular weight is obtained:

$$nHCHO \longrightarrow (CH_2O)_n$$
$$\qquad\qquad\quad \text{paraformaldehyde}$$

(c) In the presence of dilute sulphuric acid, formaldehyde polymerizes to *trioxan* which has a cyclic structure.

$$3HCHO \xrightarrow{\text{dil. } H_2SO_4} (CH_2O)_3, \text{ i.e. } \begin{matrix} O-CH_2 \\ CH_2 \quad\quad O \\ O-CH_2 \end{matrix}$$

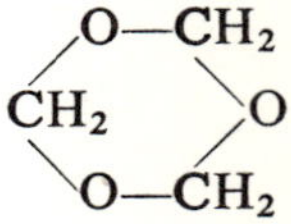

(d) If an aqueous solution of formaldehyde, to which a little slaked lime has been added, is allowed to stand for a few days a mixture of sugars, usually called *formose*, is obtained.

$$6HCHO \xrightarrow{\text{Ca(OH)}_2} C_6H_{12}O_6$$
$$\qquad\qquad\qquad \text{formose}$$

(e) Formaldehyde also differs from other aldehydes in that:

(i) It does not give the aldol condensation (inspection of the formula of aldol shows that formaldehyde cannot give an analogous product).

(ii) It does not resinify when warmed with alkalis; instead *disproportionation* occurs:

$$2HCHO + H_2O \xrightarrow[\substack{or \\ KOH}]{NaOH} CH_3OH + HCOOH$$
$$\downarrow$$
$$HCOO^-Na^+$$

This reaction is usually referred to as *Cannizzaro's reaction*, after its discoverer.

Disproportionation is, strictly, a process in which two or more new valency or oxidation states arise from one original state—a case of mutual oxidation and reduction. Here, one molecule of formaldehyde is oxidized to formic acid and another reduced to methyl alcohol.

(2) Formaldehyde does not add on ammonia, but condenses with it, water being eliminated:

$$6HCHO + 4NH_3 \longrightarrow (CH_2)_6N_4 + 6H_2O$$
hexamethylenetetramine

The reaction can be brought about by warming a mixture of formalin and concentrated ammonia. The product was at one time used as a urinary disinfectant.

The structure of hexamethylenetetramine is interesting. The molecule may be represented approximately as:

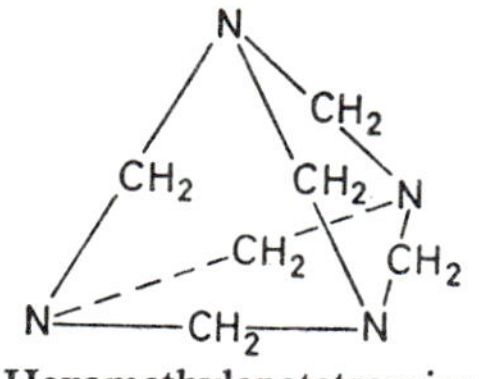

Hexamethylenetetramine

a tetrahedron with a nitrogen atom at each apex and a methylene group, —CH$_2$— in the middle of each side (the N—CH$_2$—N links are, of course, not linear as shown here).

Preparation and manufacture of formaldehyde

Formaldehyde can be obtained in the laboratory by the dry distillation of calcium formate:

$$(HCOO)_2Ca \longrightarrow CaCO_3 + HCHO$$

On the industrial scale, formaldehyde is obtained by passing a mixture of air and the vapour of methyl alcohol over a catalyst heated to a temperature of about 500°C. The catalyst may be copper, silver or platinum:

$$2CH_3OH + O_2 \xrightarrow[500°C]{Cu, \; Ag \; or \; Pt} 2HCHO + 2H_2O$$

The vapour coming over is cooled, when the steam condenses and dissolves the formaldehyde, the product being formalin.

A certain amount of formaldehyde is also manufactured in America by the controlled oxidation of natural gas and by the dehydrogenation of methyl alcohol by copper, followed by the burning of the hydrogen:

$$CH_3OH \longrightarrow HCHO + H_2$$

and then
$$2H_2 + O_2 \longrightarrow 2H_2O$$

Uses of formaldehyde

(1) As a *disinfectant* and *preservative*. Formalin is the most common preservative for biological specimens.

(2) In the manufacture of *plastics*. It has been mentioned (p. 24) that the common feature in the manufacture of both thermo- and thermo-setting plastics is the building up of large molecules by the processes of polymerization and condensation or both.

It was first known in 1872 that a resinous substance could be obtained by the condensation of formaldehyde with phenol (p. 239) but it was not until 1909 that this substance, the first synthetic plastic, was put on the market as Bakelite (named after the inventor Baekeland).

A thermo-setting plastic is also obtained by condensing formaldehyde with urea, the condensation probably taking the form of:

$$\underset{\text{urea}}{NH_2 \cdot CO \cdot NH_2} + \underset{H}{\overset{H}{\diagdown}} C{=}O \longrightarrow NH_2 \cdot CO \cdot NH \cdot CH_2OH$$

This reaction can repeat itself. In this way, huge ladder-like molecules, consisting of chains of atoms joined by cross links, are built up.

A mixture of formalin and urea is warmed in the presence of ammonia or some other alkali (as a catalyst), when a syrupy liquid is obtained. This is then mixed with a filler, e.g. wood pulp, dried under reduced pressure, and ground to a fine powder. The powder is pressed into the required mould and heated, when further condensation occurs, the powder setting to an infusible mass. The urea-formaldehyde plastics are colourless but colour can be added. They are not so strong as the urea-phenol plastics, but the fact that they are light-coloured is an advantage for some purposes. They are put to a variety of uses including the making of lacquers and cements and, more recently, as an adhesive in the manufacture of plywood. The adhesive has an advantage over glue for this purpose in that it is waterproof.

Chloral, $CCl_3 \cdot CHO$

This trichloro-derivative of acetaldehyde (first prepared by Liebig) is obtained by the action of chlorine on ethyl alcohol; the chlorine acts first as an oxidizing agent:

$$C_2H_5OH \xrightarrow{Cl_2} CH_3 \cdot CHO \xrightarrow{Cl_2} CCl_3 \cdot CHO$$

It is a colourless oily liquid which, on shaking with water, evolves much heat to form a crystalline hydrate of formula $CCl_3 \cdot CHO \cdot H_2O$, known as *chloral hydrate*.

Both chloral and chloral hydrate show most of the general properties of aldehydes. Both are hydrolysed by alkalis to chloroform:

$$CCl_3 \cdot CHO + NaOH \longrightarrow CHCl_3 + HCOONa$$

Chloral, however, gives a colour with Schiff's reagent but chloral hydrate does not. It is believed that chloral hydrate has the constitutional formula

$$CCl_3 \cdot C \underset{\diagdown OH}{\overset{\diagup H}{-}} OH$$

If so, chloral hydrate is one of the few fairly stable organic compounds (cf. formaldehyde hydrate, p. 103) in which there are two hydroxyl groups attached to one and the same carbon atom.

Chloral hydrate is used as a hypnotic or soporific (i.e. sleep inducing).

SUMMARY OF THE CHEMISTRY OF THE ALIPHATIC ALDEHYDES

The general formula is $C_nH_{2n+1} \cdot CHO$

The characteristic or functional group is: $-C \overset{\diagup H}{\underset{\diagdown O}{}}$

The group $\diagup C = O$ is called the *carbonyl* group.

The first member is HCHO, formaldehyde.

The representative member is $CH_3 \cdot CHO$, acetaldehyde.

Acetaldehyde, $CH_3 \cdot CHO$

(1) *Preparation*

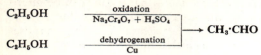

(2) *Manufacture*

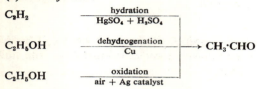

(3) *Properties*

(*a*) Acetaldehyde acts as a reducing agent (due to the hydrogen atom of the functional group); in each case, acetic acid is the oxidation product:

$$CH_3 \cdot CHO \xrightarrow[\text{silver oxide}]{\text{ammoniacal}} CH_3 \cdot COOH + Ag\downarrow$$
$$\text{(mirror)}$$

$$\xrightarrow[\substack{\text{(CuSO}_4 + \text{Rochelle salt} \\ + \text{NaOH)}}]{\text{Fehling's solution}} CH_3 \cdot COOH + Cu_2O\downarrow$$
$$\text{(hydrated, orange-red)}$$

$$\xrightarrow[\text{H}_2\text{SO}_4]{\text{K}_2\text{Cr}_2\text{O}_7} CH_3 \cdot COOH + Cr^{3+}$$
$$\text{green}$$

$$\xrightarrow[\text{H}_2\text{SO}_4]{\text{KMnO}_4} CH_3 \cdot COOH + Mn^{2+}$$
$$\text{colourless}$$

(*b*) Acetaldehyde is reduced to ethyl alcohol (by $LiAlH_4$, $NaBH_4$, catalytic hydrogenation)

$$CH_3 \cdot CHO \longrightarrow CH_3 \cdot CH_2 \cdot OH$$

(*c*) Additive reactions (due to the carbonyl, $\diagdown C{=}O$, group):

$$CH_3 \cdot C{\underset{O}{\overset{H}{\diagup}}} \xrightarrow{\text{NaHSO}_3} CH_3 \cdot C{\underset{\overset{|}{\bar{S}O_3Na^+}}{\overset{H}{-}}}OH$$

acetaldehyde sodium bisulphite

$$CH_3 \cdot C \underset{O}{\overset{H}{<}} \xrightarrow{\text{NH}_3} CH_3 \cdot C \underset{NH_2}{\overset{H}{\underset{\diagdown}{\diagup}}} OH$$

acetaldehyde ammonia

$$\xrightarrow{\text{HCN}} CH_3 \cdot C \underset{CN}{\overset{H}{\underset{\diagdown}{\diagup}}} OH$$

acetaldehyde cyanohydrin

$$\xrightarrow{\text{RMgX}} CH_3 \cdot C \underset{R}{\overset{H}{\underset{\diagdown}{\diagup}}} OMgX$$

The first step in all these additions is:

$$CH_3 \cdot C \underset{H}{\overset{O}{\diagup}} :N \longrightarrow CH_3 \cdot C \underset{H}{\overset{O^-}{\diagdown}} N^+$$

where :N represents a *nucleophile* (nucleophilic reagent). *Nucleophiles* are reagents which provide an electron pair for the formation of a new bond to another nucleus (e.g. $:CN^-$, $:NH_3$, $:NH_2OH$ etc.).

Organic 'ionic' reactions involve reactions between *nucleophiles* and *electrophiles* (not necessarily between ions!).

Electrophiles are reagents which share the electron pair provided by nucleophiles.

The carbon atom of the carbonyl group is an *electrophilic* centre:

$$\overset{\delta+}{\underset{\diagup}{\diagdown}} \overset{\delta-}{C=O} \quad \text{or} \quad \underset{\diagup}{\diagdown} C=O \longleftrightarrow \underset{\diagup}{\diagdown} \overset{+}{C} - \overset{-}{O}$$

$\longleftrightarrow$ implies an 'in-between' state (*not* an equilibrium mixture of two different forms).

(*d*) Addition-elimination reactions (formation of 'condensation' products):

$$\underset{\text{hydroxylamine}}{\overset{H}{\underset{|}{CH_3 \cdot C}} = O + H_2N \cdot OH} \longrightarrow \underset{\text{acetaldehyde oxime}}{\overset{H}{\underset{|}{CH_3 \cdot C}} = NOH + H_2O}$$

$$CH_3 \cdot \overset{\overset{\displaystyle H}{|}}{C}{=}O + H_2N \cdot NH_2 \longrightarrow CH_3 \cdot \overset{\overset{\displaystyle H}{|}}{C}{=}N \cdot NH_2 + H_2O$$
hydrazine   acetaldehyde hydrazone

or

$$CH_3 \cdot CHO + H_2N \cdot NH \cdot C_6H_5 \longrightarrow CH_3 \cdot \overset{\overset{\displaystyle H}{|}}{C}{=}N \cdot NH \cdot C_6H_5 + H_2O$$
phenylhydrazine   acetaldehyde phenylhydrazone

(e) Reactions with halogens (hydrogen atoms of methyl group replaceable by chlorine and other halogens):

$$CH_3 \cdot CHO \xrightarrow[\text{(CaCO}_3\text{)}]{\text{Cl}_2} CCl_3 \cdot CHO$$
chloral

(f) Other reactions:

(i) Polymerizations to paraldehyde (trimer), metaldehyde (tetramer) and also to aldol, $CH_3 \cdot CH(OH) \cdot CH_2 \cdot CHO$.

(ii) Acetal formation:

$$CH_3 \cdot CHO + 2C_2H_5OH \xrightarrow[\text{HCl}]{\text{dry}} CH_3 \cdot CH(OC_2H_5)_2 + H_2O$$
acetal

(iii) Replacement of oxygen by chlorine:

$$CH_3 \cdot CHO + PCl_5 \longrightarrow CH_3 \cdot CHCl_2 + POCl_3$$

(g) Tests:

(i) *Iodoform test* (also given by ethanol, acetone, etc.):

$$CH_3 \cdot CHO \xrightarrow[\text{(Na}_2\text{CO}_3\text{)}]{\text{I}_2} CHI_3$$
(smell or pale yellow ppt.)

(ii) *Schiff's test* (also given by most aldehydes):

$$\text{Magenta} \xrightarrow{\text{SO}_2} \text{decolorized} \xrightarrow{\text{CH}_3 \cdot \text{CHO}} \text{purple colour}$$

(iii) Ammoniacal silver nitrate and Fehling's solution are also used as test-reagents (see p. 94).

Formaldehyde, HCHO

The only gaseous member of the series.

Has some properties in common with other aldehydes, but differs from them in a number of respects:

(a) Does not give aldol condensation but undergoes ' disproportionation ' with sodium hydroxide:

$$2HCHO \longrightarrow HCO_2H + CH_3OH$$
(as the salt)

(*b*) With weak alkali (e.g. lime water) gives ' formose ' (mixture of sugars).

(*c*) With ammonia gives ' hexamethylenetetramine '.

Chloral, $CCl_3 \cdot CHO$

(1) *Preparation*

Action of chlorine on ethyl alcohol (p. 107):

$$C_2H_5OH \longrightarrow CH_3 \cdot CHO \longrightarrow CCl_3 \cdot CHO$$

Shows most of the properties of aldehydes. Important differences are:

(*a*) Hydrolysis by alkalis to chloroform:

$$CCl_3 \cdot CHO + NaOH \longrightarrow CHCl_3 + HCOONa$$

(*b*) Formation of crystalline hydrate, $CCl_3 \cdot CHO \cdot H_2O$, which does not show all aldehyde properties because of probable constitution:

$$CCl_3 \cdot C \overset{H}{\underset{OH}{\diagdown}} OH$$

EXAMINATION QUESTIONS

(1) A compound A, of low boiling point, contained carbon, hydrogen and oxygen. After 10 c.c. of the vapour A. was exploded with 50 c.c. of oxygen (excess), the volume was 45 c.c. and this was reduced to 25 c.c. by treatment with potash, all volumes being measured at 25°C and 760 mm. When 0·88 gm. of A. was dissolved in water, oxidized and distilled, the distillate required 20 c.c. N sodium hydroxide for neutralization. What is the structural formula of A. and what happens when it is oxidized? (H = 1, C = 12, O = 16). (C. S.)

(2) What do you understand by the following terms:
   (*a*) a displacement reaction,
   (*b*) an elimination reaction,
   (*c*) an addition-elimination reaction?

(3) Describe, with as many examples as you can, what is meant by (*a*) a substitution reaction and (*b*) an elimination reaction. (Liverpool, B.Sc. Inter.)

9

ALIPHATIC KETONES

Functional Group $\ce{>C=O}$

THE aliphatic ketones are compounds containing a carbonyl group to which are attached two alkyl groups. The general formula for the series is therefore $C_nH_{2n+1}\cdot CO\cdot C_mH_{2m+1}$, where n and m can be the same or different. If $n=m=0$, the formula obtained is HCHO, which is that of formaldehyde. If in one group $n = 0$ and in the other $m = 1$, then the formula obtained is that of acetaldehyde, $CH_3\cdot CHO$. The first member of the ketone series is thus that in which $n = m = 1$ is substituted in both alkyl groups, viz. $CH_3\cdot CO\cdot CH_3$. This is *dimethyl ketone*, commonly called *acetone*, the best known and most important ketone. $CH_3\cdot CO\cdot C_2H_5$ is methyl ethyl ketone, which is an example of an *unsymmetrical ketone*. Isomerism is possible not only among higher members of the series owing to different arrangements within the alkyl groups, but also with the aldehydes. Thus, propionaldehyde, $CH_3\cdot CH_2\cdot CHO$, is isomeric with acetone, $CH_3\cdot CO\cdot CH_3$.

Acetone, $CH_3\cdot CO\cdot CH_3$

Preparation of Acetone

(1) By the oxidation of isopropyl alcohol (analogous to the oxidation of ethyl alcohol to acetaldehyde):

$$\underset{\displaystyle \overset{CH_3}{\diagdown}\ \underset{CH_3}{\diagup}}{C}\overset{H}{\diagup}\overset{}{\diagdown}OH \longrightarrow \underset{CH_3}{\overset{CH_3}{\diagdown}}C=O$$

The oxidation may be carried out by oxidizing agents such as acidified dichromate and permanganate, or by passing a mixture of air and the vapour of the isopropyl alcohol over heated catalysts, e.g. copper, platinum black or zinc oxide.

Oxidation of secondary alcohols is a very general method for the preparation of ketones.

(2) By heating dry calcium acetate (p. 129):

$$(CH_3CO_2)_2Ca \longrightarrow CaCO_3 + CH_3{\cdot}CO{\cdot}CH_3$$

Manufacture of acetone

At one time acetone was obtained from the aqueous layer collected during the destructive distillation of wood. With the increasing demand for acetone during recent years, other methods have been worked out.

(1) By the catalytic dehydrogenation of isopropyl alcohol.

The isopropyl alcohol is obtained in large quantities from propylene, a by-product of petroleum cracking:

$$\begin{array}{c} CH_3 \\ {\diagdown} \\ CHOH \\ {\diagup} \\ CH_3 \end{array} \xrightarrow{\text{brass}} \begin{array}{c} CH_3 \\ {\diagdown} \\ C{=}O + H_2 \\ {\diagup} \\ CH_3 \end{array}$$

The vapour of the alcohol is passed over the catalyst heated to a high temperature.

Liquid isopropyl alcohol may be oxidized by pure oxygen at 90–140°C under pressure, to give acetone and hydrogen peroxide:

$$(CH_3)_2CHOH + O_2 \longrightarrow (CH_3)_2CO + H_2O_2$$

(2) By the interaction of water gas and ethylene.

The mixture, under a pressure of 150 atmospheres, and a temperature between 150° and 180°C, is passed over a catalyst consisting of cobalt (II) oxide, CoO, mixed with a little of the oxides of thorium and magnesium:

$$C_2H_4 + CO + H_2 \longrightarrow \begin{cases} \begin{array}{c} CH_3 \\ {\diagdown} \\ C{=}O \\ {\diagup} \\ CH_3 \end{array} \text{ acetone} \\[2em] CH_3{\cdot}CH_2{\cdot}C{\Big\langle}\begin{array}{c} H \\ O \end{array} \text{ propionaldehyde} \end{cases}$$

Thirty per cent of the product is acetone, the remaining 70 per cent being the isomeric propionaldehyde. The process is a German one and is an example of the Fischer–Tropsch synthesis (p. 195).

(3) By the fermentation of carbohydrates (from maize or molasses) by a culture of a bacterium known as *clostridium acetobutylicum.*

A mixture of n-butyl alcohol, acetone and ethyl alcohol is obtained.

(4) Acetone is obtained as a by-product in the conversion of benzene to phenol.

Properties of acetone

Acetone is a colourless liquid (b.p. 56°C) with a sweetish rather pleasant smell. It mixes readily with water, ethyl alcohol and ether. It is a good solvent.

Since its molecule contains the carbonyl group, acetone will show those properties of aldehydes connected with this group. Since no hydrogen atoms flank the carbonyl group as in aldehydes, the reducing properties of aldehydes are not shown.

(1) Like most simple organic compounds, acetone burns in air to form carbon dioxide and water.

It is not a reducing agent and with mild oxidizing agents undergoes no reaction. Vigorous oxidation breaks up the molecule, giving a carboxylic acid containing a smaller number of carbon atoms and carbon dioxide (cf. oxidation of an aldehyde).

$$CH_3 \cdot CO \cdot CH_3 \xrightarrow[\text{oxidation}]{\text{vigorous}} CH_3 \cdot COOH + CO_2$$

(2) Acetone forms addition compounds.

(*a*) Addition of *hydrogen*: on reduction by lithium aluminium hydride, sodium borohydride or by catalytic methods, acetone gives a *secondary alcohol*:

$$CH_3 \cdot CO \cdot CH_3 \longrightarrow \begin{matrix} CH_3 \\ {>}CHOH \\ CH_3 \end{matrix}$$

secondary propyl alcohol
(isopropylalcohol)

(*b*) Addition of *sodium bisulphite*: acetone gives the bisulphite compound:

$$\begin{matrix} CH_3 \\ {>}C{=}O \\ CH_3 \end{matrix} + Na^+HSO_3^- \longrightarrow \begin{matrix} CH_3 \quad OH \\ {>}C{<} \\ CH_3 \quad SO_3^-Na^+ \end{matrix}$$

acetone sodium bisulphite

Ketones, other than methyl ketones (i.e. containing the grouping $CH_3 \cdot CO$), usually fail to give this reaction.

(c) Addition of *hydrogen cyanide*: acetone forms a cyanhydrin:

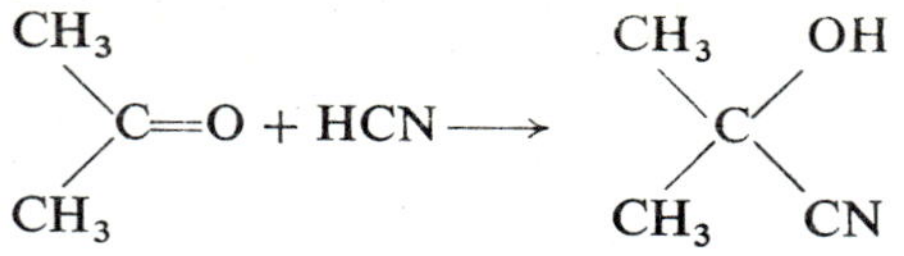

$$\underset{CH_3}{\overset{CH_3}{>}}C{=}O + HCN \longrightarrow \underset{CH_3}{\overset{CH_3}{>}}C\underset{CN}{\overset{OH}{<}}$$

acetone cyanhydrin*

(3) Acetone forms ' condensation' products with elimination of water (i.e. undergoes addition-elimination reactions).

(a) With *hydroxylamine*, acetone forms a *ketoxime*:

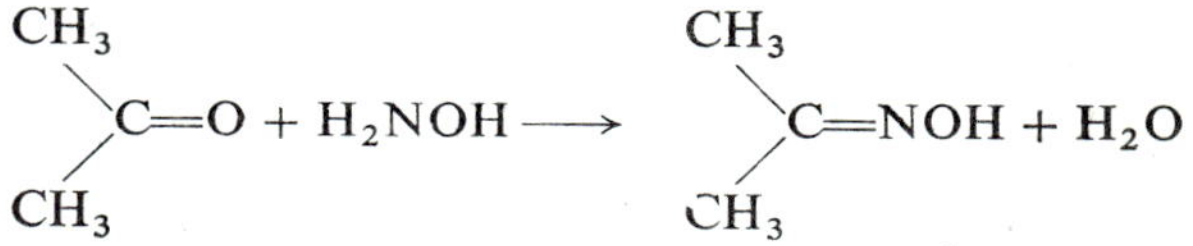

$$\underset{CH_3}{\overset{CH_3}{>}}C{=}O + H_2NOH \longrightarrow \underset{CH_3}{\overset{CH_3}{>}}C{=}NOH + H_2O$$

acetoxime

(b) With *hydrazine* and *substituted hydrazines*, acetone gives *hydrazones*:

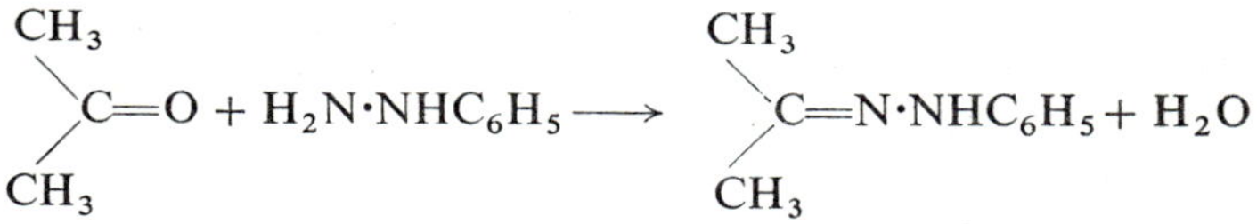

$$\underset{CH_3}{\overset{CH_3}{>}}C{=}O + H_2N{\cdot}NHC_6H_5 \longrightarrow \underset{CH_3}{\overset{CH_3}{>}}C{=}N{\cdot}NHC_6H_5 + H_2O$$

acetone phenylhydrazone

* This compound is an intermediate product in the manufacture of Perspex. It is treated with a mixture of methyl alcohol and concentrated sulphuric acid, when the methyl ester of methyl acrylic acid is obtained. The stages in its formation may be represented as follows:

(a) Hydrolysis of the nitrile group:

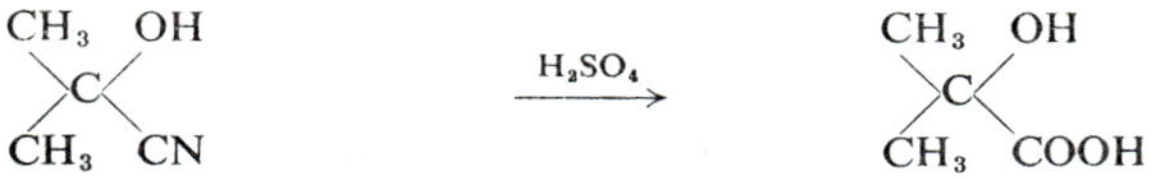

$$\underset{CH_3}{\overset{CH_3}{>}}C\underset{CN}{\overset{OH}{<}} \xrightarrow{H_2SO_4} \underset{CH_3}{\overset{CH_3}{>}}C\underset{COOH}{\overset{OH}{<}}$$

(b) Dehydration:

$$\underset{CH_3}{\overset{CH_3}{>}}C\underset{COOH}{\overset{OH}{<}} \xrightarrow{H_2SO_4} \underset{CH_3}{\overset{CH_2}{>}}C{-}COOH$$

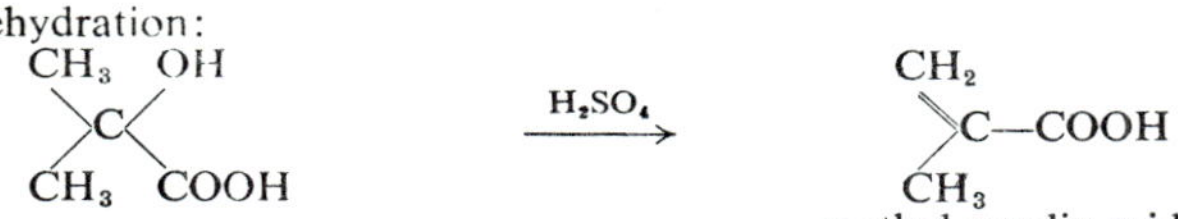

methyl acrylic acid

(c) Esterification:

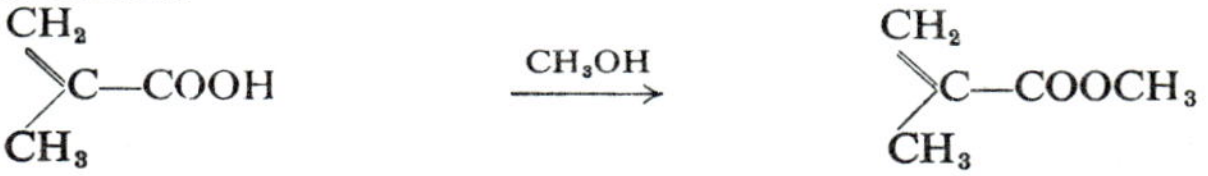

$$\underset{CH_3}{\overset{CH_2}{>}}C{-}COOH \xrightarrow{CH_3OH} \underset{CH_3}{\overset{CH_2}{>}}C{-}COOCH_3$$

methyl ester of methyl acrylic acid

Perspex is the polymer of this ester.

The mechanisms of the above addition reactions and addition-elimination (condensation) reactions are the same as those of the analogous reactions with acetaldehyde. In ketones, as well as in aldehydes, the carbon atom of the carbonyl group is electrophilic in character; thus reactions occur with 'electron-rich' nucleophiles like $:CN^-$, $:NH_2 \cdot NH \cdot C_6H_5$.

(c) A different type of condensation reaction occurs when acetone is treated with moderately concentrated sulphuric acid:

mesitylene

The reaction is of interest since it is an example of a conversion from an aliphatic compound to an aromatic compound (a benzene derivative, p. 200).

(4) Acetone forms substitution compounds.

(a) Chlorine (and also bromine) replaces successively the hydrogen atoms of the methyl group to give finally *hexachloracetone:*

$$CH_3 \cdot CO \cdot CH_3 + 6Cl_2 \longrightarrow CCl_3 \cdot CO \cdot CCl_3 + 6HCl$$
hexachloracetone

(b) Acetone gives the iodoform reaction, i.e. a substitution followed by a hydrolysis (p. 101):

$$CH_3 \cdot CO \cdot CH_3 \xrightarrow{I_2} CI_3 \cdot CO \cdot CH_3 \xrightarrow{NaOH} CHI_3 + CH_3 \cdot COONa$$

(c) Acetone gives chloroform on treatment with *bleaching powder* (p. 101). As above, this is a substitution followed by a hydrolysis:

$$CH_3 \cdot CO \cdot CH_3 \xrightarrow[powder]{bleaching} CCl_3 \cdot CO \cdot CH_3 \xrightarrow{NaOH} CHCl_3 + CH_3 \cdot COONa$$

(5) Phosphorus pentachloride replaces the oxygen atom of the carbonyl group:

$$CH_3 \cdot CO \cdot CH_3 + PCl_5 \longrightarrow CH_3 \cdot CCl_2 \cdot CH_3 + POCl_3$$

Uses of acetone

Acetone is used extensively as a solvent for a variety of purposes, e.g. acetylene, fine varnishes and lacquers, aeroplane dope, resins to make moulding powders, cellulose nitrate in the manufacture of photographic films, cellulose acetate in the manufacture of acetate rayon, gun cotton and nitroglycerine in the making of cordite. It is used in the manufacture of Perspex (p. 115).

Comparison between aldehydes and ketones

In general, ketones are less chemically active than are aldehydes.

Aldehydes and ketones show the following similarities:

(a) They can be reduced, aldehydes to primary alcohols and ketones to secondary alcohols.

(b) They add on sodium bisulphite (methyl ketones react; other ketones do not).

(c) They add on hydrogen cyanide.

(d) With hydroxylamine, they form oximes.

(e) With hydrazine and substituted hydrazines, they form hydrazones.

(f) The hydrogen atoms of the alkyl groups can be replaced by chlorine (strictly, this applies only to the CH_3 or CH_2 groups immediately adjacent to the carbonyl group).

(g) The oxygen atom of the carbonyl group is replaceable by two chlorine atoms on treatment with phosphorus pentachloride.

Ketones *differ* from aldehydes in that:

(a) They are not reducing agents (no effect on Fehling's solution or ammoniacal silver nitrate).

(b) They do not readily polymerize, and do not form resins with alkalis.

(c) They do not give Schiff's test.

SUMMARY OF THE CHEMISTRY OF KETONES

The general formula of ketones is $C_nH_{2n+1} \cdot CO \cdot C_mH_{2m+1}$; i.e. the two alkyl groups may be different, e.g. $CH_3 \cdot CO \cdot C_2H_5$, methyl ethyl ketone.

The first and representative member is $(CH_3)_2CO$, dimethyl ketone or acetone.

(1) *Preparation*

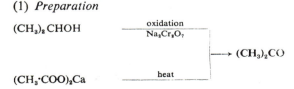

(2) *Manufacture*

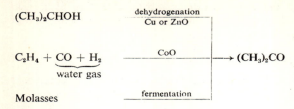

(CH$_3$)$_2$CHOH

C$_2$H$_4$ + $\underbrace{CO + H_2}_{\text{water gas}}$

Molasses

dehydrogenation
Cu or ZnO

CoO

fermentation

$\longrightarrow$ (CH$_3$)$_2$CO

(3) *Properties*

Show most of the properties of aldehydes connected with the carbonyl group:

(*a*) Addition reactions:

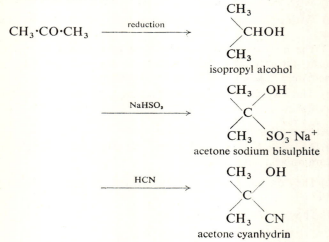

CH$_3$·CO·CH$_3$ $\xrightarrow{\text{reduction}}$

CH$_3$
$\diagdown$
$\quad$CHOH
$\diagup$
CH$_3$
isopropyl alcohol

$\xrightarrow{\text{NaHSO}_3}$

CH$_3$ $\diagup$ OH
$\quad\quad$ C
CH$_3$ $\diagdown$ SO$_3^-$ Na$^+$
acetone sodium bisulphite

$\xrightarrow{\text{HCN}}$

CH$_3$ $\quad$ OH
$\quad\quad$ C
CH$_3$ $\quad$ CN
acetone cyanhydrin

(*b*) Addition-elimination reactions

CH$_3$·CO·CH$_3$ $\xrightarrow{\text{NH}_2\text{OH}}$

CH$_3$
$\diagdown$
$\quad$C=NOH
$\diagup$
CH$_3$
acetoxime

$\xrightarrow{\text{NH}_2\cdot\text{NH}_2}$

CH$_3$
$\diagdown$
$\quad$C=N·NH$_2$
$\diagup$
CH$_3$
acetone hydrazone

(c) ' Self-condensation '

$$CH_3 \cdot CO \cdot CH_3 \xrightarrow{\text{H}_2\text{SO}_4}$$

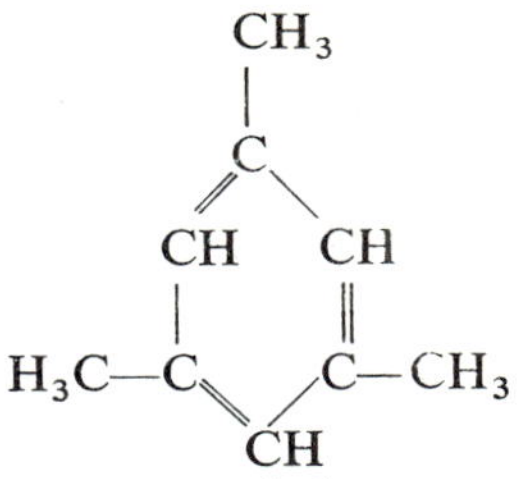

mesitylene

(d) Substitution reactions

$$CH_3 \cdot CO \cdot CH_3 \xrightarrow{\text{Cl}_2} CH_3 \cdot CO \cdot CCl_3$$
trichloracetone

$$\xrightarrow[\text{powder}]{\text{bleaching}} CHCl_3 \quad \begin{array}{l}\text{(reaction also} \\ \text{involves} \\ \text{hydrolysis)}\end{array}$$
chloroform

$$\xrightarrow{\text{PCl}_5} CH_3 \cdot CCl_2 \cdot CH_3$$
2,2-dichloropropane

EXAMINATION QUESTIONS

(1) Name the compounds with the formulae $CH_3 \cdot CHO$ and $(CH_3)_2 \cdot CO$. State **one** method of preparing each compound, and give their reactions with (a) oxidizing agents, (b) hydrogen, (c) sodium bisulphite, (d) hydroxylamine. Name the substances formed and give their structural formulae. (O. A.)

(2) Explain what is meant by (a) addition, (b) condensation, (c) polymerization, (d) substitution, and give **one** example of each from the reactions of acetaldehyde.
Describe **two** chemical tests that would enable you to distinguish between aqueous solutions of acetaldehyde and acetone. (C. A.)

(3) Describe the laboratory preparation of crude acetaldehyde from ethyl alcohol.
How does acetaldehyde react with (a) sodium hydrogen sulphite, (b) hydroxylamine, (c) phosphorus pentachloride? Write equations for the reactions and name the organic products.
Give **two** simple chemical tests which would enable you to distinguish between acetaldehyde and acetone. (A. A.)

10

THE ALIPHATIC CARBOXYLIC ACIDS

Functional Group $-C{\Large\langle}^{O}_{OH}$

THE aliphatic carboxylic acids form a homologous series of general formula $C_nH_{2n+1}COOH$. The group —COOH, the constitution of which is usually represented as $-C{\Large\langle}^{O}_{OH,}$ is known as the *carboxyl group*. Thus, the carboxyl group contains both the carbonyl group and the hydroxyl group, but because these groups are directly linked together, they influence one another and as a result carboxylic acids differ in distinctive ways from aldehydes and ketones, and from alcohols.

The first member (where $n = 0$) and having the formula of HCOOH, possesses some peculiar properties. It differs, of course, from other members in that it contains no alkyl group. The member characteristic of the series is that where $n = 1$, having the formula $CH_3 \cdot COOH$. This is *acetic acid*, which will be studied as being representative of the group.

Formic acid	HCOOH	
Acetic acid	$CH_3 \cdot COOH$	
Propionic acid	$C_2H_5 \cdot COOH$	
Butyric acid	$C_3H_7 \cdot COOH$	n-butyric acid $CH_3 \cdot CH_2 \cdot CH_2 \cdot COOH$
		iso-butyric acid ${CH_3 \atop CH_3}{\Large\rangle}CH \cdot COOH$
Palmitic acid	$C_{15}H_{31} \cdot COOH$	
Stearic acid	$C_{17}H_{35} \cdot COOH$	

The term 'fatty', also applied to this series of acids, originates from the fact that the different naturally-occurring vegetable and animal fats and oils are *esters** of these acids. The names of the

---

* Esters are compounds derived from acids (including carboxylic acids) and alcohols:

$$RC{\Large\langle}^{O}_{OH} + R'OH \longrightarrow RC{\Large\langle}^{O}_{OR'} + H_2O$$

$\quad\quad\quad\quad\quad\quad$ alcohol $\quad\quad\quad\quad$ ester

[*Footnote continued opposite*]

better-known individual acids are connected with their occurrence in nature. *Formic acid* is present in ants (Latin *formica* = ant). *Acetic acid* is the acid present in vinegar (Latin *acetum* = vinegar). The term *propionic* comes from the Greek (*proto* = first and *pion* = fat). The glyceryl ester of n-butyric acid is present to the extent of about 7 per cent in butter (Latin, *butyrum* = butter). The free acid is a constituent, in small amounts, of animal perspiration. Most of the natural oils and fats contain large amounts of the glyceryl esters of higher acids, e.g. palmitic and stearic (Greek *stear* = fat). Tallow contains a large amount of the stearin ester, and palm oil up to about 30 per cent of the glyceryl ester of palmitic acid. It is a point of interest that only those acids containing an even number of carbon atoms in the molecule occur in natural oils and fats.

The lowest members of the aliphatic carboxylic acids are colourless liquids with a pungent smell. They are soluble in water giving solutions which are acid to litmus, indicating formation of the hydroxonium ion. Some of the middle members of the series have disagreeable odours, e.g. the unpleasant smell of rancid butter and strong cheese is due to n-butyric acid set free from its ester by bacterial action.

As the carbon chain length increases, the solubility in water decreases. Those acids containing more than nine atoms of carbon in the molecule are oily or waxy solids, which are insoluble in water, and because of their low volatility possess practically no smell.

Acetic acid, $CH_3 \cdot C \overset{\displaystyle O}{\underset{\displaystyle OH}{}}$

This was one of the earliest known acids, being formed when beer and light wines turn sour. When pure, it is a colourless liquid with a strong pungent smell but since it freezes at a temperature of 16·7°C, it often solidifies in winter to an ice-like solid. For this reason, the pure anhydrous acetic acid is usually known as *glacial* acetic acid. The acid was first isolated in 1720 by Stahl and first obtained in crystalline form in 1789 by Lowitz, by the distillation of vinegar.

Many natural fats and oils are esters derived from the trihydric alcohol *glycerol* (p. 78), e.g.:

$$CH_2OH \atop CHOH \atop CH_2OH + 3RC \overset{O}{\underset{OH}{}} \longrightarrow CH_2OC \atop CHOC \atop CH_2OC + 3H_2O$$

Preparation of acetic acid

Acetic acid may be prepared in the laboratory by any one of the following methods, which are examples of general methods applicable to all members of the series.

(1) By the oxidation of ethyl alcohol or acetaldehyde, by treatment with acidified sodium or potassium dichromate or potassium permanganate (p. 63), or by using air in presence of platinum as the oxidizing agent:

$$CH_3 \cdot CH_2OH \longrightarrow \underset{\text{acetaldehyde}}{CH_3 \cdot CHO} \longrightarrow CH_3 \cdot COOH$$

(2) By the hydrolysis of methyl cyanide (p. 156).

(3) By the addition of carbon dioxide to the corresponding Grignard compound.

By passing carbon dioxide at low temperature into the methyl compound, we get:

$$O{=}C{=}O \xrightarrow{CH_3 \cdot MgI} O{=}C{\Big\langle}\begin{array}{l} CH_3 \\ OMgI \end{array}$$

and then:

$$O{=}C{\Big\langle}\begin{array}{l} CH_3 \\ OMgI \end{array} + H_2O \longrightarrow MgI(OH) + CH_3 \cdot COOH$$

Manufacture of acetic acid

(1) By bacterial oxidation of ethyl alcohol.

In the manufacture of *vinegar*, acetic acid is formed by the atmospheric oxidation of ethyl alcohol in the presence of various species of acetobacter.

Malt vinegar is obtained from wort and on the Continent is also made from wines of poor quality. In the old process, the alcoholic liquor was put into casks already containing a culture of the bacteria which breed and collect as a slimy film on the surface. When oxidation had proceeded to an acid concentration of 4–5 per cent, most of the vinegar was withdrawn and the cask then filled up again with the liquor. The amount left in the cask contained sufficient bacteria for the oxidation of the refill.

In the modern quick process, the alcoholic liquor is brought into greater contact with air by allowing it to trickle over wood shavings impregnated with a culture of bacteria. Warm air is blown through the container.

The bacteria cannot work in a solution containing a greater alcohol content than 15 per cent. Spirits and strong wines therefore do not turn sour on exposure to air (in which the spores of the oxidizing bacteria are always present).

(2) **By the atmospheric oxidation of acetaldehyde or ethyl alcohol.**

Most of the acetic acid of industry is obtained in this way. The process is carried out under a pressure of about five atmospheres and in presence of a manganese salt as a catalyst. The acetaldehyde used is obtained from acetylene (p. 91).

N.B. At one time, acetic acid was derived from the aqueous liquor obtained by the destructive distillation of wood. Lime was added to the distillate (which also contains methyl alcohol and acetone) to convert the acetic acid into calcium acetate, from which the acid was set free by distillation with sulphuric acid.

Properties of acetic acid

(1) With a melting point of 16·7°C, acetic acid may exist as a colourless liquid or a crystalline solid. It is readily soluble in water and is, in fact, hygroscopic. On adding water to glacial acetic acid, there is a contraction in volume and a consequent increase in specific gravity.

(2) Although a *weak acid* by comparison with the mineral acids, acetic acid gives distinctly acidic solutions in water, due to the dissociation

$$CH_3 \cdot C \overset{O}{\underset{OH}{<}} + H_2O \rightleftharpoons CH_3 \cdot C \overset{O}{\underset{O^-}{<}} + H_3O^+$$

whereas alcohols give neutral solutions in water (the difference is easily shown by the use of indicators). Thus the acidity of carboxylic acids shows the effect of an attached carbonyl group on the properties of a hydroxyl group. Clearly, the carbonyl group makes the fission of the O—H bond easier, or alternatively, the carbonyl group makes the anion, R—COO⁻, relatively more stable (we know that the acetate ion, $CH_3 \cdot COO^-$, can exist in water, but alcoholate anions, RO⁻, are decomposed by water giving the alcohol ROH and hydroxide anions).

It is instructive to consider the electronic structure of the acetate anion. For one particular anion, we see at once that two alternative structures can be written:

$$CH_3 \cdot C \overset{O}{\underset{O^-}{<}} \quad \text{and} \quad CH_3 \cdot C \overset{O^-}{\underset{O}{<}}$$

The two structures are exactly equivalent in stability (energy) and there is no reason why one oxygen atom, rather than the other one, should be linked to the carbon atom by a double bond. In fact, the true state of affairs—the actual structure—is believed to be intermediate between the two ' classical ' structures, i.e.

$$CH_3 \cdot C{\overset{\textstyle O}{\underset{\textstyle O^-}{\big<}}} \longleftrightarrow CH_3 \cdot C{\overset{\textstyle O^-}{\underset{\textstyle O}{\big<}}}$$

($\longleftrightarrow$ again represents an ' in-between ' state, p. 98)

or

$$\left. CH_3 C{\overset{\textstyle O}{\underset{\textstyle O}{\lessgtr}}} \right\} -$$

In the latter representation, the negative charge is shared between the two oxygen atoms and each C—O bond has some partial double bond character. The electron pair of the second bond in the carbonyl group is shared out over the system $-C{\overset{\textstyle O}{\underset{\textstyle O}{\lessgtr}}}$ as indicated by the dotted line. Thus this pair of electrons is not localized between two nuclei—the electrons are said to be *delocalized*, such delocalization being associated with *stabilization*.

In contrast, in the alkoxide anion, e.g. $CH_3 \cdot CH_2 \cdot O^-$ (present in sodium ethoxide), there is no similar process whereby the negative charge can be shared; it is, in effect, concentrated on the oxygen atom, explaining why alkoxide ions (relative to acetate anions) are very strong bases, decomposed by water:

$$CH_3 \cdot CH_2 O^- + H_2 O \longrightarrow CH_3 \cdot CH_2 OH + OH^-$$

It should be noted that, if the carbonyl group in aldehydes and ketones is correctly represented (p. 98) by

$$\overset{\delta+}{\underset{}{>}}C\overset{\delta-}{=}O \quad \text{or} \quad >C=O \longleftrightarrow >\overset{+}{C}\!\!-\!O^-$$

then attachment of a negatively charged oxygen atom to the somewhat positive carbon atom (forming the —COO⁻ anion) would be expected to produce an interaction leading to a sharing-out of the negative charge. Similarly, attachment of the electron-rich hydroxyl group to the carbonyl group (forming the —COOH group) effectively destroys the electrophilic centre; consequently, the characteristic addition or addition-elimination reactions of aldehydes and ketones are not shown by carboxylic acids. We can represent the interaction between hydroxyl and carbonyl groups as:

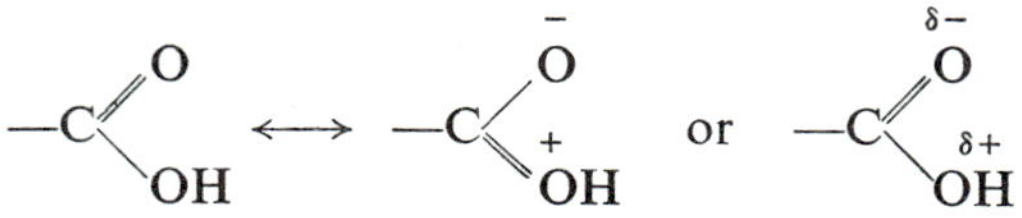

$$-C\begin{subarray}{l}\diagup O \\ \diagdown OH\end{subarray} \longleftrightarrow -C\begin{subarray}{l}\diagup O^{-} \\ \diagdown \overset{+}{OH}\end{subarray} \quad or \quad -C\begin{subarray}{l}\diagup O^{\delta-} \\ \diagdown OH^{\delta+}\end{subarray}$$

Note that the two structures (called *limiting forms* or *limiting structures*) separated by the double-headed arrow are not exactly equivalent. The actual or true state of the —COOH group is undoubtedly nearer to that represented by the classical structure $-C\begin{subarray}{l}\diagup O \\ \diagdown OH\end{subarray}$ than to the dipolar structure $-C\begin{subarray}{l}\diagup O^{-} \\ \diagdown OH^{+}\end{subarray}$. When the limiting forms are exactly equivalent, as with the acetate anion, the true state will be exactly midway between them, and the stabilization due to delocalization will be at a maximum*.

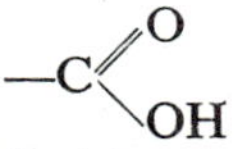

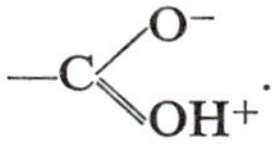

The acidity of acetic acid is shown not only by its action on indicators but also by the following reactions:

(a) With alkalis, acetic acid gives salts and water only:

$$CH_3 \cdot CO\overset{+}{O}\overset{-}{H} + Na\overset{-}{O}\overset{+}{H} \longrightarrow CH_3 \cdot CO\overset{-}{O}\overset{+}{Na} + H_2O$$

Since the acid is weak, aqueous solutions of its salts with the alkali metals are alkaline to litmus owing to hydrolysis. Thus, in titrating acetic acid with caustic soda, it is necessary to use phenolphthalein to indicate the end-point.

(b) The more electropositive metals dissolve in acetic acid evolving hydrogen:

$$2CH_3 \cdot COOH + Mg \longrightarrow (CH_3 \cdot CO\overset{-}{O})_2 Mg^{2+} + H_2$$

* It is important to understand the relation between the limiting forms (they must differ only in the distribution of electrons) and the reasons for the appearance of the charges in the dipolar form. Consider conversion of one form into the other; the process $-C$ leads to $-C$ and the upper oxygen atom acquires two electrons which were previously shared (i.e. net gain of one electron and negative charge); the lower oxygen atom shares two electrons previously unshared (i.e. net loss of one electron and hence appearance of positive charge). Note that the lower oxygen atom in the dipolar form is positively charged and trivalent, as it is in H_3O^+.

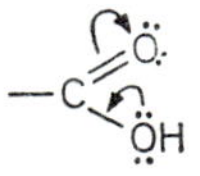

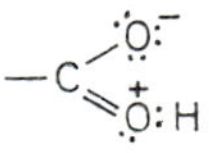

(*c*) With aqueous sodium bicarbonate, carbon dioxide is given off:

$$CH_3{\cdot}COOH + \overset{+}{Na}HCO_3^- \longrightarrow CH_3{\cdot}CO\overset{-}{O}\overset{+}{Na} + H_2O + CO_2$$

This reaction is a convenient laboratory test for a carboxylic acid (but note that mineral acids, and some organic acids other than carboxylic acids will give a positive test, e.g. sulphonic acids) (p. 206).

(3) In addition to forming salts with alkalis, acetic acid reacts with alcohols to form neutral compounds known as *esters* (Chapter 12). These two types of reaction are, in fact, very dissimilar; the reaction with (e.g.) sodium hydroxide (effectively OH^- ions) is a proton transfer—it is virtually instantaneous and complete and so can be used quantitatively (i.e. in a titration). On the other hand, reaction of acetic acid with ethyl alcohol is relatively slow (extremely slow in the absence of anhydrous mineral acid as a catalyst), requires heat, and does not go to completion, a position of equilibrium being reached:

$$CH_3{\cdot}CO_2H + CH_3{\cdot}CH_2OH \rightleftharpoons \underset{\text{ethyl acetate}}{CH_3{\cdot}C\begin{cases} O \\ OCH_2{\cdot}CH_3 \end{cases}} + H_2O$$

The first step involved is thought to be:

$$CH_3{\cdot}C\begin{cases} O \\ OH \end{cases} + \underset{\text{(catalyst)}}{HCl} \rightleftharpoons CH_3{\cdot}C\begin{cases} \overset{+}{O}H \\ OH \end{cases} \longleftrightarrow CH_3{\cdot}\overset{+}{C}\begin{cases} OH \\ OH \end{cases} + Cl^-$$

Protonation of the carboxyl group produces a reactive electrophilic centre at the carbon atom which is then attacked by the oxygen atom of the alcohol molecule (acting as a nucleophile):

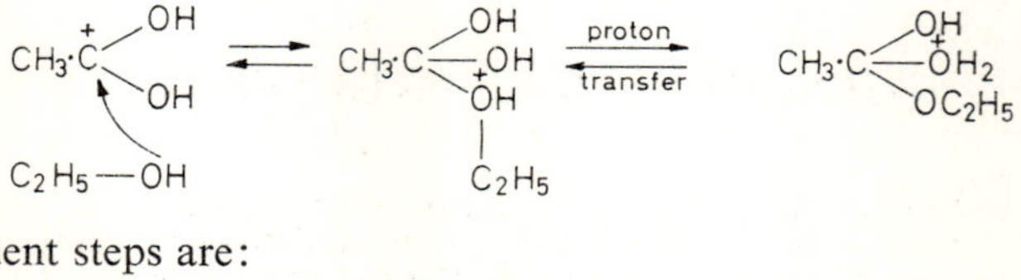

Subsequent steps are:

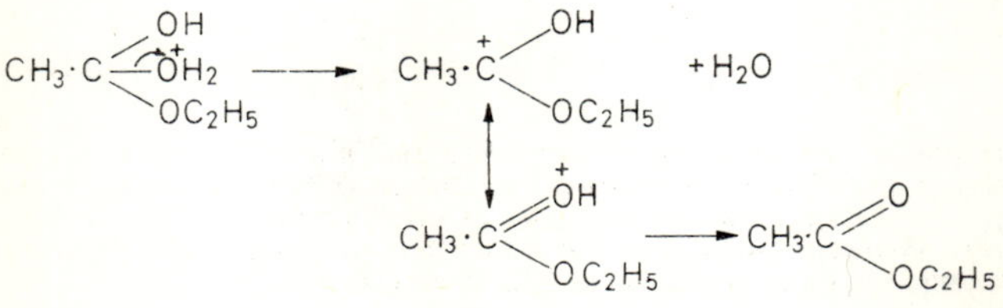

Thus the water produced in the reaction contains oxygen from the carboxylic acid group; the $-OC_2H_5$ group in the ester (ethyl acetate) comes from the alcohol. This can be proved by labelling the oxygen of the alcohol with the heavy isotope ^{18}O; all of the ^{18}O appears in the ester and none in the water.

(4) The hydroxyl group of acetic acid can be replaced by a chlorine atom to form an *acid chloride*. This can be done by treatment of the glacial acetic acid with phosphorus pentachloride or trichloride, or by thionyl chloride, $SOCl_2$ (p. 60):

$$CH_3 \cdot COOH + PCl_5 \longrightarrow CH_3 \cdot C{\overset{\displaystyle O}{\underset{\displaystyle Cl}{<}}} + HCl + POCl_3$$

acetyl chloride

$$3CH_3 \cdot COOH + PCl_3 \longrightarrow 3CH_3 \cdot COCl + P(OH)_3$$

orthophosphorous acid

$$CH_3 \cdot COOH + SOCl_2 \longrightarrow CH_3 \cdot COCl + SO_2 + HCl$$

thionyl chloride

In these reactions, the hydroxyl group in acetic acid behaves similarly to the hydroxyl group in alcohols.

(5) The hydrogen atoms of the methyl group of acetic acid can be replaced successively by chlorine atoms, to form the mono-, di- and tri-chloracetic acids. This is done by passing chlorine through boiling acetic acid in sunlight. The process is speeded up by the presence of a ' halogen carrier ', e.g. iodine, sulphur or red phosphorus.

$$CH_3 \cdot COOH + Cl_2 \longrightarrow CH_2Cl \cdot COOH + HCl$$

monochloracetic acid

$$CH_2Cl \cdot COOH + Cl_2 \longrightarrow CHCl_2 \cdot COOH + HCl$$

dichloracetic acid

$$CHCl_2 \cdot COOH + Cl_2 \longrightarrow CCl_3 \cdot COOH + HCl$$

trichloracetic acid

The strength of these acids increases with the increase in chlorine content (p. 296). The fluorine compound $CF_3 \cdot COOH$ is even stronger than $CCl_3 \cdot COOH$. Note that, in these reactions, an analogy can be seen to the chemistry of aldehydes and ketones.

(6) Reduction of acetic acid to a primary alcohol, formerly difficult, is now easy by means of lithium aluminium hydride:

$$CH_3 \cdot COOH \xrightarrow{\text{LiAlH}_4} CH_3 \cdot CH_2OH$$

The reduction probably proceeds *via* the aldehyde, which is easily reduced to ethyl alcohol:

$$CH_3 \cdot C \overset{O}{\underset{OH}{\big<}} \longrightarrow \left(CH_3 \cdot CH \overset{OH}{\underset{OH}{\big<}} \right) \longrightarrow CH_3 \cdot C \overset{O}{\underset{H}{\big<}} \longrightarrow CH_3 \cdot CH_2OH$$

(7) On heating the acid in a basic solvent (e.g. quinoline), in the presence of a trace of finely divided copper, carbon dioxide is removed to give a hydrocarbon:

$$CH_3 \cdot COOH \overset{Cu}{\longrightarrow} CH_4 + CO_2$$

This removal of the carboxyl group, or *decarboxylation*, can also be done by heating the sodium salt of the acid with an alkali, e.g. soda-lime.

Uses of acetic acid

(1) In the manufacture of acetate esters, e.g. cellulose acetate (Chapter 25).

(2) In the manufacture of metal acetates, e.g. aluminium acetate and the acetates of lead (p. 130).

(3) In the manufacture of white lead (basic lead carbonate) used in making paint.

(4) As a valuable solvent.

(5) As a precipitating agent of casein from milk and latex from rubber.

(6) In the manufacture of acetone and acetic anhydride.

Salts of acetic acid—the acetates

Many metallic acetates are soluble in water. A solution of lead(II) acetate $(CH_3 \cdot COO)_2 Pb$ (most lead salts are insoluble in water) is commonly used in the laboratory as a test for hydrogen sulphide, and also to confirm the presence of the sulphate ion.

Many of the chemical properties of acetates have been given already. They may be summarized as follows:

(1) Mineral acids liberate acetic acid from acetates. Glacial acetic acid is obtained by distilling a mixture of concentrated sulphuric acid and anhydrous sodium acetate:

$$CH_3 \cdot \overset{-}{C}O\overset{+}{O}Na + H_2SO_4 \longrightarrow NaHSO_4 + CH_3 \cdot COOH$$

(2) On heating, the acetates of the alkali and alkaline earth metals evolve acetone.

$$(CH_3 \cdot \overline{COO})_2 Ca^{2+} \longrightarrow CaCO_3 + CH_3 \cdot CO \cdot CH_3$$

This reaction provides a simple laboratory method for the preparation of acetone (and other ketones), used when the corresponding acid is readily available.

(3) If mixed with soda-lime and heated, most acetates (and also acetic acid itself) evolve methane:

$$CH_3 \cdot CO\overset{+}{O}\overset{-}{N}a + NaOH \longrightarrow CH_4 + Na_2CO_3$$

(4) Sodium acetate and acetyl chloride, $CH_3 \cdot COCl$ (p. 136) react together to form acetic anhydride (p. 139):

acetic anhydride

(5) Electrolysis of aqueous solutions of sodium or potassium acetate produces ethane, a phenomenon discovered by Kolbe early in the development of organic chemistry.

The acetate anions, discharging at the anode, probably form transient methyl radicals, which combine to form ethane:

$$2CH_3 \cdot COO^- \longrightarrow \begin{array}{c} CH_3 \\ | \\ CH_3 \end{array} + 2CO_2 + 2e^-$$

Unlike the Wurtz synthesis of hydrocarbons (p. 51), the Kolbe synthesis is a general method of practical utility which can be extended to compounds other than hydrocarbons, e.g.

(6) A neutral aqueous solution of an acetate with a neutral solution of iron(III) chloride gives a deep red coloration. (Formic acid and its salts give a similar reaction, as do many other carboxylic acids.)

(7) On adding silver nitrate to a neutral aqueous solution of an acetate, there is a white precipitate of silver acetate, which is one of the very few sparingly soluble acetates:

$$CH_3 \cdot COO^- + Ag^+ \longrightarrow CH_3 \cdot COOAg\downarrow$$
$$\text{white}$$

The precipitate turns black on exposure to light due to separation of silver, but it may be dried in the dark without any decomposition occurring.

Uses of acetates

Many acetates have important uses in industry. *Verdigris* (basic copper acetate) is used as a green pigment. It is made by covering copper sheets with cloths soaked in vinegar.

Iron(III) and *aluminium acetates* are used in the dyeing of cotton as *mordants*. The cloth to be dyed is dipped into an aqueous solution of the acetate. On subsequent heating with steam, the acetate is hydrolysed to the hydroxide which is left on the threads. In the dye bath, the dye attaches itself strongly to this compound.

Aluminium acetate is also used in waterproofing, the process being similar. The deposit of the aluminium hydroxide renders the material waterproof by filling up the pores.

Lead acetate, $(CH_3 \cdot COO)_2Pb \cdot 3H_2O$, or ' sugar of lead ', is used in medicine. The *tetra-acetate*, $(CH_3 \cdot COO)_4Pb$, is a useful oxidizing agent and, being soluble in organic solvents, is particularly valuable in organic chemistry.

Formic acid, $HC\overset{\displaystyle O}{\underset{\displaystyle OH}{\diagdown}}$

Formic acid was first obtained by Marggraf in 1749 by distilling red ants. The compound is also present in honey bees and in stinging nettles. It can be prepared by catalytic oxidation of methyl alcohol or formaldehyde, or by combination of carbon monoxide and steam at high pressures:

$$CO + H_2O \longrightarrow HCOOH$$

and by other methods which are not of general interest.

Properties of formic acid

Formic acid is a colourless corrosive liquid, with a pungent smell. It is a much stronger acid than acetic.

Writing the structural formula as $H-C\overset{\displaystyle O}{\underset{\displaystyle OH}{\diagdown}}$ it can be seen that the molecule contains the functional group —CHO as well as the —COOH group. Formic acid and its derivatives do show some resemblance to aldehydes, particularly in being reducing agents.

The following properties of formic acid (and formates) are peculiar to the acid and in marked contrast to acetic acid and the series of carboxylic acids in general:

(1) Formic acid and formates show most of the reducing properties of aldehydes (Chapter 8). Carbon dioxide and carbonates, respectively, and water are the oxidation products:

$$HCOOH \xrightarrow{\text{oxidation}} CO_2 + H_2O$$

Formic acid reduces ammoniacal silver oxide, potassium permanganate, potassium and sodium dichromates. It also reduces mercury(II) chloride, first to the white insoluble mercury(I) chloride and then to grey metallic mercury. It does *not* reduce Fehling's solution, although its methyl and ethyl esters do to some slight extent.

(2) With concentrated sulphuric acid, formic acid and formates evolve carbon monoxide (and provide a good laboratory method of preparing this gas):

$$HCOONa + H_2SO_4 \longrightarrow NaHSO_4 + CO + H_2O$$

(3) With phosphorus pentachloride, formic acid forms carbon monoxide and hydrogen chloride (not the acid chloride):

$$HCOOH + PCl_5 \longrightarrow POCl_3 + CO + 2HCl$$

(4) The action of heat on formates is as follows:

(*a*) The formates of the alkali metals yield oxalates and hydrogen:

$$\begin{matrix} H \vert COONa \\ + \\ H \vert COONa \end{matrix} \longrightarrow Na_2C_2O_4 + H_2$$
sodium oxalate (p. 174)

(This is a useful laboratory method of obtaining pure hydrogen, and, in particular, hydrogen free from the hydrides of elements, e.g. C_2H_2, PH_3, AsH_3, such as may be present in hydrogen obtained by the action of an acid on a metal.)

(*b*) The formates of the alkaline earth metals yield formaldehyde:

$$(HCOO)_2Ca \longrightarrow CaCO_3 + HC\begin{matrix} \diagup H \\ \diagdown O \end{matrix}$$

This reaction resembles that of calcium acetate, which gives acetone.

SUMMARY OF THE CHEMISTRY OF ALIPHATIC CARBOXYLIC ACIDS

The general formula is $C_nH_{2n+1}·COOH$
The first member is HCOOH, formic acid.
The representative member is $CH_3·COOH$, acetic acid.

Acetic acid,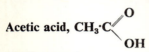

(1) *Preparation*

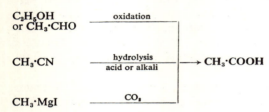

(2) *Manufacture*

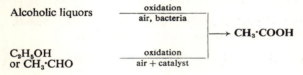

(3) *Properties*

(a) Weak *monobasic* acid, partially dissociated in aqueous solution: Forms salts with alkalis:

$$CH_3·COOH + Na\overset{+}{O}\overset{-}{H} \longrightarrow CH_3·CO\overset{-}{O}\overset{+}{Na} + H_2O$$

Gives hydrogen with strongly electropositive metals:

$$2CH_3·COOH + Mg \longrightarrow (CH_3·C\overset{-}{O}O)_2Mg^{2+} + H_2$$

Reacts with sodium bicarbonate giving carbon dioxide:

$$2CH_3COOH + Na\overset{+}{H}C\overset{-}{O_3} \longrightarrow CH_3CO\overset{-}{O}\overset{+}{Na} + CO_2 + H_2O$$

(b) Forms esters with alcohols, in equilibrium mixture:

$$CH_3·COOH + C_2H_5OH \rightleftharpoons CH_3·COOC_2H_5 + H_2O$$
ethyl acetate

(c) Forms the acid chloride with phosphorus trichloride and penta-chloride and with thionyl chloride ($SOCl_2$):

$$CH_3 \cdot COOH \longrightarrow CH_3 \cdot COCl$$
$$\text{acetyl chloride}$$

(d) Hydrogen atoms of methyl group replaceable by chlorine:

$$CH_3 \cdot COOH \rightarrow CH_2Cl \cdot COOH \longrightarrow CHCl_2 \cdot COOH \rightarrow CCl_3 \cdot COOH$$
$$\text{mono-} \qquad\qquad\qquad \text{di-} \qquad\qquad \text{tri-chloracetic acid}$$

(e) Undergoes decarboxylation:

$$CH_3 \cdot COOH \xrightarrow[\text{heat}]{\text{Cu}} CH_4 + CO_2$$

(f) Reduction by lithium aluminium hydride gives ethyl alcohol:

$$CH_3 \cdot COOH \xrightarrow{\text{LiAlH}_4} CH_3 \cdot CH_2OH$$

(g) The —COOH group can be represented as

explaining lack of aldehyde-ketone properties (the $\diagdown C{=}O$ group is modified by the attached —OH group).

Sodium acetate, $CH_3 \cdot COONa$

Properties

$$CH_3 \cdot \overset{-}{C}OO\overset{+}{N}a \xrightarrow{\text{conc. H}_2\text{SO}_4} CH_3 \cdot COOH$$

$$\xrightarrow{\text{heat}} (CH_3)_2CO$$
$$\text{acetone}$$

$$\xrightarrow[\text{heat}]{\text{NaOH}} CH_4$$

$$\xrightarrow{\text{CH}_3 \cdot \text{COCl}} CH_3 \cdot CO \cdot O \cdot CO \cdot CH_3$$
$$\text{acetic anhydride}$$

$$\xrightarrow[\text{(Kolbe)}]{\text{electrolysis}} C_2H_6 + 2CO_2 \quad \text{(at anode)}$$

The acetate anion is stabilized by 'delocalization of electrons', i.e.

No similar delocalization can stabilize $CH_3 \cdot CH_2 \cdot O^-$ which is therefore a much stronger base than the acetate anion (i.e. acetic acid is a stronger ' acid ' than ethyl alcohol).

Formic acid, $HC \big\langle \begin{smallmatrix} O \\ OH \end{smallmatrix}$

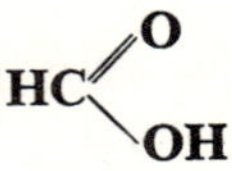

Properties

The following are properties peculiar to formic acid (the only member of the series containing the —CHO group).

(*a*) Shows reducing properties, being oxidized to carbon dioxide and water; it reduces:

(*i*) Ammoniacal silver oxide $\longrightarrow Ag\downarrow$

(*ii*) $MnO_4^- \longrightarrow Mn^{2+}$

(*iii*) $Cr_2O_7^{2-} \longrightarrow Cr^{3+}$

(*iv*) $HgCl_2 \longrightarrow Hg_2Cl_2\downarrow \longrightarrow 2Hg\downarrow$

It does *not* reduce Fehling's solution.

(*b*) The acid and also formates give carbon monoxide with concentrated sulphuric acid:

$$\overset{-}{H}COO\overset{+}{N}a + H_2SO_4 \longrightarrow \overset{+}{N}aHSO_4^{-} + CO + H_2O$$

(*c*) With phosphorus pentachloride, gives carbon monoxide and hydrogen chloride, *not* formyl chloride:

$$HCOOH + PCl_5 \longrightarrow POCl_3 + CO + HCl$$

The formates

Sodium formate on heating, gives the oxalate and hydrogen:

$$2HCOONa \longrightarrow Na_2C_2O_4 + H_2$$

Calcium formate, on heating, gives formaldehyde:

$$(HCOO)_2Ca \longrightarrow CaCO_3 + HCHO$$

EXAMINATION QUESTIONS

(1) The first member of a homologous series frequently differs in its reactions from the general chemical behaviour of the series. Illustrate this statement by reference to (*a*) the monohydric alcohols, (*b*) the aldehydes, (*c*) the monocarboxylic acids.

You are given three test-tubes containing respectively an alcohol, an aldehyde and formic acid. Describe the positive tests (**one** in each case) which you would carry out in the laboratory to distinguish each of these from the other two. State clearly what you would do and what you would see. (J. M. B. A.)

(2) If provided with a ten per cent solution of acetic acid, describe how you would obtain a specimen of glacial acetic acid.

How, and under what conditions, does acetic acid react with (*a*) chlorine, (*b*) phosphorus trichloride, (*c*) ethyl alcohol?

Give **two** chemical properties of formic acid in which it differs from acetic acid. (O. A.)

11

DERIVATIVES OF THE ALIPHATIC CARBOXYLIC ACIDS

Functional Groups

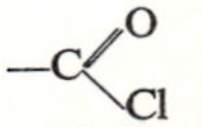

acid chlorides

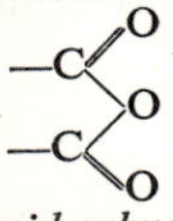

acid anhydrides

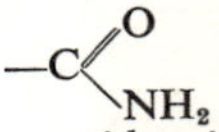

acid amides

ACETIC acid derivatives of these types have been mentioned already. They will be considered as representative members.

Acetyl chloride, $CH_3 \cdot C \overset{O}{\underset{Cl}{\Large\diagdown}}$

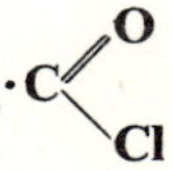

The monovalent group $CH_3 \cdot CO—$ is known as the *acetyl* group. In acetic acid it may be considered as being combined with a hydroxyl group. Acetyl chloride is thus the product of the replacement of the hydroxyl group of acetic acid by a chlorine atom. The following are the standard methods of carrying out this replacement and any one of them may be used in the laboratory as a method of preparation.

Preparation of acetyl chloride

(1) By the action of thionyl chloride on glacial acetic acid:

$$CH_3 \cdot COOH + SOCl_2 \longrightarrow CH_3 \cdot COCl + HCl + SO_2$$

Ten grammes of glacial acetic acid are cautiously mixed with 20 g of thionyl chloride in a 100 ml flask fitted with a reflux condenser as shown in *Figure 11.1*. The flask is warmed in a water bath* until evolution of gases stops. The anhydrous calcium chloride (or silica gel) tube prevents the ingress of moisture.

The above quantities are equimolar proportions and since the by-products of the reaction are gaseous, the residue in the flask should be pure acetyl chloride. It is, however, distilled from a water bath and the fraction coming over between 53° and 56°C is collected (the b.p. of acetyl chloride is 55°C).

* The preparation should be carried out in a fume cupboard.

(2) By the action of phosphorus trichloride on glacial acetic acid:

$$3CH_3 \cdot COOH + PCl_3 \longrightarrow 3CH_3 \cdot COCl + P(OH)_3$$

The phosphorus trichloride is dropped slowly on to the acetic acid contained in a flask kept cool by surrounding it with cold water. The flask is then warmed for some time to complete the

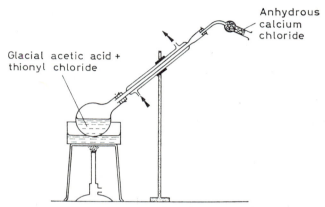

Figure 11.1. Preparation of acetyl chloride.

reaction and the contents fractionally distilled to obtain the acetyl chloride.

(3) By the action of phosphorus pentachloride on glacial acetic acid:

$$CH_3 \cdot COOH + PCl_5 \longrightarrow CH_3 \cdot COCl + POCl_3 + HCl$$

Properties of acetyl chloride

Acetyl chloride is a chemically reactive, colourless liquid (b.p. 55°C). It fumes strongly in moist air and because of this is an unpleasant liquid to work with. It must be stored in bottles fitted with ground glass stoppers. Further, as a precaution after use, the stopper is sealed into the neck by coating with molten paraffin wax.

(1) Acetyl chloride is hydrolysed extremely vigorously by water, the acid being regenerated. Care should be taken when carrying out this reaction and a very small quantity of acetyl chloride should be used:

$$CH_3 \cdot COCl + H_2O \longrightarrow CH_3 \cdot COOH + HCl$$

The corresponding hydrolysis by alkali is dangerously vigorous.

(2) Acetyl chloride reacts readily with alcohols to form esters:

$$CH_3 \cdot COCl + C_2H_5OH \longrightarrow CH_3 \cdot COOC_2H_5 + HCl$$
ethyl acetate

(3) Acetyl chloride reacts with ammonia to form acetamide:

$$CH_3 \cdot COCl + NH_3 \longrightarrow CH_3 \cdot CO \cdot NH_2 + HCl$$
acetamide $\downarrow$

$$(NH_4Cl)$$

This reaction is dangerously violent. It is moderated by using ammonium carbonate as a source of ammonia.

Primary amines (p. 160), e.g. ethylamine, $C_2H_5NH_2$, give a substituted amide, but here, as with ammonia, the reactions are most vigorous:

$$CH_3 \cdot COCl + C_2H_5NH_2 \longrightarrow CH_3 \cdot CO \cdot NHC_2H_5 + HCl$$
N-ethylacetamide

(4) With acetyl chloride and anhydrous sodium acetate, the acid anhydride is formed (p. 139):

$$CH_3 \cdot COCl + Na\overset{+}{O}\overset{-}{O}C \cdot CH_3 \longrightarrow NaCl + (CH_3CO)_2O$$
acetic anhydride

A similar reaction occurs with other metal acetates and can be used to prepare anhydrous chlorides of metals, since the acetic anhydride formed removes any water present in the metal acetate.

(5) In the presence of aluminium chloride, acetyl chloride reacts with the aromatic hydrocarbon *benzene*, to form a ' mixed ' *aliphatic-aromatic ketone*. The reaction is an example of the *Friedel–Crafts reaction* described more fully on p. 207:

$$CH_3 \cdot COCl + C_6H_6 \xrightarrow{\ AlCl_3\ } HCl + CH_3 \cdot CO \cdot C_6H_5$$
benzene   acetophenone

The carbonyl group in acetyl chloride

As we have seen, the properties of the carbonyl group, exemplified by the reactions of aldehydes and ketones, are modified in the carboxyl group, —COOH, by the attached hydroxyl group. Effectively, the electron-rich hydroxyl group ' neutralizes ' the electrophilic centre:

In contrast, when the carbonyl group is directly attached to a strongly electronegative chlorine atom, as in an acid chloride, the electrophilic character of the carbonyl carbon atom is strengthened:

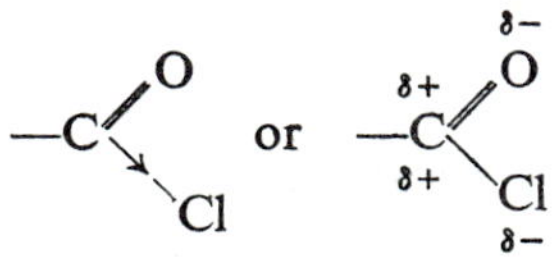

Thus acid chlorides are *more* reactive towards nucleophiles than are aldehydes and ketones; acid chlorides will react even with relatively ' weak ' nucleophiles such as water or alcohols:

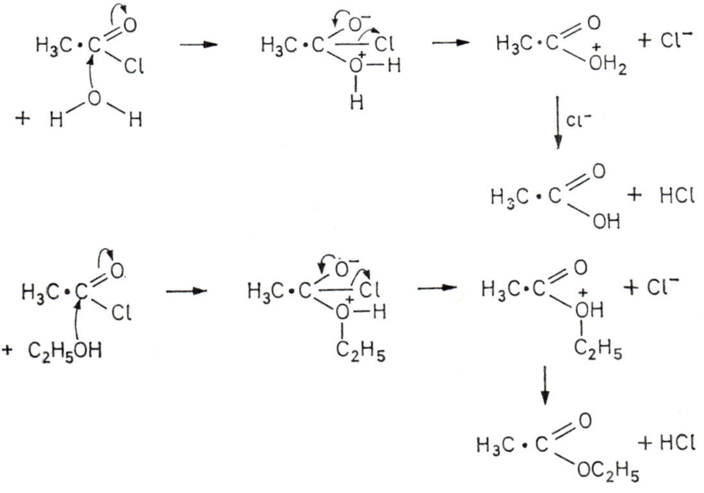

The other reactions of acetyl chloride given above can be formulated in a similar way, e.g.

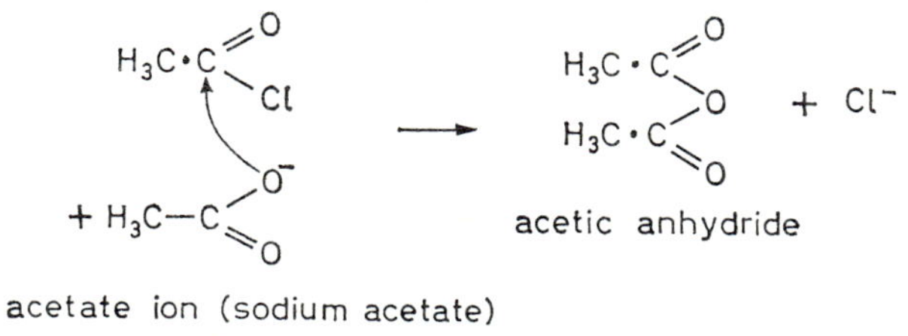

acetic anhydride

acetate ion (sodium acetate)

Acetic anhydride, (CH₃CO)₂O

Acetic anhydride is the product formed by the removal of one molecule of water from two molecules of the acid.

Preparation of acetic anhydride

The anhydride is prepared by the interaction of acetyl chloride and sodium acetate (see previous section).

This method of preparation is used on the large scale to manufacture the compound. In another industrial-scale method, acetic acid vapour is ' cracked ' at 700°C in presence of a catalyst to give the substance *ketene*, a very reactive compound:

$$CH_3 \cdot COOH \longrightarrow \underset{\text{ketene}}{CH_2{=}C{=}O} + H_2O$$

The ketene reacts with acetic acid to give the anhydride:

$$CH_3 \cdot COOH + CH_2{=}C{=}O \longrightarrow (CH_3CO)_2O$$

Properties of acetic anhydride

Acetic anhydride is a colourless liquid, b.p. 137°C. It is not very soluble in cold water, but dissolves readily in warm water, when it is hydrolysed. Although not so active chemically as acetyl chloride, it resembles it closely in chemical behaviour. Both reagents provide a means of introducing the $CH_3 \cdot CO{-}$ or acetyl group. The introduction of the acetyl group into a compound is known as *acetylation* and both acetyl chloride and acetic anhydride are thus *acetylating agents*. Acetic anhydride is more commonly used, having the advantage that it does not fume in air, and is not so vigorous in its action as is acetyl chloride.

(1) Acetic anhydride is hydrolysed by warm water and more quickly by alkalis to give acetic acid and an acetate respectively:

$$(CH_3 \cdot CO)_2O + H_2O \longrightarrow 2CH_3 \cdot COOH$$

$$(CH_3 \cdot CO)_2O + 2NaOH \longrightarrow 2CH_3 \cdot COONa + H_2O$$

(2) With alcohols, acetic anhydride forms esters and the free acid:

(3) Acetic anhydride reacts with ammonia to form acetamide and ammonium acetate:

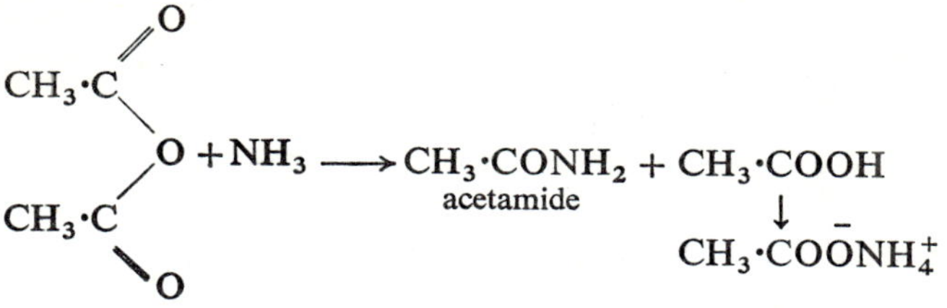

$$CH_3 \cdot C \overset{O}{\underset{}{\diagup}} \quad O + NH_3 \longrightarrow CH_3 \cdot CONH_2 + CH_3 \cdot COOH$$

$$CH_3 \cdot C \overset{}{\underset{O}{\diagdown}}$$

acetamide

$$\downarrow$$

$$CH_3 \cdot CO\overline{O}NH_4^+$$

Amines are acetylated by a similar process to give substituted amides (p. 165):

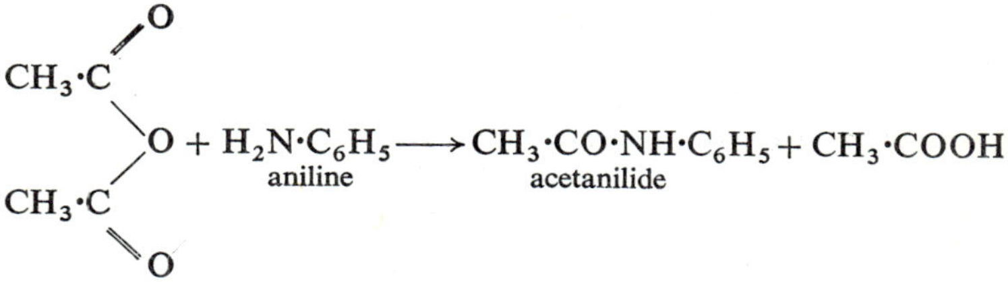

$$CH_3 \cdot C \overset{O}{\diagup}$$
$$\qquad O + H_2N \cdot C_6H_5 \longrightarrow CH_3 \cdot CO \cdot NH \cdot C_6H_5 + CH_3 \cdot COOH$$
$$CH_3 \cdot C \overset{}{\underset{O}{\diagdown}}$$

aniline $\qquad$ acetanilide

Note the similarity of the above reactions to those of acetyl chloride:

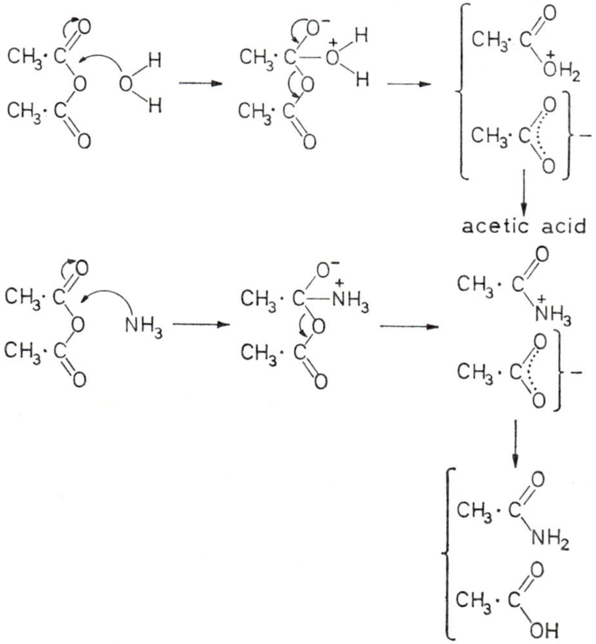

acetic acid

Acetamide, $CH_3 \cdot C \underset{NH_2}{\overset{O}{<}}$

Acetamide is representative of the class of compounds known as *acid amides*, the characteristic group of which is $-C \underset{NH_2}{\overset{O}{<}}$. The group is derived from the carboxyl group by replacing its hydroxyl group by the monovalent $-NH_2$ or amino group.

Preparation of acetamide

Acetamide can be obtained by the action of ammonia on acetyl chloride or acetic anhydride (pp. 138 and 140).

In the laboratory, acetamide is conveniently prepared from ammonium acetate, a quantity of which is refluxed for about two hours with glacial acetic acid as solvent:

$$CH_3 \cdot CO\overset{-}{O}\overset{+}{N}H_4 \xrightarrow{CH_3 \cdot COOH} CH_3 \cdot C \underset{NH_2}{\overset{O}{<}} + H_2O$$

The acetamide is recovered from the reaction mixture by distillation in the apparatus shown below.

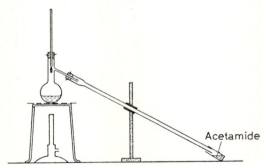

Figure 11.2. Preparation of acetamide.

Since the b.p. of acetamide is 222°C it is essential to use an air condenser. The fraction collecting between 213° and 224°C is collected. The melting point of acetamide is 82°C, the compound may solidify in the condenser tube, in which case the tube must be warmed to get the acetamide into the receiver.

As prepared above, acetamide smells strongly of mice. This smell is due to an impurity which can be removed by crystallizing from acetone, when an odourless specimen of acetamide is obtained.

Properties of acetamide

Acetamide is a white crystalline solid, as are most amides. Formamide, $HCONH_2$, is exceptional in being a liquid at room temperature (a liquid which incidentally has a high dielectric constant of 84).

(1) Acetamide and amides in general are essentially neutral, in contrast to amines (e.g. methylamine, $CH_3 \cdot NH_2$) which are decidedly basic. This difference between amides and amines is obviously due to the presence in amides of the carbonyl group which, when attached to an amino-group, reduces the electron density on the nitrogen atom, hence reducing the basicity.

(2) Acetamide is hydrolysed slowly by boiling with water to give the ammonium salt:

$$CH_3 \cdot CONH_2 + H_2O \xrightarrow{\text{boil}} CH_3 \cdot COO\overset{-}{N}\overset{+}{H_4}$$

Hydrolysis proceeds much more rapidly in the presence of a mineral acid or an alkali; with alkali, ammonia is liberated and can be detected by the usual tests.

N.B. Ammonium acetate also gives ammonia and sodium acetate on treatment with caustic soda, but this reaction will take place in the cold, whereas that with acetamide requires heating. This different behaviour serves to distinguish between acetamide and ammonium acetate.

(3) Acetamide is dehydrated to a nitrile (p. 156) by distillation with phosphorus pentoxide:

$$CH_3 \cdot C \underset{NH_2}{\overset{O}{\diagup}} \xrightarrow{P_2O_5} CH_3 \cdot CN + H_2O$$

acetonitrile
or
methyl cyanide

(4) On treatment with bromine and aqueous or alcoholic potash, acetamide is converted into methylamine:

$$CH_3 \cdot CONH_2 \xrightarrow[KOH]{Br_2} CH_3 \cdot NH_2$$
methylamine

This change, known as *Hofmann's reaction*, is obviously not a straightforward one since it involves a *rearrangement* of the molecule (the nitrogen atom originally attached to the carbonyl group becomes attached to the methyl group). Although complicated in mechanism (a discussion of which can be found in more advanced books) the reaction is of practical utility (see p. 163).

(5) With lithium aluminium hydride the acid amide is reduced to a primary amine:

$$CH_3 \cdot CO \cdot NH_2 \xrightarrow{\text{LiAlH}_4} CH_3 \cdot CH_2 NH_2$$
$$\text{ethylamine}$$

It should be clear from the properties given above, that acetamide differs considerably from acetyl chloride and acetic anhydride; it is much more stable and, unlike the other two compounds, is not used as an acetylating agent. The alkaline hydrolysis of acetamide can, however, be formulated in similar terms to the reactions of acetyl chloride (p. 139), i.e.

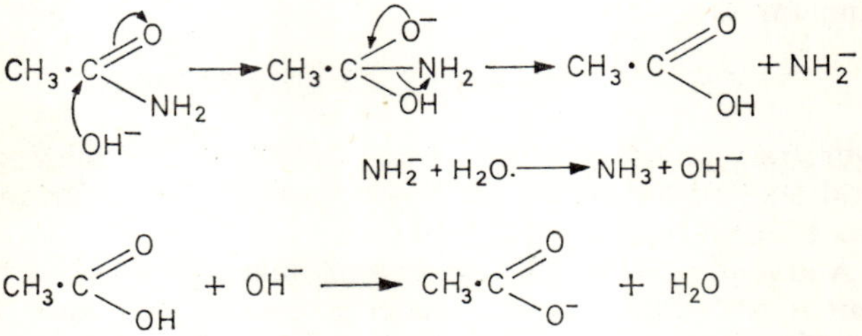

$$NH_2^- + H_2O \longrightarrow NH_3 + OH^-$$

SUMMARY OF THE CHEMISTRY OF ACID CHLORIDES, ACID ANHYDRIDES AND ACID AMIDES

Acid chlorides

The general formula is $C_nH_{2n+1} \cdot C \overset{\displaystyle O}{\underset{\displaystyle Cl}{\diagdown}}$

The first and representative member is $CH_3 \cdot COCl$, acetyl chloride (formyl chloride is unstable).

(1) *Preparation*

$$CH_3 \cdot COOH \xrightarrow[\substack{\text{or PCl}_3 \\ \text{or PCl}_5}]{\text{SOCl}_2} CH_3 \cdot COCl$$

(2) *Properties*

Acid anhydrides

The general formula is
$$C_nH_{2n+1}C \underset{\displaystyle C_nH_{2n+1}C}{\overset{\displaystyle O}{\diagdown}} O$$

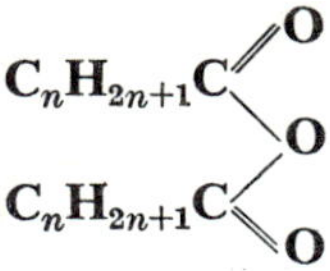

The first and representative member is $CH_3 \cdot C \diagup^O_{\diagdown} O$, acetic anhydride.

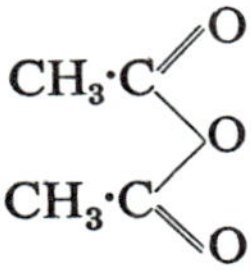

(No corresponding anhydride of formic acid is known.)

(1) *Preparation*

$$CH_3 \overset{-}{COO}\overset{+}{Na} + CH_3 \cdot COCl \longrightarrow \begin{array}{c} CH_3 \cdot C \\ CH_3 \cdot C \end{array}$$

(2) *Properties*

$$CH_3 \cdot C \diagup^O_{\diagdown} O \begin{cases} \xrightarrow{H_2O} CH_3 \cdot COOH \\ \xrightarrow{NaOH} CH_3 \cdot COONa \\ \xrightarrow{C_2H_5OH} CH_3 \cdot COOC_2H_5 \\ \qquad\quad + CH_3 \cdot COOH \\ \xrightarrow{NH_3} CH_3 \cdot CO \cdot NH_2 \\ \qquad\quad + CH_3 \cdot \overset{-}{COO}\overset{+}{NH_4} \end{cases}$$

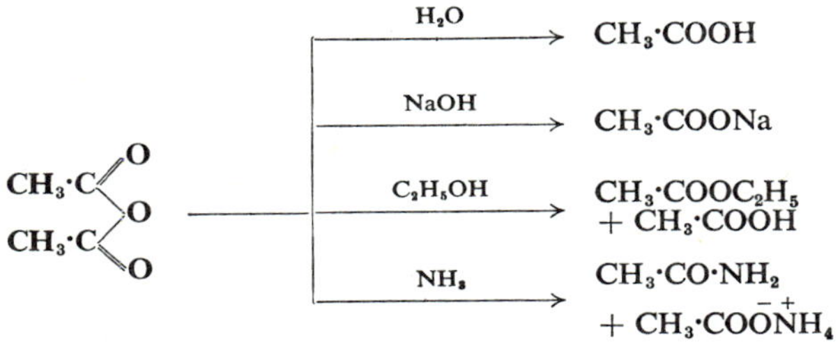

Acid amides

The general formula is $C_nH_{2n+1} \cdot C \diagup^O_{\diagdown NH_2}$

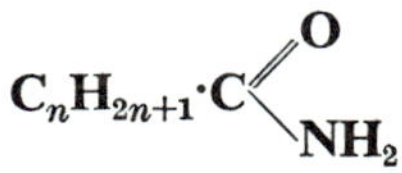

The first member is $H{-}C \diagup^O_{\diagdown NH_2}$, formamide.

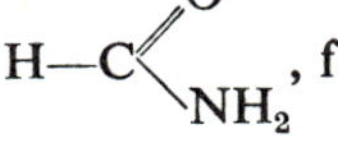

The representative member is $CH_3 \cdot C \diagup^O_{\diagdown NH_2}$, acetamide.

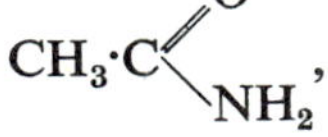

(1) *Preparation*

$$CH_3 \cdot \overset{-}{COO}\overset{+}{NH_4} \xrightarrow[\text{glacial acetic acid}]{\text{dehydration}} CH_3 \cdot CO \cdot NH_2$$

(2) *Properties*

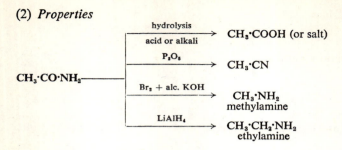

EXAMINATION QUESTIONS

(1) Describe fully **one** method of preparation, the physical properties and **five** important chemical properties of acetyl chloride. Give **one** simple test to distinguish it from acetic anhydride.   (S. A.)

(2) The chemical structures assigned to aldehydes, ketones, amides and carboxylic acids all contain the carbonyl group. To what extent is the structural similarity reflected in the chemical behaviour of these compounds? (Liverpool, B.Sc., Part I)

12

ESTERS

Functional Group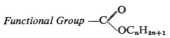

THE esters of the aliphatic carboxylic acids of general formula $C_nH_{2n+1} \cdot COOC_mH_{2m+1}$ have been referred to already. Isomerism is common in the series, even among the low members. Ethyl formate, $HCOOC_2H_5$, is isomeric with methyl acetate $CH_3 \cdot COOCH_3$ and these are isomeric with propionic acid, $C_2H_5 \cdot COOH$.

The lower members are volatile liquids with pleasant odours; as the molecular weight increases, volatility decreases, and the compounds become viscous, oily liquids or waxy solids. Many esters occur naturally as the sweet-smelling constituents of plants.

Ethyl acetate, $CH_3 \cdot COOC_2H_5$

One of the commonest esters is ethyl acetate, $CH_3 \cdot COOC_2H_5$, and this is taken as the representative one for detailed study.

Preparation of ethyl acetate

Esters were studied as early as 1780 by Scheele who discovered methods of preparation which are the same in principle as those in use to-day.

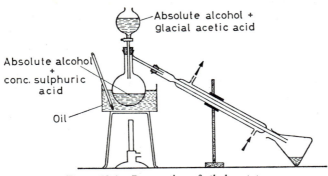

Figure 12.1. Preparation of ethyl acetate.

(1) By the interaction of ethyl alcohol and glacial acetic acid in the presence of concentrated sulphuric acid:

$$CH_3 \cdot COOH + C_2H_5OH \rightleftharpoons CH_3 \cdot COOC_2H_5 + H_2O$$

This is the usual laboratory preparation.

The reaction from left to right is known as *esterification*. Although this equation seems equivalent to the neutralization reaction of inorganic chemistry, i.e.

$$acid + base \longrightarrow salt + water$$

there is a marked contrast, in that these organic reactions reach an equilibrium position, and never go to completion; moreover they are relatively slow.

The Fischer–Speier method of preparing ethyl acetate consists in passing hydrogen chloride through the mixture of alcohol and acetic acid while being refluxed. Here, the acid is a catalyst only and the yield of ester is the equilibrium yield. There is some advantage in the use of hydrogen chloride since no olefin is formed from the alcohol, as does occur to some extent with concentrated sulphuric acid. Frequently only a small proportion of hydrogen chloride is used.

A mixture of equal volumes of concentrated sulphuric acid and absolute alcohol (50 ml of each) is made in a 500 ml distillation flask fitted with a tap funnel and attached to a Liebig condenser. A mixture of equal volumes of glacial acetic acid and absolute alcohol (100 ml of each) is put into the tap funnel (*Figure 12.1*).

The flask is heated in an oil bath at a temperature of 140°C. Then the mixture is run in from the tap funnel at the same rate as the distillate comes over.

In addition to ethyl acetate, the distillate contains ethyl alcohol, ether and sulphur dioxide. The distillate is transferred to a tap funnel. It is purified by shaking with aqueous sodium carbonate until there is no further effervescence, indicating removal of the acids originally present. The lower aqueous layer is drawn off. The remaining layer, containing the ethyl acetate, is now shaken with a concentrated solution of calcium chloride which removes any ethyl alcohol by combining with it (p. 60). The lower layer is again drawn off. The remainder, consisting of ethyl acetate and ether and a small amount of water, is allowed to stand for some short time over a few pieces of calcium chloride and then distilled over a water bath. Ether first comes over at a temperature range of 35–40°C. Ethyl acetate (b.p. 77°C) is collected between 74° and 79°C.

The mechanism of the acid-catalysed esterification of carboxylic acids has been discussed previously (p. 126).

(2) By the action of ethyl alcohol on acetyl chloride or acetic anhydride (pp. 138, 140):

$$C_2H_5OH + CH_3 \cdot COCl \longrightarrow CH_3 \cdot COOC_2H_5 + HCl$$
$$C_2H_5OH + (CH_3 \cdot CO)_2O \longrightarrow CH_3 \cdot COOC_2H_5 + CH_3 \cdot COOH$$

(3) By the interaction of silver acetate and an ethyl halide:

$$CH_3{\cdot}COOAg + C_2H_5I \longrightarrow CH_3{\cdot}COOC_2H_5 + AgI$$

By contrast with method (1), the reactions in methods (2) and (3) proceed to completion.

Properties of ethyl acetate

(1) Ethyl acetate is hydrolysed very slowly on warming with water, an equilibrium being established after a long time:

$$CH_3{\cdot}COOC_2H_5 + H_2O \rightleftharpoons CH_3{\cdot}COOH + C_2H_5OH$$

The time taken to reach equilibrium is cut down considerably by the presence of mineral acids; effectively, the reactions involved in esterification (p. 126) are reversed in the presence of a large excess of water.

Complete hydrolysis of an ester is usually achieved by heating it with sodium or potassium hydroxide dissolved in water, or frequently, in aqueous ethanol (provided that the addition of ethanol does not complicate the isolation of the products!). Alkaline hydrolysis of esters is similar in mechanism to reactions of other acid derivatives previously discussed:

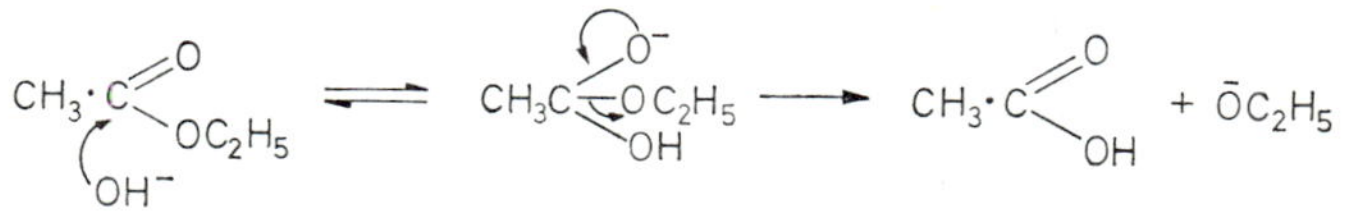

followed by an irreversible step which drives the reaction to completion:

$$CH_3{\cdot}C\begin{matrix}\nearrow O\\ \searrow OH\end{matrix} + \bar{O}C_2H_5 \longrightarrow CH_3{\cdot}C\begin{matrix}\nearrow O\\ \searrow \bar{O}\end{matrix} + HOC_2H_5$$

This hydrolysis with an alkali is sometimes referred to as *saponification,* a term originating in the soap industry (p. 152) where a similar hydrolysis is carried out.

An ethyl ester can be identified by boiling a little of it with caustic soda and collecting, in this case, the ethyl alcohol distilling over. This can be identified by the iodoform test.

(2) Ethyl acetate reacts slowly with ammonia to form acetamide and ethyl alcohol:

$$CH_3 \cdot COOC_2H_5 + NH_3 \longrightarrow CH_3 \cdot CONH_2 + C_2H_5OH$$

Mechanism:

(3) Ethyl acetate is reduced to ethyl alcohol by lithium aluminium hydride:

$$CH_3 \cdot COOC_2H_5 \xrightarrow{\text{LiAlH}_4} CH_3 \cdot CH_2OH$$

Uses of ethyl acetate and other esters

The following are some of the more important uses of acetates and the esters of the lower aliphatic carboxylic acids. A reference to the uses of the esters of the higher acids is made on p. 152.

(1) As solvents in the making of lacquers.
(2) In the manufacture of artificial silk.
(3) In the manufacture of some plastics.
(4) In the manufacture of synthetic perfumes and flavouring essences.

Waxes, fats and oils

These compounds which are widely distributed throughout the animal and vegetable kingdoms are, in general, high molecular weight esters derived from long chain 'fatty' acids, e.g. palmitic acid, $C_{15}H_{31} \cdot COOH$, stearic acid, $C_{17}H_{35} \cdot COOH$ and oleic acid $C_{17}H_{33} \cdot COOH$. The first two are members of the series of carboxylic acids of general formula $C_nH_{2n+1} \cdot COOH$ and are saturated compounds. Oleic acid is a member of a similar series of acids derived from ethylene and, like ethylene, is unsaturated containing a double bond. These acids are combined with alcohols, the commonest of which is glycerol, the trihydric alcohol of formula $CH_2OH \cdot CHOH \cdot CH_2OH$, but high members of the monohydric alcohols of general formula $C_nH_{2n+1}OH$ are also found.

The nature of these substances has been known for some time. Scheele heated olive oil with litharge and obtained glycerine. The Frenchman Chevreul, at the beginning of the nineteenth century, showed that fatty substances were compounds of glycerol and acids of high molecular weight. Half way through the century, Berthelot actually synthesized some simple fats by heating mixtures of glycerol and fatty acids in closed vessels to temperatures of around 300°C.

In Nature, fatty material is a reserve food. It is stored in various parts of the animal body, sometimes just under the skin where its non-conducting power as well as its high calorific value is particularly valuable during the winter of the temperate zones. Many of our hibernating animals rely on a fat store, built up in the summer months, to keep them alive during the hibernation period. In the plant world, it is a common constituent of the food store of the seed, enabling it to grow until it can develop roots and leaves to obtain food from its environment.

The classification of fatty materials into waxes, fats and oils is based on melting point. *Waxes* are solids with a melting point above the body temperature. *Fats* are solids at room temperature, and *oils*, at the same temperature, are liquids. Beeswax, the material from which the honey bee makes its comb, is a mixture of esters of saturated acids containing 26 to 28 carbon atoms, and saturated monohydric alcohols. Spermaceti from the head of the sperm whale, is chiefly the palmitate of a monohydric alcohol of formula $C_{16}H_{33}OH$ (cetyl alcohol). Most fats and oils are glycerol esters. Oils contain a higher proportion of esters from unsaturated acids. They are more common in the vegetable than in the animal kingdom, except in the case of fish oils which actually contain a greater proportion of unsaturated acids than do vegetable oils.

Animal fats and oils are separated from the fatty tissue by direct heating—the process of *rendering*—or by means of steam. Vegetable oils are extracted from the seeds or fruits by pressure, either in the cold or at a higher temperature. In the latter case, a higher yield is obtained but the oil is sometimes mixed with undesirable impurities. Recently methods of extraction by solvents have been used.

Uses of fats and oils

(1) As a food. Fats and oils rank with carbohydrates and proteins as the three essential food-stuffs of animals. They have a higher calorific value than have carbohydrates. Cotton seed, soya-bean seed and linseed of the flax plant are used extensively in the making of cattle foods.

(2) In the manufacture of *margarine* (p. 152).

(3) In the manufacture of *soap* (p. 152).

(4) As *drying oils* in paints. Highly unsaturated oils absorb oxygen from the air and then polymerize to form an elastic waterproof substance which, in thinly-spread paint, acts as a protective film. Linseed oil is one of the commonest drying oils used in the manufacture of paints.

(5) The *free acids* obtained from fats are mixed with paraffin wax to make candles and to soften rubber.

(6) The *aluminium, calcium and lead salts* of these acids are heated with petroleum oils to make lubricating greases.

(7) *Magnesium and zinc stearates* are used in face powders and as moulding powders in the plastic industry.

Manufacture of margarine

Half way through the last century, the increasing population in Europe along with the development of the Americas and Australia, brought an inevitable food shortage. To meet the growing demand for fat, the French scientist Mège Mouriés experimented to find a substitute for butter. He extracted the softer fat from beef tallow and churned it with milk. Production started about 1870. As the industry developed, animal fats became in short supply and vegetable oils were used to mix with them.

In 1899, Sabatier and Senderens had found that hydrogen could be added to the double bond of an ethylene compound in presence of nickel as a catalyst, making it into a saturated compound. This process of *hydrogenation* was applied to vegetable oils in about 1910 and revolutionized the margarine industry. Hydrogen under pressure is bubbled through the vegetable oil heated to about 200°C and containing finely divided nickel in suspension. The oil is 'hardened' to a fat, for the oleate in it is converted into the stearate. Hydrogenation, however, can be stopped at any point when the oil has reached the required consistency, and the product can be used directly for making margarine without adding any animal fat. Oils from groundnuts, cotton seed, sunflower seeds, soya-beans, coconut fruits, palm kernels and whale oil are the raw materials now in use. After hardening to a semi-solid state, they are churned with milk and vitamins A and D added. Before use, the milk is treated with certain micro-organisms to produce in margarine the flavour of butter.

Manufacture of soap

Soap consists of the sodium or potassium salts (according to the variety of soap) of the complex acids present in fats and oils. The fat is heated by means of steam with an aqueous solution of caustic soda, when hydrolysis (or saponification to use the common term) takes place, to form the sodium salts of the acids in the fat and set free the glycerol. Much of the soap remains in solution and to precipitate it, common salt is added. This is the process known as 'salting out', caused by the addition of sodium ions ('common ion effect'). The glycerol remains in aqueous solution from which it is recovered by distillation.

The soap is removed, boiled again with dilute alkali to complete saponification, and yet again with water to make it ' smooth '.

The product is *hard soap*, the best varieties of which are made from beef and mutton tallow. Cheaper kinds are made from those same vegetable oils as are used in the margarine industry. Hydrogenation is used in the soap industry as in the manufacture of margarine.

In a modern method, the fats are hydrolysed by superheated steam to the free acids and glycerol. The free acids are then converted into soap by the addition of caustic soda. The salting out process is avoided.

Soft soap is made by using caustic potash instead of caustic soda and the glycerol is left in. Oils containing more oleate are used here. Since, in general, potassium salts are more soluble than the corresponding sodium salts, soft soap is more soluble in water than is hard soap.

Some toilet soaps contain a certain proportion of potassium salts. The solubility of a soap depends also on the nature of the acids present. Liquid soap is an aqueous solution of the potassium salts of the acids in coconut oil. Shaving soap, also fairly soluble, consists of the potassium salts of palmitic and stearic acids. A soap suitable for washing in salt water (which contains the magnesium ion) is made from coconut oil and contains the sodium salts of some of the lower fatty acids. The magnesium salts of these acids are soluble in water, whereas those of palmitic and stearic acids are thrown out of solution as a scum when we attempt to wash in hard water (containing calcium and magnesium ions) with ordinary toilet soap.

Transparent soap is made by dissolving hard soap in industrial spirit, evaporating off the alcohol and pouring the molten residue into moulds to ' crystallize '.

In all kinds of soap, disinfectants, skin tonics (e.g. sulphur), scent and colouring matter may be added.

The different soap powders consist of powdered soap mixed with washing soda.

The cleansing action of soap

The cleansing power of soap is believed to be due to the fact that the anion consists of $RCOO^-$, where R represents a long hydrocarbon chain. This part dissolves in an oil, with the carboxyl group remaining soluble in water; hence when a mixture of oil and water is treated with soap, the interfacial tension is decreased and the oil readily breaks up into tiny droplets.

Dirt usually sticks to a surface by means of a film of grease. A detergent loosens the dirt because its oil-soluble part breaks up the grease, i.e. it emulsifies it. A detergent must therefore contain in its molecule a water-soluble group and an oil-soluble group. The oil-soluble group must contain between 10 and 18 carbon atoms. Below this number, a group possesses no emulsifying power. With more than 18, the salt is not sufficiently soluble in water.

Compounds other than soap having these two essential characteristics also possess cleansing power. The earliest of these detergents were made from vegetable oils which were highly unsaturated. They were treated with sulphuric acid which added on to the double bond as it adds on to ethylene. The sodium salts of these compounds provided the necessary water-soluble group.

During relatively recent years, however, a variety of new detergents has been made. Hydrocarbons containing the requisite number of carbon atoms have been used as the starting material. They may be aliphatic or aromatic, saturated or unsaturated. Many of these have been obtained from natural petroleum. They are converted into alcohols which are then treated with sulphuric acid and then sodium carbonate to make a sodium derivative. In most cases, sodium phosphate, Na_3PO_4, and foaming agents are added. Some of the detergents in use today have the advantage over soap that their calcium and magnesium salts are soluble in water. Their cleansing power is not therefore affected by the calcium and magnesium ions of hard water, although, like soaps, they contain long-chain anions. Certain detergents are *non-ionic*, and are made up of groups which are *hydrophilic* (water-loving, and imparting solubility) and other groups which are *hydrophobic* (water-repellant) but with the power of dissolving oils and fats. The hydrophilic groups in these non-ionic detergents are usually derived from ethylene oxide (p. 78).

SUMMARY OF THE CHEMISTRY OF ESTERS

The general formula is $C_nH_{2n+1}\cdot COOC_mH_{2m+1}$
Isomerism is very common in the series, e.g.:

$$CH_3\cdot COOC_2H_5 \qquad \text{and} \qquad C_2H_5\cdot COOCH_3$$
$$\text{ethyl acetate} \qquad\qquad\qquad \text{methyl propionate}$$

The representative member is $CH_3\cdot COOC_2H_5$, ethyl acetate.

(1) *Preparation*

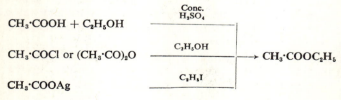

(2) *Properties*

$$CH_3 \cdot COOC_2H_5 \quad
\begin{cases}
\xrightarrow[\text{(acid or alkali)}]{\text{hydrolysis}} & CH_3 \cdot COOH + C_2H_5OH \\
\xrightarrow{NH_3} & CH_3 \cdot CONH_2 + C_2H_5OH \\
\xrightarrow{LiAlH_4} & CH_3 \cdot CH_2OH
\end{cases}$$

EXAMINATION QUESTIONS

(1) Starting from a monocarboxylic acid, give in each case, **one** reaction or series of reactions by which you could prepare (*a*) an acid chloride, (*b*) an acid anhydride, (*c*) an amide, (*d*) an ester. Indicate the necessary reagents and conditions.

Name (*i*) **one** *type* of reaction which is common to (*a*), (*b*), (*c*) and (*d*), (*ii*) **one** *type* of reaction which is common to (*a*) and (*b*) but is not given by the others, (*iii*) **one** *type* of reaction which is given only by (*c*). Write equations for **one** example in each case. (J. M. B. A.)

(2) A liquid *A*, formula $C_9H_{10}O_2$, was refluxed with aqueous potassium hydroxide solution until the two liquids merged. On distilling the product, a solution of *B* giving the iodoform test was obtained. The solution remaining in the distilling flask when acidified with hydrochloric acid gave a white precipitate of *C*. Identify *A*, *B* and *C*, and write the equation for the reaction between *A* and potassium hydroxide.

Describe exactly how (*a*) the iodoform test is performed, (*b*) *C* is prepared from benzyl chloride.

How does *C* react with (*i*) soda-lime, (*ii*) phosphorus pentachloride? (O. A.)

(3) Describe *briefly* the difference in chemical character between a carbon–carbon double bond and a carbon–oxygen double bond. Write an account of the chemistry of a simple compound containing a carbonyl group. (Liverpool, B.Sc. Inter.)

(4) An organic compound was found to have the composition C = $33 \cdot 18\%$, H = $4 \cdot 60\%$, Cl = $32 \cdot 72\%$. When boiled with aqueous sodium hydroxide it gave a mixture of sodium chloride, sodium carbonate and ethyl alcohol. When the original substance was treated with ammonia a substance $C_3H_7O_2N$ was first produced, but by the prolonged action of ammonia a substance CH_4ON_2 was obtained. Suggest structural formulae for these compounds and explain the charges involved.

[H = $1 \cdot 00$; C = $12 \cdot 0$; O = $16 \cdot 0$; Cl = $35 \cdot 5$] (L. S.)

(5) The compounds below all contain the group $CH_3 \cdot CO$. To what extent is this structural similarity reflected in their behaviour?

$$CH_3 \cdot CO \cdot CH_3 \quad CH_3 \cdot CO \cdot OH \quad CH_3 \cdot CO \cdot OCH_3$$
$$CH_3 \cdot CO \cdot NH_2 \quad CH_3 \cdot CO \cdot Cl \quad CH_3 \cdot CO \cdot CCl_3$$

(Liverpool B.Sc., Part I)

13

ALIPHATIC NITRILES AND ISONITRILES
(CYANIDES AND ISOCYANIDES)

Functional Groups —CN, —NC

NITRILES OR CYANIDES

THESE compounds, which are derivatives of hydrocyanic or prussic acid, HCN, have the general formula of $C_nH_{2n+1}\cdot CN$, the constitution of the cyanide group being $-C\equiv N$. The first member, i.e. where $n = 1$, is *methyl cyanide*, $CH_3\cdot CN$. It is also called *acetonitrile* from the fact that it is readily converted into acetic acid.

Preparation of acetonitrile

Acetonitrile may be prepared by either of the following methods: (1) By the action of sodium or potassium cyanide in aqueous ethanolic solution on a methyl halide (p. 49).

(N.B. Since methyl chloride at room temperature is a gas, it would be difficult to use this halide.)

$$CH_3I + KCN \longrightarrow CH_3\cdot CN + KI$$
$$\text{methyl cyanide}$$
$$\text{or acetonitrile}$$

(2) By the dehydration of acetamide by distilling it with phosphorus pentoxide or thionyl chloride:

$$CH_3\cdot CONH_2 \xrightarrow{P_2O_5} CH_3\cdot CN$$
$$CH_3\cdot CONH_2 + SOCl_2 \longrightarrow CH_3\cdot CN + 2HCl\uparrow + SO_2\uparrow$$

This method gives the pure nitrile; the product from method (1) contains some isonitrile, $CH_3\cdot NC$.

Properties of acetonitrile

Acetonitrile is a colourless liquid of b.p. 82°C, readily soluble in water. It is itself a useful solvent for a wide variety of substances. Higher members of the series are only slightly soluble in water. By contrast with hydrocyanic acid, and also with isonitriles (below), nitriles have a rather pleasant odour and are only slightly toxic.

The important chemical properties are given below.

(1) Acetonitrile is readily hydrolysed by mineral acids or alkalis:

$$CH_3\cdot CN \xrightarrow{H_2O} CH_3\cdot CO\cdot NH_2 \xrightarrow{H_2O} CH_3\cdot CO\overset{-}{O}\overset{+}{N}H_4$$

Using an acid as the hydrolysing agent, the final product is acetic acid, and with an alkali the acetate of the alkali metal. (2) On reduction of acetonitrile by sodium and absolute alcohol, a *primary amine* (p. 160) is obtained. This is the *Mendius reaction*. The process is an additive one, the triple bond between the carbon and nitrogen atoms being filled up:

$$CH_3 \cdot CN \xrightarrow{2H_2} CH_3 \cdot CH_2 \cdot NH_2$$
ethylamine

The Mendius reaction has now been largely replaced by catalytic reduction, using Raney nickel and hydrogen.

If lithium aluminium hydride is added to acetonitrile, and the product subsequently hydrolysed, acetaldehyde is obtained:

$$4CH_3 \cdot CN + LiAlH_4 \longrightarrow [\text{Derivative of } CH_3 \cdot CH{=}NH]$$
$$\downarrow H_2O$$
$$4CH_3 \cdot CHO$$

Nitriles are extremely useful in organic syntheses, as illustrated below:

$$ROH \longrightarrow RBr \xrightarrow{KCN} RCN \xrightarrow{H_2(Ni)} RCH_2 NH_2$$
$$\downarrow$$
$$R \cdot CO_2H \xrightarrow{LiAlH_4} R \cdot CH_2OH$$

These reactions constitute a general method for lengthening a carbon chain. In older books, the method given often involves conversion of the primary amine, $RCH_2 \cdot NH_2$, to the primary alcohol, $R \cdot CH_2OH$, with nitrous acid (p. 165). In practice, this is not an effective method, since rearrangements occur and more than one alcohol is formed.

ISONITRILES

Isonitriles, *isocyanides* or *carbylamines* of general formula $C_nH_{2n+1}NC$ are isomeric with nitriles, the nitrogen atom being positioned between the carbon atom and the alkyl group.

Preparation of isonitriles

(1) By the action of silver cyanide on an alkyl halide in aqueous ethanolic solution:

$$CH_3I + AgCN \longrightarrow CH_3 \cdot NC + AgI$$
methyl isocyanide,
or methyl carbylamine

Here as in the preparation of the nitrile using potassium cyanide, the product contains a little of the isomeric compound. A pure product is given by the Hofmann method below.

(This different action with silver cyanide may be due to crystals of potassium cyanide being ionic K^+CN^- while silver cyanide is covalent, having the composition —Ag—CN—Ag—CN—Ag— i.e. it contains both Ag—CN and Ag—NC linkages.)

(2) By the action of chloroform and alcoholic potash on a primary amine (p. 165). This is *Hofmann's carbylamine reaction*:

$$CH_3NH_2 + CHCl_3 + 3KOH \longrightarrow CH_3 \cdot NC + 3KCl + 3H_2O$$

Properties of isonitriles

The boiling points of isonitriles are below those of the corresponding isomeric nitriles, e.g. methyl carbylamine boils at 59°C, acetonitrile at 83°C. By contrast with nitriles, they possess a most objectionable odour and are highly toxic. *Hofmann's reaction* serves as a very sensitive test for either a *primary amine* or for *chloroform*.

The behaviour of isonitriles on hydrolysis and reduction is in marked contrast to that of nitriles.

(1) Isonitriles are hydrolysed by mineral acids but not by alkalis, a primary amine and formic acid being formed:

$$CH_3 \cdot NC \xrightarrow[HCl]{H_2O} CH_3 \cdot NH_2 + HCOOH$$
$$\downarrow$$
$$[CH_3 \cdot NH_3]^+ Cl^-$$
methylamine hydrochloride

(2) On reduction of an isonitrile, a secondary amine is obtained:

$$CH_3 \cdot NC \xrightarrow{2H_2} CH_3 \cdot NH \cdot CH_3$$
dimethylamine

SUMMARY OF THE CHEMISTRY OF NITRILES AND ISONITRILES

Nitriles (or cyanides)

The general formula is $C_nH_{2n+1} \cdot CN$

The first and representative member is $CH_3 \cdot CN$, methyl cyanide, or acetonitrile.

Preparation and properties

Isonitriles (or isocyanides or carbylamines)

The general formula is $C_nH_{2n+1} \cdot NC$

The first and representative member is $CH_3 \cdot NC$, methyl isocyanide, methyl carbylamine.

Preparation and properties

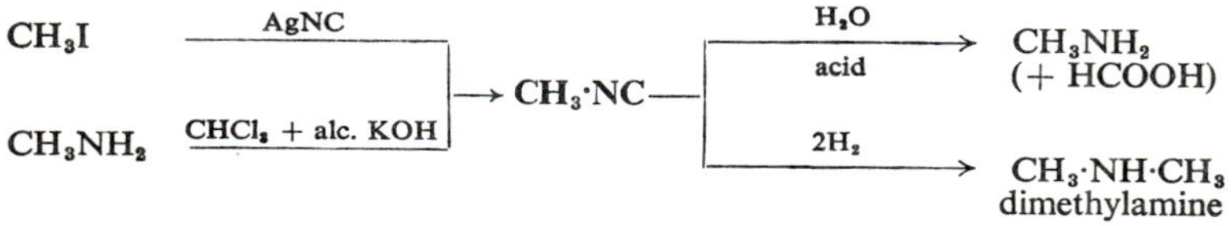

CH_3I $\xrightarrow{\text{AgNC}}$

$\rightarrow CH_3 \cdot NC$

CH_3NH_2 $\xrightarrow{\text{CHCl}_3 + \text{alc. KOH}}$

$\xrightarrow[\text{acid}]{\text{H}_2\text{O}} CH_3NH_2$ (+ HCOOH)

$\xrightarrow{2H_2} CH_3 \cdot NH \cdot CH_3$ dimethylamine

EXAMINATION QUESTION

(1) What is meant by the term unsaturation? Give two tests which can be used to determine whether an organic substance is unsaturated.

State, giving your reasons, whether the following substances may be regarded as saturated or unsaturated: ethylene, ethylene dibromide, ethyl alcohol, ethyl cyanide, acetaldehyde. (O. and C.A.)

14

ALIPHATIC AMINES

Functional Groups $R—N\underset{R}{\overset{R}{\diagup}}$

(R = hydrogen or alkyl)

THESE compounds may be looked upon as derivatives of ammonia in which the hydrogen atoms are replaced by alkyl groups. Three classes of such compounds are obtainable according to the number of hydrogen atoms replaced. They are called respectively *primary*, *secondary* and *tertiary* amines. The constitutional formulae of the methyl compounds of each class, the general formulae (using R to represent the alkyl group), the characteristic group and finally (for comparison purposes) examples of primary, secondary and tertiary alcohols, are given below:

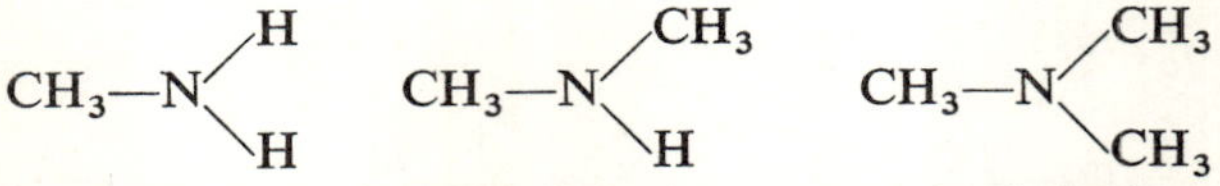

$CH_3—N\underset{H}{\overset{H}{\diagup}}$ $CH_3—N\underset{H}{\overset{CH_3}{\diagup}}$ $CH_3—N\underset{CH_3}{\overset{CH_3}{\diagup}}$

methylamine   dimethylamine   trimethylamine
(primary)   (secondary)   (tertiary)

RNH_2 R_2NH R_3N

$—NH_2$ $\diagdown NH$ $\diagdown N$
                  $\diagup$ $\diagup$

$\underset{OH}{\overset{H}{CH_3—C—H}}$ $\underset{OH}{\overset{CH_3}{CH_3—C—H}}$ $\underset{OH}{\overset{CH_3}{CH_3—C—CH_3}}$

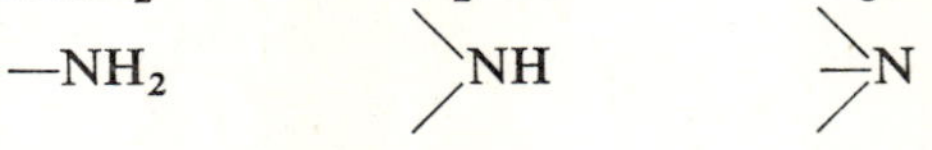

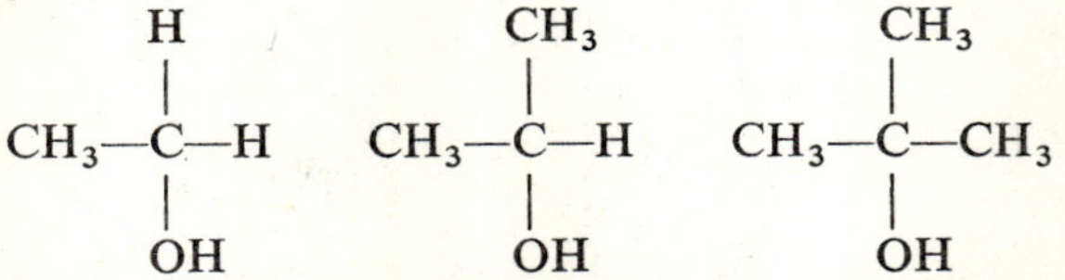

a primary alcohol   a secondary alcohol   a tertiary alcohol

It should be noted that the usage of the terms primary, secondary and tertiary in the amine series differs from that in the alcohol series. Thus $CH_3 \cdot CH(OH) \cdot CH_3$, isopropyl alcohol, is a secondary alcohol, but the corresponding amine, $CH_3 \cdot CH(NH_2) \cdot CH_3$, is a primary amine because it contains the $—NH_2$ group.

Isomerism is common in the amine series. Thus isomerism may occur within the alkyl group:

e.g. $CH_3 \cdot CH_2 \cdot CH_2 \cdot NH_2$ and $CH_3 \cdot CH(NH_2) \cdot CH_3$
n-propylamine iso-propylamine

or the different types of amine may be isomeric:

e.g. $CH_3 \cdot CH_2 \cdot CH_2 \cdot NH_2$, $CH_3 \cdot CH_2 \cdot NH \cdot CH_3$
n-propylamine N-methylethylamine

and $(CH_3)_3N$
trimethylamine

A fourth class of compounds corresponding to ammonium salts is known, e.g. $N(CH_3)_4I$. These compounds are called *quaternary ammonium compounds* and the example given is *quaternary methyl ammonium iodide*.

In 1849, Wurtz boiled methyl isocyanate with caustic potash and obtained a gas with a strong ammoniacal smell. This was methyl-amine, CH_3NH_2, the first amine to be prepared.

$$CH_3 - N = C = O + H_2O \xrightarrow{KOH} CH_3NH_2 + CO_2$$
$$\downarrow$$
$$K_2CO_3$$

In the same year, Hofmann heated an alcoholic solution of methyl iodide with ammonia in a sealed tube and obtained a mixture of the three amines and also the quaternary methyl ammonium iodide.

The three methylamines (mono-, di- and tri-) of formula CH_3NH_2, $(CH_3)_2NH$, and $(CH_3)_3N$ are all gases at room temperature and hence not easy to work with. In the subsequent study of amines, the ethyl derivatives, all of which are liquids, are for convenience taken as the representative compounds, although the corresponding methyl derivatives show similar reactions.

The three methylamines are present in the brine in which herrings have been cured, being products of the decay of fish.

These three amines are *manufactured* by passing ammonia and the vapour of methyl alcohol over alumina heated to about 400°C:

$$CH_3OH + NH_3 \xrightarrow[400°C]{Al_2O_3} CH_3NH_2 + H_2O$$

similarly,

$$CH_3NH_2 \xrightarrow{CH_3OH} (CH_3)_2NH \xrightarrow{CH_3OH} (CH_3)_3N$$

Action of ammonia on ethyl iodide (Hofmann's reaction)

On heating an alcoholic solution of ethyl iodide with ammonia in a sealed vessel, the following reactions occur successively:

$$C_2H_5I + NH_3 \longrightarrow C_2H_5 \cdot \overset{+}{N}H_3 \ \overset{-}{I}$$

ethyl ammonium iodide or
ethylamine hydriodide

$$C_2H_5 \cdot NH_2 + C_2H_5I \longrightarrow (C_2H_5)_2 \overset{+}{N}H_2 \ \overset{-}{I}$$

diethyl ammonium iodide

$$(C_2H_5)_2NH + C_2H_5I \longrightarrow (C_2H_5)_3 \overset{+}{N}H \ \overset{-}{I}$$

triethyl ammonium iodide

$$(C_2H_5)_3N + C_2H_5I \longrightarrow (C_2H_5)_4 \overset{+}{N} \ \overset{-}{I}$$

quaternary ethyl ammonium iodide

(The iodide is chosen because it is the most reactive. The order of reactivity is iodide > bromide > chloride.)

The products of these reactions are all ionic compounds. Each reaction is an example of a displacement reaction (p. 48).

The free amines can be obtained from the mono-, di- and tri-hydriodides by the addition of an alkali, thus:

$$C_2H_5 \cdot \overset{+}{N}H_3 \ \overset{-}{I} + NaOH \longrightarrow C_2H_5 \cdot NH_2 + NaI + H_2O$$

ethylamine

The reaction actually involved here is simply a proton transfer:

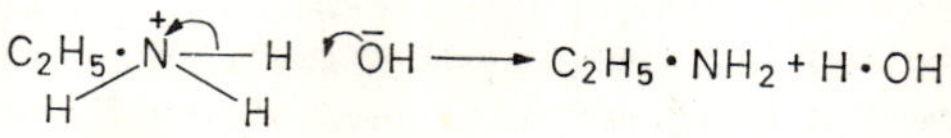

Such reactions are virtually instantaneous and do not require heat (unless the amine salt is very insoluble).

The quaternary salt is not affected by alkali.

Hofmann's alkylation reaction is not a very convenient method of preparing any particular amine, for separation of the various products is by no means easy. Some control over the relative amounts obtained is possible, e.g. excess ammonia gives, as is to be expected, a higher proportion of the primary amine.

Preparation of primary amines

The following reactions in which ethylamine is taken as the example, provide suitable methods of preparing primary amines, since no secondary or tertiary compounds are obtained during the reaction.

(1) By the reaction of bromine and potash (or sodium hypobromite or sodium hypochlorite) on an acid amide (Hofmann's amide reaction).

The reaction has already been referred to (p. 143), the change involving an unusual intramolecular rearrangement. Briefly, the steps involved are:

$$C_2H_5 \cdot CO \cdot NH_2 \xrightarrow{Br_2} C_2H_5 \cdot CO \cdot NHBr \xrightarrow{KOH} C_2H_5 \cdot NCO$$

propionamide   propiobromamide   ethyl isocyanate

$$C_2H_5 \cdot NCO \xrightarrow[(OH^-)]{H_2O} C_2H_5 \cdot NH_2 + CO_2$$

$$\downarrow$$

$$K_2CO_3$$

(2) By the hydrolysis of ethyl isocyanate (above) (Wurtz method):

$$C_2H_5 \cdot NCO \xrightarrow[KOH]{H_2O} C_2H_5 \cdot NH_2$$

(3) By the reduction of a nitrile (p. 157):

$$CH_3 \cdot CN \xrightarrow{H_2(Ni)} CH_3 \cdot CH_2 \cdot NH_2$$

acetonitrile   ethylamine

(4) By the reduction of a nitroalkane by (i) hydrogen, in presence of Raney nickel, or (ii) reduction with zinc and hydrochloric acid:

$$C_2H_5 \cdot NO_2 + 3H_2 \xrightarrow{Ni} C_2H_5 \cdot NH_2 + 2H_2O$$

This method has been available since the discovery of vapour phase nitration of alkanes (p. 11).

(5) By the reduction of an amide with lithium aluminium hydride:

$$CH_3 \cdot CO \cdot NH_2 \xrightarrow{LiAlH_4} CH_3 \cdot CH_2 \cdot NH_2$$

Preparation of secondary and tertiary amines

The lithium aluminium hydride reduction (above) is, in fact, a very general method:

$$R \cdot CO \cdot NH \cdot R' \xrightarrow{LiAlH_4} R \cdot CH_2 \cdot NH \cdot R'$$

$$R \cdot CO \cdot N \Big\langle {R' \atop R''} \xrightarrow{LiAlH_4} R \cdot CH_2 \cdot N \Big\langle {R' \atop R''}$$

Properties of primary amines

Methylamine is a gas at room temperature (b.p. $-6°C$) and ethylamine is a volatile liquid (b.p. $19°C$). The lower members have a strongly ammoniacal smell and resemble ammonia in their chemical behaviour. The following properties are given for ethylamine as a representative amine:

(1) Ethylamine is a slighter stronger base than ammonia (p. 298). It dissolves in water giving a solution which is alkaline to litmus, showing a hydroxyl ion concentration greater than 10^{-7} g-ions/litre:

$$C_2H_5 \cdot NH_2 + H_2O \rightleftharpoons C_2H_5 \cdot NH_3{}^+ + OH^-$$

With the mineral acids, it dissolves to form salts with respective formulae of

$$[C_2H_5 \cdot NH_3]^+ Cl^-, [C_2H_5 \cdot NH_3]^+ NO_3{}^- \text{ and } [C_2H_5\overset{+}{N}H_3]_2SO_4{}^{2-}$$

The free amine is liberated from these salts by addition of an alkali, as ammonia is set free from an ammonium salt by similar treatment.

Ethylamine gives white fumes of ethyl ammonium chloride in the presence of hydrochloric acid vapour, cf. ammonia test.

Like ammonia, ethylamine in aqueous solution precipitates some metallic hydroxides from aqueous solutions of their salts.

(2) Ethylamine forms addition compounds similar to ammonia because it has the power to form a co-ordinate link (p. xxii):

(*a*) Adds on to salts, e.g. calcium chloride, as does ammonia. Hence calcium chloride cannot be used to dry amines.

(*b*) Adds on to acceptor molecules (p. 290).

(*c*) The basic property of amines (property (1) above) is really an example of the addition of the amine to the proton of an acid:

$$C_2H_5 \cdot NH_2 + HCl \longrightarrow [CH_3 \cdot NH_3]^+ Cl^-$$

or, in detail:

(3) With an alkyl halide, ethylamine gives a mixture of secondary and tertiary amines—or rather their hydrohalides (Hofmann's reaction, p. 162):

$$C_2H_5 \cdot NH_2 + CH_3I \longrightarrow [C_2H_5 \, NH_2 \cdot CH_3]^+ I^-$$
methyl ethyl ammonium iodide

$$C_2H_5 \cdot NH \cdot CH_3 + CH_3I \longrightarrow [C_2H_5 \cdot NH(CH_3)_2]^+ I^-$$
dimethyl ethyl ammonium iodide

(4) With chloroform and potash, ethylamine gives a carbylamine (p. 158):

$$C_2H_5 \cdot NH_2 + Cl_3HC + 3KOH \longrightarrow C_2H_5 \cdot NC + 3KCl + 3H_2O$$
$$\text{ethyl carbylamine}$$

(5) With nitrous acid and ethylamine, ethanol is formed and nitrogen evolved. With higher members of the primary amine series, nitrogen is again evolved but more than one alcohol is frequently obtained, often with other products. The amine is dissolved in dilute hydrochloric acid and aqueous sodium or potassium nitrite added:

$$C_2H_5 \cdot NH_2 \xrightarrow[\text{HO·NO}]{} C_2H_5OH + N_2\uparrow + H_2O$$

(This reaction compares with the method of preparing nitrogen by heating ammonium nitrite solution—the ammonia being regarded as analogous to a primary amine:

$$HNH_2 + ONOH \longrightarrow 2H_2O + N_2\uparrow.)$$

(6) With either acetyl chloride or acetic anhydride, an acetyl group is introduced into the molecule of ethylamine to form a compound known as an *N-substituted amide*. The prefix ' *N-* ' indicates that the substituting group is attached to the nitrogen atom:

N-ethylacetamide

The reaction with acetic anhydride:

$$(CH_3CO)_2O + C_2H_5 \cdot NH_2 \longrightarrow CH_3 \cdot CO \cdot NH \cdot C_2H_5 + CH_3 \cdot CO_2H$$

can be formulated similarly.

This process of *acetylation*, or (more generally) *acylation*, as it is called, is of great value in organic syntheses. The amino group is very vulnerable to attack by oxidizing agents, but the substituted group —NHCOCH$_3$ is not. If therefore a compound containing an amino group is to be treated by an oxidizing agent and the amino group is required in the product, it is first 'protected' by being acetylated. The free amino compound is readily obtained from the product by hydrolysis with aqueous acids:

$$CH_3 \cdot CO \cdot NH \cdot C_2H_5 \xrightarrow[\text{(acid)}]{H_2O} CH_3 \cdot CO_2H + C_2H_5 \cdot \overset{+}{N}H_3\overset{-}{Cl}$$
$$\downarrow OH^-$$
$$C_2H_5 \cdot NH_2$$

Properties of secondary amines

The lower members possess a fishy smell; higher members have a rather objectionable smell, but higher still, with little volatility, there is practically no smell.

(1) Secondary amines show basic properties, as do primary amines. Diethylamine dissolves in water to form a solution of the hydroxide which is alkaline to litmus:

$$(C_2H_5)_2NH + H_2O \rightleftharpoons (C_2H_5)_2NH_2^+ + OH^-$$
$$\text{diethyl ammonium hydroxide}$$

Salts are formed with mineral acids. Other additive properties are shown, e.g. with calcium chloride, similar to those given for primary amines on p. 164.

(2) With secondary amines and an alkyl halide a tertiary amine is obtained (as its hydrogen halide salt):

$$(C_2H_5)_2NH + C_2H_5I \longrightarrow (C_2H_5)_3NH^+I^-$$
$$\text{triethyl ammonium iodide}$$

(3) With nitrous acid, secondary amines give a compound called a *nitrosamine*, no nitrogen being formed:

$$(C_2H_5)_2NH + HONO \longrightarrow (C_2H_5)_2N \cdot NO + H_2O$$
$$\text{diethylnitrosamine}$$

(the group —N=O is known as the *nitroso* group).

(4) Secondary amines are acetylated on treatment with *acetyl chloride* or acetic anhydride, a di-substituted amide being obtained:

$$(C_2H_5)_2NH + ClCO \cdot CH_3 \longrightarrow CH_3 \cdot CO \cdot N(C_2H_5)_2 + HCl$$
$$\textit{N}\text{-diethylacetamide}$$

$$(C_2H_5)_2NH + (CH_3 \cdot CO)_2O \longrightarrow CH_3 \cdot CO \cdot N(C_2H_5)_2$$
$$+ CH_3 \cdot COOH$$

(5) Secondary amines do not undergo the carbylamine reaction with chloroform.

Properties of tertiary amines

The lower members have a strong fishy odour.

(1) Although less soluble in water than the corresponding primary and secondary derivatives, they do dissolve and give alkaline solutions, which with mineral acids form salts. Other *additive properties* similar to those of primary and secondary amines are also shown.

(2) Tertiary amines readily react with *alkyl halides* giving *quaternary ammonium* salts:

$$(C_2H_5)_3N + C_2H_5I \longrightarrow (C_2H_5)_4N^+I^-$$
quaternary ethyl ammonium iodide

Tertiary amines do not form amides with acylating agents (acetyl chloride or acetic anhydride), nor can they form carbylamines (isocyanides) with chloroform and alkali. Reaction can occur with nitrous acid in weakly acidic conditions, but nitrogen is not formed.

Distinctions between primary, secondary and tertiary amines

(1) A *primary amine* is distinguished from a secondary and tertiary amine by its action with chloroform and potash, when a carbylamine is formed. The carbylamine is readily recognizable by its objectionable smell. Secondary and tertiary amines do not give a positive test.

(2) Amines react with cold (0°–5°C) dilute nitrous acid as follows:

(*a*) A *primary amine* gives an (observable) effervescence of nitrogen:

$$C_2H_5 \cdot NH_2 \xrightarrow{\text{HNO}_2} N_2 + C_2H_5OH + H_2O$$

With other primary amines, more than one alcohol may be formed.

(*b*) A *secondary amine* gives a nitrosamine, which separates out as a yellow oil because it is no longer basic; the oil can be extracted with ether:

$$(C_2H_5)_2NH \xrightarrow{\text{HNO}_2} (C_2H_5)_2N \cdot NO$$

(*c*) *Tertiary amines* do not react readily with nitrous acid, if the solution is strongly acidic (pH < 3).

Quaternary ammonium compounds

In constitution, these compounds differ from the salts of amines in that the nitrogen atom has no hydrogen attached to it directly.

They are prepared by the action of an alkyl halide on a tertiary amine, e.g.:

$$(C_2H_5)_3N + C_2H_5I \longrightarrow (C_2H_5)_4N^+I^-$$

quaternary ethyl ammonium iodide
or tetraethyl ammonium iodide

They are not affected by strong bases; on shaking with aqueous 'silver hydroxide', a solution of the quaternary ammonium hydroxide is obtained:

$$(C_2H_5)_4NI + AgOH \longrightarrow (C_2H_5)_4N^+ + OH^- + AgI\downarrow$$

(This reaction is more effectively carried out by using a basic ion-exchange resin, whereby the I^- ions are replaced by OH^- ions).

Tetraethyl ammonium hydroxide is a white crystalline deliquescent solid. It is a strong base corresponding in strength with caustic soda. In aqueous solution it is completely dissociated:

$$(C_2H_5)_4N^+OH^- \longrightarrow (C_2H_5)_4N^+ + OH^-$$

Like caustic soda, it attacks glass. It absorbs carbon dioxide from the air and will liberate ammonia from ammonium salts:

$$(C_2H_5)_4N^+OH^- + \overset{+}{N}H_4\overset{-}{Cl} \longrightarrow (C_2H_5)_4N^+Cl^- + NH_3 + H_2O$$

On being heated it gives a tertiary amine and an alkene:

$$(C_2H_5)_4N^+OH^- \longrightarrow H_2O + N(C_2H_5)_3 + CH_2{=}CH_2$$

ethylene

SUMMARY OF THE CHEMISTRY OF ALIPHATIC AMINES

Amines are derivatives of ammonia obtained by replacing the hydrogen atoms in turn by alkyl groups (represented as R):

$$RNH_2 \qquad R_2NH \qquad R_3N$$

primary amine   secondary amine   tertiary amine

Mixtures of amine salts are obtained by the action of an alkyl halide on ammonia:

$$C_2H_5I + NH_3 \longrightarrow C_2H_5{\cdot}\overset{+}{N}H_3\ \overset{-}{I}$$

$$C_2H_5{\cdot}NH_2 + C_2H_5I \longrightarrow (C_2H_5)_2\overset{+}{N}H_2\ \overset{-}{I} \quad \text{etc.}$$

Primary amines

The representative member is ethylamine $C_2H_5{\cdot}NH_2$, (Methylamine reacts similarly, but is a gas).

(1) *Preparation*

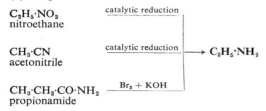

C$_2$H$_5$·NO$_2$
nitroethane   catalytic reduction

CH$_3$·CN
acetonitrile   catalytic reduction → C$_2$H$_5$·NH$_2$

CH$_3$·CH$_2$·CO·NH$_2$
propionamide   Br$_2$ + KOH

(2) *Properties*

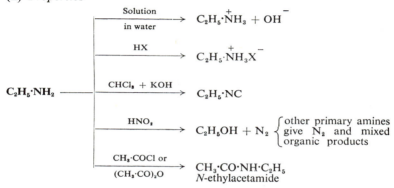

C$_2$H$_5$·NH$_2$ ——

Reagent	Product
Solution in water	C$_2$H$_5$·$\overset{+}{N}$H$_3$ + OH$^-$
HX	C$_2$H$_5$·$\overset{+}{N}$H$_3$X$^-$
CHCl$_3$ + KOH	C$_2$H$_5$·NC
HNO$_2$	C$_2$H$_5$OH + N$_2$ { other primary amines give N$_2$ and mixed organic products
CH$_3$·COCl or (CH$_3$·CO)$_2$O	CH$_3$·CO·NH·C$_2$H$_5$ *N*-ethylacetamide

Secondary amines

The representative member is (C$_2$H$_5$)$_2$NH, diethylamine.

(1) *Preparation*

$$CH_3 \cdot CO \cdot NH \cdot C_2H_5 \xrightarrow{\text{LiAlH}_4} (C_2H_5)_2NH$$

(2) *Properties*

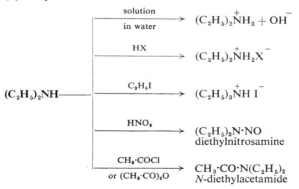

(C$_2$H$_5$)$_2$NH——

Reagent	Product
solution in water	(C$_2$H$_5$)$_2$$\overset{+}{N}H_2$ + OH$^-$
HX	(C$_2$H$_5$)$_2$$\overset{+}{N}H_2X^-$
C$_2$H$_5$I	(C$_2$H$_5$)$_3$$\overset{+}{N}$H I$^-$
HNO$_2$	(C$_2$H$_5$)$_2$N·NO diethylnitrosamine
CH$_3$·COCl or (CH$_3$·CO)$_2$O	CH$_3$·CO·N(C$_2$H$_5$)$_2$ *N*-diethylacetamide

Distinctions between primary, secondary and tertiary amines

$$\underset{\text{primary}}{RNH_2} \xrightarrow[\text{(HCl + NaNO}_2)]{HNO_2} (ROH) + N_2\uparrow + H_2O$$
alcohol and other products

$$\underset{\text{secondary}}{R_2NH} \xrightarrow{HNO_2} \underset{\text{a nitrosamine}}{R_2N \cdot NO}$$

$$\underset{\text{tertiary}}{R_3N} \xrightarrow{HNO_2} \text{no action (in strong acid)}$$

RNH_2 gives carbylamine (isocyanide) test.

R_3N is not acylated by $R \cdot COCl$ or anhydride.

EXAMINATION QUESTIONS

(1) Give names and formulae for all the products formed when methyl iodide is heated with alcoholic ammonia in a sealed tube.

What is the effect, if any, upon these products of (*a*) heat, (*b*) acetyl chloride, (*c*) nitrous acid?

What structural feature has acetamide in common with an aliphatic primary amine? Give **three** points of chemical behaviour in which acetamide differs from compounds of this type. (J. M. B. A.)

(2) *A* is a crystalline solid, containing carbon, hydrogen, nitrogen and chlorine only. Its molecular weight is 67·5. When warmed with sodium hydroxide solution, *A* produces a colourless gas, having a characteristic odour. When dissolved in hydrochloric acid and treated with sodium nitrite solution, *A* produces an odourless gas.

Identify *A* and explain the reactions given above. Give **one** other test you would apply to *A* to support your answer. Describe how you would make *A*, starting with an organic compound containing carbon, hydrogen and oxygen only and using only inorganic reagents. Outline **one** alternative synthesis of *A*, starting with the **same** organic compound. (S. A.)

(3) Starting with methyl iodide describe how (*a*) acetamide, (*b*) ethylamine, may be prepared.

Compare the properties of acetamide and ethylamine. (O. and C.A.)

(4) By means of formulae and brief explanatory notes indicate how acetic acid could be transformed into the following compounds, using only inorganic reagents: (*a*) ethyl acetate, (*b*) methylamine, (*c*) ethylamine, (*d*) acetone, (*e*) ethane. (Liverpool, B.Sc., Part I)

15

DIBASIC ACIDS AND THEIR DERIVATIVES

Carbonic acid, H_2CO_3

THIS acid, the simplest of the dibasic acids, is known in solution only, for when attempts are made to isolate the pure compound from its aqueous solution, decomposition to carbon dioxide and water occurs. Carbon dioxide dissolved in water behaves as a weak acid, being only partially dissociated:

$$CO_2 + H_2O \rightleftharpoons [H_2CO_3] \rightleftharpoons 2H^+ + CO_3^{2-}$$

The salts of carbonic acid with metals (other than sodium or potassium) are rather unstable, the volatile carbon dioxide being evolved from them when heated. Carbonates are usually studied as a part of inorganic chemistry.

Carbonic acid is of interest in organic chemistry for the following reasons:

(a) it plays an important part in both plant and animal metabolism.

(b) some of its derivatives, notably urea, are of theoretical interest and of great commercial importance.

Carbonyl chloride or phosgene, $COCl_2$

This extremely poisonous gas (it was used as a poison gas in World War I) is the acid chloride of carbonic acid. It is prepared by the direct addition of chlorine to carbon monoxide in sunlight.

$$CO + Cl_2 \longrightarrow COCl_2$$

It is formed in small quantities by the atmospheric oxidation of chloroform (p. 53) in presence of sunlight.

The gas is only slightly soluble in water; hydrolysis occurs slowly in aqueous solution:

$$COCl_2 + 2H_2O \longrightarrow H_2CO_3 + 2HCl$$

A similar reaction occurs with alcohol, the final product being ethyl carbonate (the reaction is analogous to that of acetyl chloride with ethanol):

$$COCl_2 + 2C_2H_5OH \rightleftharpoons CO(OC_2H_5)_2 + 2HCl$$
ethyl carbonate

171

This ester can also be obtained by the distillation of a mixture of silver carbonate and ethyl iodide:

$$2C_2H_5I + Ag_2CO_3 \longrightarrow (C_2H_5)_2CO_3 + 2AgI$$

Urea or carbamide, $O{=}C\begin{smallmatrix}\diagup NH_2 \\ \diagdown NH_2\end{smallmatrix}$

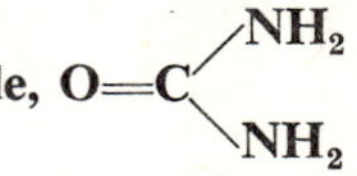

Urea is the diamide of carbonic acid—the monamide would have a formula

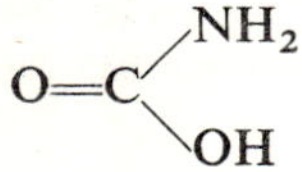

$$O{=}C\begin{smallmatrix}\diagup NH_2 \\ \diagdown OH\end{smallmatrix}$$

This latter compound, known as *carbamic acid*, has not been isolated but its ammonium salt is known. The relationships between carbonic acid and its nitrogen derivatives are seen by comparing the structural formulae of these compounds:

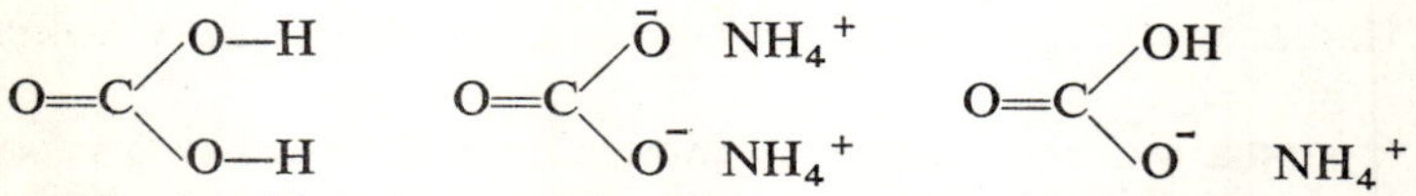

$O{=}C\begin{smallmatrix}\diagup O{-}H \\ \diagdown O{-}H\end{smallmatrix}$ $O{=}C\begin{smallmatrix}\diagup \bar{O}\ NH_4{}^+ \\ \diagdown O^-\ NH_4{}^+\end{smallmatrix}$ $O{=}C\begin{smallmatrix}\diagup OH \\ \diagdown O^-\ NH_4{}^+\end{smallmatrix}$

carbonic acid   ammonium carbonate*   ammonium bicarbonate

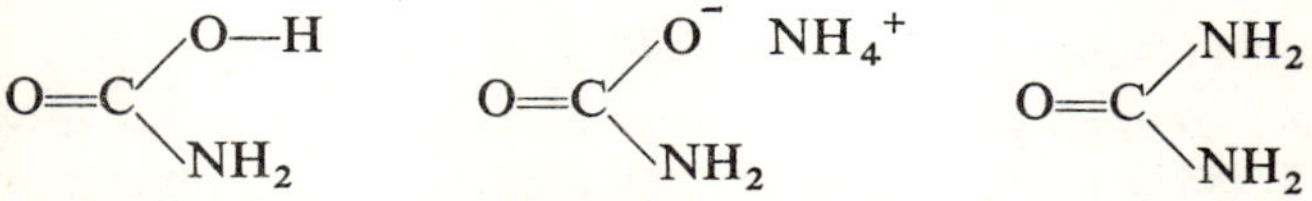

$O{=}C\begin{smallmatrix}\diagup O{-}H \\ \diagdown NH_2\end{smallmatrix}$ $O{=}C\begin{smallmatrix}\diagup O^-\ NH_4{}^+ \\ \diagdown NH_2\end{smallmatrix}$ $O{=}C\begin{smallmatrix}\diagup NH_2 \\ \diagdown NH_2\end{smallmatrix}$

carbamic acid   ammonium carbamate   urea

Urea is an excretion product of fishes, amphibia and all other animals. Proteins are digested by these animals to amino-acids (Chapter 25). In the liver, these compounds are broken up still further into ammonia which is converted into urea. The urea is removed in aqueous solution by the kidneys.

Preparation of urea

The synthesis of this essentially ' organic ' compound by Wöhler is referred to on p. xv, a synthesis depending on the change which takes place when an aqueous solution of ammonium cyanate is heated:

$$NH_4{}^+\ OCN^- \rightleftharpoons CO(NH_2)_2$$

* ' Commercial ' ammonium carbonate contains also ammonium bicarbonate and ammonium carbamate.

Urea is also obtained by the reaction of ammonia on phosgene, an application of a general method of making acid amides, and a reaction which throws light on the constitution of urea:

$$COCl_2 + 4NH_3 \longrightarrow CO(NH_2)_2 + 2NH_4Cl$$

Manufacture of urea

Urea is obtained on the commercial scale by heating together carbon dioxide and ammonia (the latter in excess of the amount required) to a temperature of about 150°C and under a pressure of 40 atmospheres or higher:

$$CO_2 + 2NH_3 \longrightarrow CO(NH_2)_2 + H_2O$$

Ammonium carbamate is an intermediate product.

Properties of urea

Urea is a colourless crystalline substance readily soluble in water and alcohol but not in ether.

(1) Urea is more basic than simple amides like acetamide, $CH_3 \cdot CO \cdot NH_2$. This is not surprising since urea contains two amino–groups but only one carbonyl group. Urea nitrate, a salt-like compound, is obtained as a white precipitate when concentrated nitric acid is added to a concentrated aqueous solution of urea:

$$\begin{matrix} NH_2 \\ \\ NH_2 \end{matrix}\!\!\!\!\!\! C{=}O + HO \cdot NO_2 \longrightarrow$$

$$\left[\begin{matrix} NH_2 \\ \\ NH_2 \end{matrix}\!\!\!\!\!\! \overset{+}{C}{=}OH \longleftrightarrow \begin{matrix} \overset{+}{NH_2} \\ \\ NH_2 \end{matrix}\!\!\!\!\!\! C{-}OH \right] \overset{-}{NO_3}$$

(2) Urea gives ammonia when heated with aqueous alkali (a typical amide reaction):

$$CO(NH_2)_2 + 2OH^- \longrightarrow 2NH_3 + CO_3^{2-}$$

This hydrolysis of urea occurs in the soil and the ammonia obtained is later oxidized and made available for plant life.

(3) Nitrous acid reacts with the amino groups of urea evolving nitrogen. Other reactions take place also for the yield of nitrogen is not quantitative:

$$CO(NH_2)_2 + 2HNO_2 \longrightarrow H_2CO_3 + 2N_2\uparrow + 2H_2O$$
$$\downarrow$$
$$CO_2 + H_2O$$

(4) Urea gives Hofmann's reaction for amides (p. 143) on addition of bromine and an alkali, or sodium hypobromite. Here, however, hydrazine, N_2H_4, is the first product but this is immediately oxidized to nitrogen:

$$CO(NH_2)_2 + 3NaOBr \longrightarrow N_2 + 3NaBr + CO_2 + 2H_2O$$

The reaction is quantitative and provides a means of estimating the amount of urea in a given solution (e.g. in urine), the nitrogen evolved being collected and measured.

(5) On being heated, urea melts, evolves ammonia and then re-solidifies. Condensation occurs to give the compound *biuret*:

$$NH_2 \cdot CO \cdot NH_2 + HNH \cdot CO \cdot NH_2$$
$$\longrightarrow NH_2 \cdot CO \cdot NH \cdot CO \cdot NH_2 + NH_3$$
$$\text{biuret}$$

If a little dilute caustic soda is added to the residue and then one drop of very dilute copper (II) sulphate solution, a deep pink (to purple) coloration appears. This is the *biuret test* for *proteins*. The test is given by any compound containing the linkage —CO·NH—, a grouping which is present in proteins (Chapter 25).

Uses of urea

In recent years, urea has been used in rapidly increasing amounts for the following purposes:

(1) In the manufacture of plastics.
(2) As a fertilizer.
(3) In the manufacture of pig and cattle foods.
(4) In medicine.

Oxalic acid, $H_2C_2O_4$

Oxalic acid is the first member of a series of dibasic acids each of which contains two carboxyl groups per molecule. The structural formula of oxalic acid is:

$$\begin{array}{c} HO-C \diagup^{O} \\ | \\ HO-C \diagdown_{O} \end{array}$$

i.e. the molecule consists of two carboxyl groups only.

The second member of the series is *malonic acid* of formula

$$H_2C \begin{array}{c} \diagup COOH \\ \diagdown COOH \end{array}$$

and the third is *succinic acid* of formula

$$\begin{array}{c} H_2C\!-\!COOH \\ | \\ H_2C\!-\!COOH \end{array}$$

Many of these dibasic acids and their derivatives are widely distributed throughout the vegetable kingdom. Potassium hydrogen oxalate, $KOOC \cdot COOH$, is found in the wood sorrels (*oxalis*), in the leaves and roots of rhubarb and other plants. Calcium oxalate is found in crystalline form in many plant cells.

Preparation of oxalic acid

(1) By heating cane sugar (sucrose) with concentrated nitric acid, which oxidizes the sugar*.

This was the method by which the acid was first prepared by Scheele in 1776. It is still a laboratory method of preparation. It should be carried out in a fume cupboard, for large amounts of nitrous fumes (reduction products of the nitric acid) are evolved.

Almost any carbohydrate, not only sucrose, can be used.

(2) By the oxidation of ethylene glycol by nitric acid (fume cupboard!)*.

Oxalic acid is obtained as the final oxidation product. The oxidation follows the usual course for primary alcohol groups, viz. through the aldehyde to the carboxylic acid:

$$\begin{array}{c} CH_2OH \\ | \\ CH_2OH \end{array} \xrightarrow[\text{HNO}_3]{\text{conc.}} \begin{array}{c} COOH \\ | \\ COOH \end{array}$$

Manufacture of oxalic acid

Oxalic acid is manufactured by heating sodium formate:

$$2HCOONa \longrightarrow (COONa)_2 + H_2$$

The free oxalic acid is obtained from the resulting sodium oxalate by treatment with sulphuric acid.

At one time, oxalic acid was manufactured by fusing sawdust with caustic soda in a current of air. The sodium oxalate was

---

* Potentially dangerous reactions, which should only be attempted under adequate supervision.

extracted from the residue by water treatment and the free acid obtained from it. The method is comparable to that of Scheele, for the process is essentially an oxidation by the oxygen of the air of the carbohydrate cellulose in the sawdust.

Properties of oxalic acid

Oxalic acid is a colourless crystalline solid which separates from aqueous solution as the hydrate of formula $H_2C_2O_4 \cdot 2H_2O$.

(1) Oxalic acid is the strongest of the dicarboxylic acids. The nearer together the two carboxyl groups in such acids, the greater is the degree of dissociation. Succinic acid is thus weaker than malonic.

Two series of salts, e.g. $K_2C_2O_4$ and KHC_2O_4 are obtainable. A compound of the formula $KHC_2O_4 \cdot H_2C_2O_4 \cdot 2H_2O$ called *potassium quadroxalate* or commonly ' salts of lemon ' is also known. It is used to remove ink stains and iron mould, by forming a soluble potassium iron(II) oxalate.

(2) On heating hydrated oxalic acid, water of crystallization is first evolved. If the heating is continued slowly, the anhydrous solid sublimes, some decomposition to formic acid with loss of carbon dioxide occurring:

$$HOOC \cdot COOH \longrightarrow CO_2 + HCOOH$$

Carbon monoxide is also formed:

$$HCOOH \longrightarrow H_2O + CO$$

Decomposition of oxalic acid to carbon monoxide occurs readily if the solid is warmed with concentrated sulphuric acid which acts as a dehydrating agent:

$$H_2C_2O_4 \xrightarrow{\substack{\text{conc.} \\ \text{H}_2\text{SO}_4}} H_2O + CO + CO_2$$

This is the common laboratory method of preparing carbon monoxide. Oxalates give a similar action for they are first converted to the free acid by the sulphuric acid.

(3) Oxalic acid forms esters. Diethyl oxalate is obtained by heating together anhydrous oxalic acid and ethyl alcohol. It is not necessary to have a mineral acid present as catalyst (oxalic acid is a sufficiently strong acid to catalyse the reaction):

$$(COOH)_2 + 2C_2H_5OH \longrightarrow (COOC_2H_5)_2 + 2H_2O$$

(4) Oxalic acid forms an acid chloride, $(COCl)_2$, *oxalyl chloride*. This is obtained by the reaction of phosphorus pentachloride or thionyl chloride on the anhydrous acid:

$$(COOH)_2 \xrightarrow[\substack{\text{or} \\ \text{SOCl}_2}]{\text{PCl}_5} (COCl)_2 \quad \text{oxalyl chloride}$$

Like the acid chlorides of most carboxylic acids, this is hydrolysed by water or more rapidly by alkalis, reacts with alcohols to form esters and with ammonia to form an amide (below).

(5) Oxalic acid forms an amide, $(CONH_2)_2$, *oxamide*. This compound can be prepared:

(*a*) By distilling ammonium oxalate:

$$(COONH_4)_2 \longrightarrow (CONH_2)_2 + 2H_2O$$
$$\text{oxamide}$$

(*b*) By the reaction of ammonia on the acid chloride or an ester:

$$(COCl)_2 + 4NH_3 \longrightarrow (CONH_2)_2 + 2NH_4Cl$$

Like the amides of other carboxylic acids, oxamide is dehydrated by heating with phosphorus pentoxide, the product being cyanogen:

$$(CONH_2)_2 \xrightarrow{\text{P}_2\text{O}_5} C_2N_2 + 2H_2O$$

The free acid is regenerated by treatment with nitrous acid, nitrogen being evolved:

$$(CONH_2)_2 \xrightarrow{\text{HNO}_2} (COOH)_2 + N_2$$

(6) Oxalic acid is oxidized by warm acidified potassium permanganate to carbon dioxide:

$$5(COOH)_2 + 2MnO_4^- + 6H^+ \longrightarrow 10CO_2 + 2Mn^{2+} + 8H_2O$$

or

$$5(COOH)_2 + 2KMnO_4 + 3H_2SO_4$$
$$\longrightarrow 10CO_2 + K_2SO_4 + 2MnSO_4 + 8H_2O$$

Uses of oxalic acid

Oxalic acid is used in the dye industry and in the manufacture of inks.

Iron(II) oxalate (which incidentally is yellow in contrast to the usual green colour of an iron(II) salt) is used as a developer. The oxalate ion has a very strong tendency to form complex ions with many metal ions.

Malonic acid, $H_2C\underset{\displaystyle \diagdown COOH}{\overset{\displaystyle \diagup COOH}{}}$

This second member of the series of dicarboxylic acids may be synthesized from acetic acid, using reactions already met in previous chapters:

$$CH_3 \cdot COOH \xrightarrow{Cl_2} \underset{\text{monochloroacetic acid}}{CH_2Cl \cdot COOH} \xrightarrow{KCN} \underset{\text{cyanoacetic acid}}{CH_2(CN) \cdot COOH}$$

$$\xrightarrow{\text{hydrolysis}} CH_2(COOH)_2$$

(The monochloroacetic acid is converted to the potassium salt before treatment with KCN, otherwise it would yield hydrogen cyanide, HCN.)

Malonic acid has typical carboxylic acid properties; in addition, it has the special property of decomposing when heated to its melting point (136°), losing carbon dioxide and yielding acetic acid:

$$H_2C\underset{\displaystyle \diagdown COOH}{\overset{\displaystyle \diagup COOH}{}} \xrightarrow{\text{heat}} CH_3 \cdot COOH + CO_2$$

Malonic acid, or rather its ethyl ester, is extremely valuable in organic syntheses, since one or both of the hydrogen atoms of the methylene group are successively replaceable by sodium atoms, the replacing agent being sodium ethoxide:

$$\underset{\text{diethyl malonate}}{H_2C(COOC_2H_5)_2} \xrightarrow{C_2H_5ONa} Na\overset{+}{H}\overset{-}{C}(COOC_2H_5)_2$$

This sodium compound reacts with an alkyl halide, for example

$$CH_3I + Na\overset{+}{H}\overset{-}{C}(COOC_2H_5)_2 \longrightarrow CH_3 \cdot CH(COOC_2H_5)_2 + NaI$$

Like malonic acid, the free acid obtainable from this ester contains two carboxyl groups attached to the same carbon atom and is decomposed on heating, thus:

$$CH_3 \cdot CH(COOH)_2 \longrightarrow \underset{\text{propionic acid}}{CH_3 \cdot CH_2COOH} + CO_2$$

Such reactions provide a *means of synthesizing the higher mono-carboxylic acids.*

Succinic acid, $\begin{array}{l} CH_2{\cdot}COOH \\ | \\ CH_2{\cdot}COOH \end{array}$

This acid was known in the sixteenth century. It is obtained as a product of the distillation of amber. It may be synthesized from ethyl alcohol or ethylene by steps shown below:

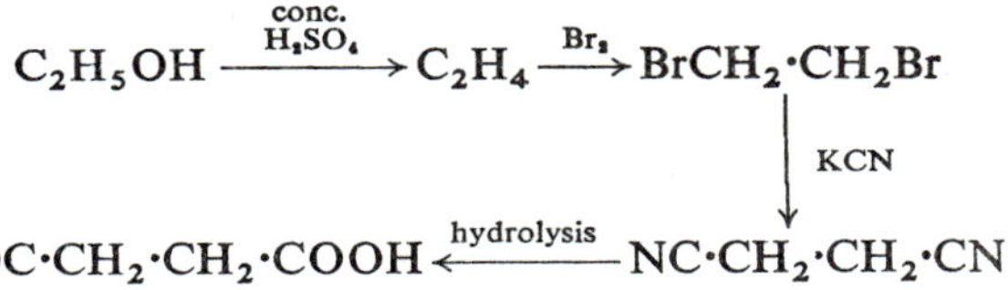
conc. H₂SO₄
C₂H₅OH
C₂H₄
Br₂
BrCH₂·CH₂Br
KCN
HOOC·CH₂·CH₂·COOH
hydrolysis
NC·CH₂·CH₂·CN

One of its most interesting properties is its power to form an intramolecular acid anhydride, a cyclic compound:

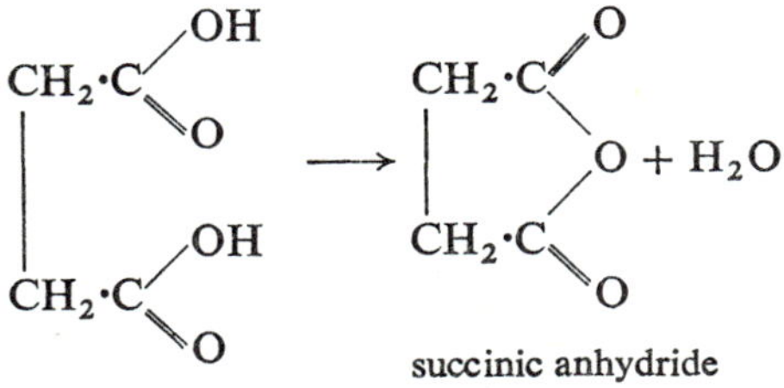
OH
CH₂·C
O
OH
CH₂·C
O
O
CH₂·C
O + H₂O
CH₂·C
O

succinic anhydride

Neither oxalic nor malonic acid forms such an anhydride.

Adipic acid, $\begin{array}{l} CH_2{\cdot}CH_2{\cdot}COOH \\ | \\ CH_2{\cdot}CH_2{\cdot}COOH \end{array}$

Adipic acid was first obtained by oxidation of fats (*adeps*, fat). It is now synthesized, and used extensively to form polymers, of which 'Nylon' is an example, by reaction with hexamethylene diamine, $H_2N(CH_2)_6NH_2$:

$$\ldots\overset{\overset{\displaystyle O}{\|}}{C}(CH_2)_4\overset{\overset{\displaystyle O}{\|}}{C}NH(CH_2)_6NH\overset{\overset{\displaystyle O}{\|}}{C}(CH_2)_4\overset{\overset{\displaystyle O}{\|}}{C}NH\ldots$$

$$\text{Tartaric acid,} \quad \begin{array}{l} \text{CH(OH)·COOH} \\ | \\ \text{CH(OH)·COOH} \end{array}$$

Tartaric acid is perhaps the most widely distributed throughout the vegetable kingdom of all acids, occurring in many berries and fruits. Its potassium hydrogen salt was known from the earliest times. It is present in grape juice and is deposited during fermentation in impure form as a crystalline substance known as *argol* or *wine lees*. The pure salt, of formula

$$\text{HOOC·CH(OH)·CH(OH)·CO}\overset{-}{\text{O}}\overset{+}{\text{K}}$$

is known as *cream of tartar* and is the acid component of baking powder. On neutralization with caustic soda, the sodium potassium salt is obtained. This compound is known as *Rochelle salt* and is used in the preparation of Fehling's solution (p. 94).

The free tartaric acid was first isolated by Scheele in 1769. The acid has great historical interest because of the brilliant work done on it by Pasteur (p. 333).

It may be synthesized from succinic acid as shown:

$$\begin{array}{l} \text{CH}_2\text{·COOH} \\ | \\ \text{CH}_2\text{·COOH} \end{array} \xrightarrow{\text{Br}_2} \begin{array}{l} \text{CHBr·COOH} \\ | \\ \text{CHBr·COOH} \end{array} \xrightarrow[\text{Ag}_2\text{O}]{\text{moist}} \begin{array}{l} \text{CH(OH)·COOH} \\ | \\ \text{CH(OH)·COOH} \end{array}$$

SUMMARY OF THE CHEMISTRY OF DIBASIC ACIDS AND THEIR DERIVATIVES

Carbonic acid

carbonic acid	ammonium bicarbonate	ammonium carbonate

carbamic acid (unstable)	ammonium carbamate	urea

Urea: preparation

$$\overset{+}{\text{NH}}_4(\text{OCN})^- \rightleftharpoons \text{O}=\text{C}\begin{array}{l} \diagup \text{NH}_2 \\ \diagdown \text{NH}_2 \end{array}$$

Urea: properties

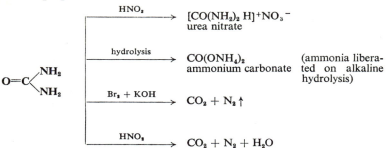

Dicarboxylic acids

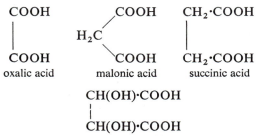

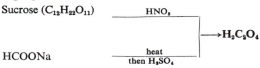

Oxalic acid: preparation

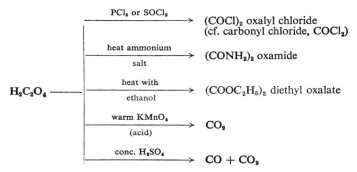

Oxalic acid: properties

Malonic acid: synthesis

$$CH_3 \cdot COOH \xrightarrow{Cl_2} CH_2Cl \cdot COOH \xrightarrow[KCN]{KOH} CH_2 \cdot CN \cdot COOH$$

$$\downarrow \text{hydrolysis (dilute acid)}$$

$$CH_2 \Big\langle \begin{matrix} COOH \\ COOH \end{matrix}$$

Two peculiar properties:

(*a*) Evolves carbon dioxide on heating. This is characteristic of dicarboxylic acids in which both carboxyl groups are attached to the same carbon atom:

$$H_2C \Big\langle \begin{matrix} COOH \\ COOH \end{matrix} \xrightarrow{\text{heat}} CH_3 \cdot COOH + CO_2$$

(*b*) Hydrogen atoms of methylene group in esters are acidic (react with strong bases):

$$H_2C(COOC_2H_5)_2 \xrightarrow{CH_3ONa} Na\overset{+}{H}\overset{-}{C}(COOC_2H_5)_2$$
diethyl malonate

This sodium compound is very useful in synthesis, e.g.:

$$Na\overset{+}{H}\overset{-}{C}(COOC_2H_5)_2 \xrightarrow{CH_3I} CH_3 \cdot CH(COOC_2H_5)_2$$

$$\downarrow \text{hydrolysis}$$

$$CH_3 \cdot CH_2 \cdot COOH \xleftarrow{\text{heat}} CH_3 \cdot CH(COOH)_2$$
propionic acid

Succinic acid: synthesis

$$C_2H_5OH \xrightarrow[\text{H}_2\text{SO}_4]{\text{conc.}} C_2H_4 \xrightarrow{Br_2} CH_2Br \cdot CH_2Br$$

$$\downarrow KCN$$

$$\underset{\begin{matrix} | \\ CH_2 \cdot COOH \end{matrix}}{CH_2 \cdot COOH} \xleftarrow{\text{hydrolysis}} NC \cdot CH_2 \cdot CH_2 \cdot CN$$

Peculiar property:—forms an internal anhydride which is a ring compound:

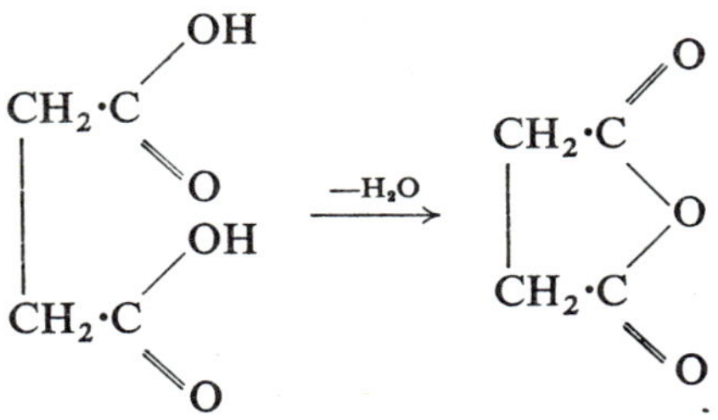

$$CH_2 \cdot C \diagup OH \diagdown O \quad | \quad CH_2 \cdot C \diagup OH \diagdown O \xrightarrow{-H_2O} CH_2 \cdot C \diagup O \diagdown O \quad | \quad CH_2 \cdot C \diagdown O$$

succinic anhydride

Tartaric acid: synthesis

$$\begin{array}{c} CH_2 \cdot COOH \\ | \\ CH_2 \cdot COOH \end{array} \xrightarrow{Br_2} \begin{array}{c} CHBr \cdot COOH \\ | \\ CHBr \cdot COOH \end{array} \xrightarrow[Ag_2O]{moist} \begin{array}{c} CH(OH) \cdot COOH \\ | \\ CH(OH) \cdot COOH \end{array}$$

Two commonly known derivatives:

Cream of tartar (in baking powder):

$$\begin{array}{c} CH(OH) \cdot CO\overset{-}{O}K^{+} \\ | \\ CH(OH) \cdot COOH \end{array}$$

Rochelle salt (in Fehling's solution):

$$\begin{array}{c} CH(OH) \cdot CO\overset{-}{O}K^{+} \\ | \\ CH(OH) \cdot CO\overset{-}{O}Na^{+} \end{array}$$

EXAMINATION QUESTIONS

(1) Indicate how you could obtain a pure specimen of urea from ammonium sulphate and potassium cyanate.

How does urea react with (*a*) nitric acid, (*b*) sodium hydroxide, (*c*) nitrous acid?

Describe the biuret test for urea, and indicate **one** method of determining urea in solution. (O. A.)

(2) How can urea be obtained from (*a*) potassium cyanate, (*b*) carbon dioxide? Describe the action of heat on urea. Give **two** examples of the uses of urea. Which of the reactions of urea could be used for its estimation? (O. and C. A.)

16

PREPARATIVE METHODS AND ORGANIC SYNTHESIS

PREPARATIVE METHODS

IN previous chapters of this book emphasis has been laid on the properties and reactions of the more important aliphatic compounds. Methods of preparation have not been treated systematically, for the very good reason that the various reactions by which a particular type of compound may be prepared are often unrelated to one another (except that they give the same product). Nevertheless, the student of organic chemistry needs to know how to prepare compounds just as much as he or she needs to know how compounds behave, and in this chapter an attempt is made to summarize the preparative methods commonly used in aliphatic chemistry.*

Alkanes†

(1) $$RCO_2^- Na^+ + NaOH \xrightarrow{\text{heat}} RH + Na_2CO_3$$
(soda lime)

In this form the reaction is chiefly used for demonstration purposes (e.g. preparation of methane in the school laboratory).

(2) $$RI \text{ (or } RBr) \xrightarrow[\text{ether}]{\text{Mg}} RMgI \xrightarrow{\text{H}_2\text{O}} RH + Mg(OH)I$$
(Grignard reagent)

(3) $$2RCO_2^- \xrightarrow{\text{electrolysis}} (2R\cdot) + 2CO_2 + 2e^-$$
$$\downarrow$$
$$R-R$$

---

* The facts presented have all been encountered previously but they are now organized in a different way. A useful consequence is that study of this chapter will provide a means of revision of much of the earlier material.

† R represents a simple alkyl group, e.g. CH_3-, $CH_3\cdot CH_2-$, $CH_3\cdot CH_2\cdot CH_2-$ etc. The symbol R′ represents an alkyl group which can be the same as R or a different alkyl group. X represents Cl, Br or I.

The classical Kolbe synthesis of (symmetrical) hydrocarbons, which can be extended, e.g.

$$2\,(CH_3O \cdot \underset{\underset{O}{\|}}{C} \cdot CH_2 \cdot CH_2 \cdot CH_2 \cdot CO_2^- Na^+) \longrightarrow CH_3O \cdot \underset{\underset{O}{\|}}{C}(CH_2)_6 \underset{\underset{O}{\|}}{C} \cdot OCH_3$$

(4) $\quad RCH{=}CHR' \xrightarrow[\text{catalyst}]{H_2} RCH_2 \cdot CH_2R'$

A widely used type of catalytic hydrogenation. Note the related, stepwise, reduction of acetylenes (alkynes):

$$RC{\equiv}CR' \xrightarrow[\text{catalyst}]{H_2} RCH{=}CHR' \longrightarrow RCH_2 \cdot CH_2R'$$

(5) $\quad 2RX + 2Na \longrightarrow R{-}R + 2NaX$

The classical Wurtz synthesis which is chiefly of historical importance (other products are also formed).

Alkenes

(1) $\quad ROH \xrightarrow[\text{or } H_3PO_4]{\text{heat with } H_2SO_4}$ alkene (dehydration of alcohols)

e.g. $CH_3 \cdot CH(OH)CH_3 \longrightarrow CH_3 \cdot CH{=}CH_2 + H_2O$

The ease with which an alcohol undergoes dehydration depends on its structure; in general, primary alcohols require vigorous conditions (concentrated acid, high temperature) whereas tertiary alcohols are often easily dehydrated (milder heating with diluted acids).

An alternative method is to pass the alcohol vapour over heated alumina.

(2) $\quad \underset{\text{alkyl halide}}{RX} \xrightarrow[\text{alc. KOH}]{\text{heat with}}$ alkene (dehydrohalogenation)

e.g. $CH_3 \cdot CHI \cdot CH_3 \longrightarrow CH_3 \cdot CH{=}CH_2 + HI$

(an example of an elimination reaction).

(3) 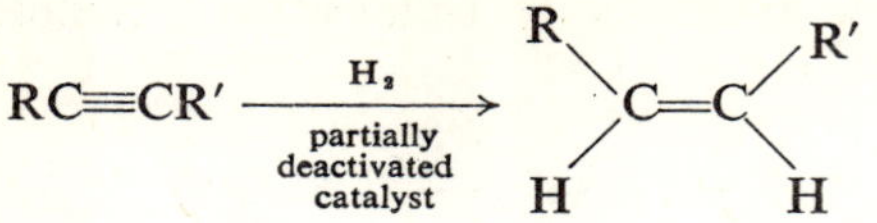 $RC\equiv CR' \xrightarrow[\substack{\text{partially}\\\text{deactivated}\\\text{catalyst}}]{H_2}$

$$\begin{array}{c} R \\ \\ H \end{array} C=C \begin{array}{c} R' \\ \\ H \end{array}$$

An important method, since mono- and di-alkyl acetylenes are readily available (see below). This partial hydrogenation normally gives the *cis* isomer of the alkene.

Alkynes (acetylenes)

(1) $HC\equiv CH \xrightarrow[\text{(liq. NH}_3\text{)}]{\overset{+}{Na} \overset{-}{NH_2}} \overset{+}{Na}\overset{-}{C}\equiv CH \xrightarrow[\text{(liq. NH}_3\text{)}]{RI} RC\equiv CH$

$RC\equiv CH \xrightarrow[\text{similarly}]{} RC\equiv CR'$

A very general method and commonly used (because acetylenes can be converted to many other types of compound). [Note that acetylene is readily available in cylinders but can be easily made from calcium carbide (calcium acetylide):

$$CaC_2 + 2H_2O \longrightarrow HC\equiv CH + Ca(OH)_2]$$

(2) $RCHBr \cdot CH_2Br \xrightarrow[\text{KOH}]{\text{hot alc.}} RC\equiv CH + 2HBr$

     (a dihalogeno-
     compound)

A dehydrohalogenation reaction related to alkene synthesis (2) above.

Alkyl halides

(1) $ROH + HX \longrightarrow RX + H_2O$

Chlorides, bromides and iodides are prepared by variations of this method:

e.g. $ROH + HBr$
        (conc. aq. acid)

$\hspace{5cm} RBr$

$ROH + KBr + H_2SO_4$
           conc.

(2) $ROH + SOCl_2 \longrightarrow RCl + SO_2 + HCl$

(3) $ROH + PCl_5 \longrightarrow RCl + POCl_3 + HCl$

Other phosphorus halides are also used (e.g. PBr_3, PCl_3) and are sometimes generated *in situ*:

$$ROH \xrightarrow[Br_2]{red\ P} RBr$$

(4) $RH + Cl_2 \xrightarrow[(or\ heat)]{light} RCl + HCl$

Reactions of this type are important in industry but are not often performed on alkanes in the laboratory.

(5) $R{\cdot}CH{=}CH_2 + HCl \longrightarrow R{\cdot}CHCl{\cdot}CH_3$
                  (anhydrous)

Monohydric alcohols

(1) $RMgX + HC\overset{O}{\underset{H}{\diagdown}}$ $\longrightarrow RCH_2{\cdot}OMgX \xrightarrow{H_2O} RCH_2OH$
   (Grignard
   reagent)
                  (formaldehyde)

$RMgX + R'{\cdot}C\overset{O}{\underset{H}{\diagdown}} \longrightarrow \underset{R'}{RCH{\cdot}OMgX} \xrightarrow{H_2O} \overset{R}{\underset{R'}{\diagdown}}CHOH$

$RMgX + \overset{R'}{\underset{R''}{\diagdown}}C{=}O \longrightarrow R'\overset{R}{\underset{R''}{-}}C{\cdot}OMgX \xrightarrow{H_2O} R'\overset{R}{\underset{R''}{-}}COH$

A very general method, applicable to the preparation of primary, secondary and tertiary alcohols.

(2) $RCHO \xrightarrow[or\ NaBH_4]{LiAlH_4} RCH_2OH$

$\overset{R}{\underset{R'}{\diagdown}}C{=}O \xrightarrow[or\ NaBH_4]{LiAlH_4} \overset{R}{\underset{R'}{\diagdown}}CHOH$

Other methods of reduction are available, e.g. with aluminium isopropoxide, $Al[OCH(CH_3)_2]_3$.

(3) $RCO_2H \text{(or } RCO_2C_2H_5) \xrightarrow{\text{LiAlH}_4} RCH_2OH$

Note that sodium borohydride is *not* effective here.

(4) $RCH{=}CH_2 \xrightarrow{\text{H}_2\text{SO}_4} \underset{\substack{|\\O{\cdot}SO_2{\cdot}OH}}{RCH{\cdot}CH_3} \xrightarrow{\text{H}_2\text{O}} \underset{\substack{|\\OH}}{RCH{\cdot}CH_3}$

Used industrially for the manufacture of simple alcohols.

(5) $RX \xrightarrow[\text{or Ag}_2\text{O/H}_2\text{O}]{\text{aq. NaOH}} ROH$

In practice, the reverse reaction $ROH \xrightarrow{\text{HX}} RX$ is more important.

(6) $RNH_2 \xrightarrow[\substack{(\text{NaNO}_2 + \text{aq.}\\ \text{acid})}]{\text{HO{\cdot}NO}} ROH + N_2$

Sometimes useful, but frequently the reaction yields other products in addition to the expected alcohol.

(7) $RC{\overset{\displaystyle O}{\underset{\displaystyle OR'}{\big<}}} \xrightarrow[\text{aq./alc. KOH}]{\text{aq. or}} RC{\overset{\displaystyle O}{\underset{\displaystyle OH}{\big<}}} + R'OH$
    (as salt)

Useful where a particular ester is easily available (e.g. from natural sources).

Ethers

(1) $RX + Na\overset{+}{O}\overset{-}{R'} \longrightarrow RO{\cdot}R' + Na^+X^-$

A very general method (the Williamson synthesis).

(2) Simple ethers are also formed, under appropriate conditions, when alcohols are heated with concentrated acids (usually H_2SO_4),

e.g. the laboratory preparation of diethyl ether. Note that this is a very limited method.

Aliphatic aldehydes

(1) $$\underset{\text{(primary alcohol)}}{RCH_2OH} \xrightarrow[\text{acid'}]{\text{'chromic}} RCHO$$

Aldehydes are themselves easily oxidized; thus the aldehyde must be removed from the reaction mixture as quickly as possible, e.g. by distillation.

(2) $$RCH_2OH \xrightarrow[\text{Cu}]{\text{heated}} RCHO + H_2$$

An example of dehydrogenation.

(3) $$HC{\equiv}CH \xrightarrow[\text{(Hg SO}_4\text{)}]{H_2SO_4/H_2O} (H_2C{=}CH{\cdot}OH) \longrightarrow CH_3{\cdot}CHO$$

A specific method for acetaldehyde only.

Aliphatic ketones

(1) $$\begin{matrix}R\\ \quad\rangle CH{\cdot}OH\\ R'\end{matrix} \xrightarrow[\text{acid'}]{\text{'chromic}} \begin{matrix}R\\ \quad\rangle C{=}O\\ R'\end{matrix}$$

(2) $$\begin{matrix}R\\ \quad\rangle CH{\cdot}OH\\ R'\end{matrix} \xrightarrow[\text{Cu}]{\text{heated}} \begin{matrix}R\\ \quad\rangle C{=}O\\ R'\end{matrix}$$

(3) $$RC{\equiv}CH \xrightarrow[\text{(Hg SO}_4\text{)}]{H_2SO_4/H_2O} (R\underset{\displaystyle OH}{\underset{|}{C}}{=}CH_2) \longrightarrow RCO{\cdot}CH_3$$

$$RC{\equiv}CR \xrightarrow{\text{similarly}} RCO{\cdot}CH_2R$$

Two examples of the usefulness of alkylacetylenes.

(4) $(RCO_2^-)_2Ca^{2+} \xrightarrow{heat} RCOR + CaCO_3$

An old method which is still useful when the carboxylic acid is readily available.

Aliphatic carboxylic acids

(1) $RCH_2 \cdot OH \xrightarrow[\text{acid'}]{\text{'chromic}} (RCHO) \xrightarrow[\text{acid'}]{\text{'chromic}} RCO_2H$

(2) $RX \longrightarrow RMgX \xrightarrow[\text{(in ether)}]{CO_2} RCO_2MgX \xrightarrow[\text{acid}]{aq.} RCO_2H$

(3) $RX \xrightarrow[\text{(aq. alc.)}]{KCN} RCN \xrightarrow[\text{(or alkali)}]{\text{hot aq. acid}} RCO_2H$

(4) $RC\overset{O}{\underset{OR'}{\big<}} \xrightarrow[\text{aq./alc. KOH}]{\text{aq., or}} RC\overset{O}{\underset{\underset{O K^+}{}}{\big<}} \xrightarrow[\text{acid}]{aq.} RCO_2H$

Acid chlorides

(1) $RCO_2H + SOCl_2 \longrightarrow RCO \cdot Cl + SO_2 + HCl$

Very frequently used, because the other products are gaseous and therefore easily removed.

(2) $3RCO_2H + PCl_3 \longrightarrow 3RCOCl + P(OH)_3$

(3) $RCO_2H + PCl_5 \longrightarrow RCOCl + POCl_3 + HCl$

Acid anhydrides

(1) $RCO \cdot Cl + RCO_2^- Na^+ \longrightarrow RCO \cdot O \cdot COR + NaCl$

Acid amides

(1) $RCOCl + NH_3 \longrightarrow RCONH_2 + HCl$

Ammonium carbonate is often used in place of ammonia, giving a less violent reaction. Note the related reactions:

$$RCOCl + NH_2R' \longrightarrow RCONHR'$$

$$RCOCl + NH\begin{array}{c}R'\\ \diagup\\ \diagdown\\ R''\end{array} \longrightarrow RCON\begin{array}{c}R'\\ \diagup\\ \diagdown\\ R''\end{array}$$

leading to N-substituted amides.

(2) $RCO \cdot O \cdot COR + NH_3 \longrightarrow RCONH_2 + RCO_2H$

Anhydrides react similarly with $R'NH_2$ and $R'NH \cdot R''$.

(3) $RCO_2^- NH_4^+ \xrightarrow[\text{in R·CO}_2\text{H}]{\text{heat}} RCO \cdot NH_2$

Used in a laboratory preparation of acetamide but not often encountered otherwise.

Esters

(1) $RCO_2H + R'OH \overset{(H+)}{\rightleftharpoons} RCOOR' + H_2O$

The acid-catalysed esterification process.

(2) $RCOCl + R'OH \longrightarrow RCOOR' + HCl$

or $RCO \cdot O \cdot COR + R'OH \longrightarrow RCOOR' + RCO_2H$

(3) $\left.\begin{array}{l}RCO_2^- K^+\\[1em]\text{or } RCO_2^- Ag^+\end{array}\right\} \xrightarrow{R'X} RCOOR' + KX$
$\qquad\qquad\qquad\qquad\qquad$ (or AgX)

Alkyl cyanides (nitriles)

(1) $RX + K^+ CN^- \xrightarrow[\text{aq. alc.}]{\text{heat in}} RCN + KX$

Usually carried out with the alkyl bromide or iodide.

(2) $RCONH_2 \xrightarrow[\text{P}_2\text{O}_5]{\text{heat with}} RCN + H_2O$

Aliphatic amines

(1) $RX + NH_3 \longrightarrow RNH_3^+ X^- \xrightarrow[\text{alkali}]{\text{aq.}} RNH_2$

Not a good method, in general, because numerous other products are formed (e.g. R_2NH, R_3N etc.).

(2) $RCONH_2 \xrightarrow[\text{aq. alkali}]{Br_2} RNH_2$ (Hofmann reaction)

Often useful in practice (but complicated in mechanism!).

(3) $RCN \xrightarrow[\text{(catalyst)}]{H_2} RCH_2 \cdot NH_2$

(4) $RNO_2 \xrightarrow[\text{(catalyst)}]{H_2} RNH_2$
(nitro alkane)

(5) $RCONH \cdot R' \xrightarrow{LiAlH_4} RCH_2 \cdot NH \cdot R'$

$$RCON \underset{R''}{\overset{R'}{<}} \xrightarrow{LiAlH_4} RCH_2 \cdot N \underset{R''}{\overset{R'}{<}}$$

Useful for the preparation of secondary and tertiary amines.

ORGANIC SYNTHESIS

The synthesis of organic compounds forms an important and indeed essential part of organic chemistry. We have seen that many organic compounds occur naturally, and that such compounds may be made up of large molecules with highly elaborate structures. But the simpler organic compounds, whether they occur naturally or not, are of the greatest importance in that they form the starting materials for syntheses which lead to materials of great industrial value. Hence, a large proportion of the vast and world-wide chemical industry is concerned with organic preparations. Many of the simpler organic compounds are manufactured on a scale of millions of tons annually, and then used, for example, as solvents, as monomers for the ever-growing plastics industry, or as starting materials in the preparation of dyestuffs, detergents, pharmaceuticals and other ' fine ' chemicals. Much research in organic chemistry is devoted to the preparation of new compounds; but a great deal of time and effort is also given to the development of better methods of making known compounds. These activities involve organic syntheses which consist, in general, of sequences of organic reactions arranged so as to lead to the desired end-products.

To the beginner in the study of organic chemistry, lists of methods of preparation such as that just given in this chapter may seem to present a bewilderingly large number of possibilities. It should be remembered, however, that the yields of a particular product obtained by various routes can show considerable differences, and also that the necessary starting materials may not be equally available (e.g. one particular compound may be available commercially at low price, or may be made easily by an established procedure). It is therefore useful to have a choice of several general methods for each type of organic compound.

Some simple examples of organic reaction sequences are outlined below.

(a) Conversion of *acetic acid* into (i) *ethylamine* and (ii) *methylamine*:

$$CH_3 \cdot CO_2H \longrightarrow CH_3 \cdot CO \cdot Cl \longrightarrow CH_3 \cdot CO \cdot NH_2$$

$$\xrightarrow{P_2O_5} \qquad \qquad \searrow Br_2/KOH$$

$$CH_3 \cdot CN \xrightarrow[\text{(Ni)}]{H_2} CH_3 \cdot CH_2 \cdot NH_2 \qquad \qquad CH_3 \cdot NH_2$$
$$\text{ethylamine} \qquad \qquad \text{methylamine}$$

(b) Conversion of *ethyl alcohol* into (i) *acetone*, (ii) *isopropyl alcohol*, (iii) *n-propyl alcohol*, (iv) *sec-butyl alcohol* (*butan-2-ol*):

$$CH_3 \cdot CH_2OH \longrightarrow CH_3 \cdot CHO \longrightarrow CH_3 \cdot CO_2H$$

$$\text{distil Ca salt}$$

$$CH_3 \cdot CO \cdot CH_3 \xrightarrow[\text{or LiAlH}_4]{H_2 \text{ (cat)}} CH_3 \cdot CH(OH) \cdot CH_3$$
$$\text{acetone} \qquad \qquad \text{isopropyl alcohol}$$

$$CH_3 \cdot CH_2OH \longrightarrow CH_3 \cdot CH_2I \longrightarrow CH_3 \cdot CH_2 \cdot CN$$

$$CH_3 \cdot CH_2 \cdot CO_2H \xrightarrow{LiAlH_4} CH_3 \cdot CH_2 \cdot CH_2 \cdot OH \text{ (n-propyl alcohol)}$$

$$CH_3 \cdot CH_2OH \longrightarrow CH_3 \cdot CH_2 \cdot MgI$$

$$\longrightarrow CH_3 \cdot CHO \qquad \rangle \longrightarrow CH_3 \cdot CH_2 \cdot CH \cdot CH_3$$
$$\qquad \qquad \qquad \qquad \qquad \qquad OH$$
$$\qquad \qquad \qquad \qquad \qquad \text{sec-butyl alcohol}$$

(c) Conversion of *ethyl alcohol* to *lactic acid*, $CH_3 \cdot CH(OH) \cdot CO_2H$.

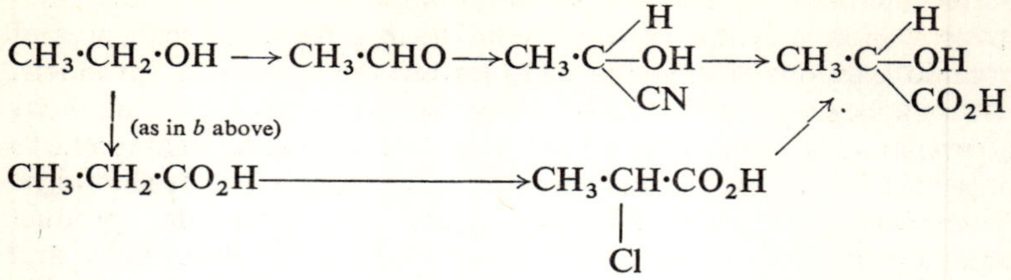

$$CH_3 \cdot CH_2 \cdot OH \longrightarrow CH_3 \cdot CHO \longrightarrow CH_3 \cdot C\!\!\underset{CN}{\overset{H}{\underset{\big|}{\big|}}}\!\!OH \longrightarrow CH_3 \cdot C\!\!\underset{CO_2H}{\overset{H}{\underset{\big|}{\big|}}}\!\!OH$$

$\downarrow$ (as in *b* above)

$$CH_3 \cdot CH_2 \cdot CO_2H \xrightarrow{\hspace{3cm}} CH_3 \cdot \underset{\underset{Cl}{|}}{CH} \cdot CO_2H$$

Note that the product contains two different functional groups, but that the two routes given involve applications of standard reactions of mono-functional compounds.

A useful exercise, which the student can pursue indefinitely (or until exhausted), is to start with a few simple organic compounds (e.g. methyl alcohol and ethyl alcohol) and to construct a chart showing the various substances which could be prepared from them (a rather large sheet of paper is desirable!).

Part of such a chart, using only a few types of reaction, is shown in *Figure 16.1*.

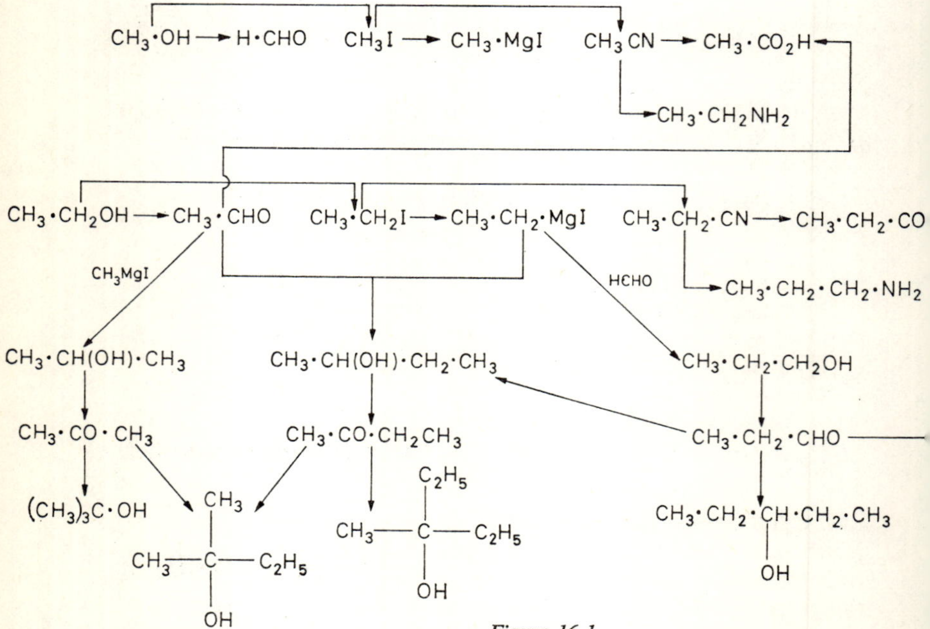

Figure 16.1.

THE FISCHER-TROPSCH SYNTHESIS

In 1902 the two French chemists Sabatier and Senderens synthesized methane from hydrogen and carbon monoxide and dioxide respectively, by passing the gaseous mixture over nickel heated to a temperature of about 250°C:

$$CO + 3H_2 \longrightarrow CH_4 + H_2O$$

$$CO_2 + 4H_2 \longrightarrow CH_4 + 2H_2O$$

In 1913, a German firm showed that high hydrocarbons and some oxy- compounds could be similarly obtained from mixtures of hydrogen and carbon monoxide.

Ten years later, the two German chemists Fischer and Tropsch started research work on the subject and very soon showed that it was possible to synthesize a great variety of organic aliphatic compounds from a mixture of hydrogen and carbon monoxide, by varying the composition of the mixture, the conditions of temperature and pressure, and the nature of the catalyst. Production plants were soon set up and extended very considerably during World War II. There has been much development since, both in America and in this country. Very recently, the synthesis of aromatic compounds by the process has been achieved; there seems, in fact, no limit to the possibilities of the process in industrial organic chemistry.

The synthesis of methyl alcohol from water gas and hydrogen by passing the mixture under pressure, over heated zinc oxide (with other oxides as promoters) has now almost completely displaced the older method of obtaining this useful compound by the destructive distillation of wood:

$$\underbrace{CO + H_2}_{\text{water gas}} + H_2 \xrightarrow[\substack{Cr_2O_3 \text{ promoter} \\ 300°-400°C \\ 200-400 \text{ atm}}]{\text{ZnO catalyst}} CH_3OH$$

Primarily, however, the Fischer–Tropsch synthesis is essentially the hydrogenation of carbon monoxide to paraffins and olefins, represented as follows:

$$nCO + (2n + 1)H_2 \longrightarrow C_nH_{2n+2} + nH_2O$$

and

$$nCO + 2nH_2 \longrightarrow C_nH_{2n} + nH_2O$$

The ratio of hydrogen to carbon monoxide, temperature and pressure of the reaction, and the nature of the catalyst determine the composition of the product. As would be expected, the greater the proportion of hydrogen, the greater the percentage of alkane obtained. With some catalysts, the process of hydrocarbon synthesis is followed by polymerization, and in such cases, the number of carbon atoms in the compound obtained may be high. High pressures tend to help the latter process.

By the end of World War II, Fischer–Tropsch plants in Germany were supplying a variety of most valuable products. Lower-boiling hydrocarbons were made and used as petrols. These were of good quality, for by combination of the alkane and the alkene (by alkylation, p. 44) a product of high octane number was obtained. Higher fractions were used as fuels. Solid paraffin waxes, most useful in electrical work, were also produced. More important still, by secondary processes these hydrocarbons were subjected to partial oxidation to yield carboxylic acids. Alkanes, of carbon values between 10 and 18, were oxidized to carboxylic acids which were converted by glycerol into edible fats or into sodium or potassium salts for use as soap. Salts of the carboxylic acids from alkanes of carbon values between 18 and 24 made valuable lubricants.

Further, by an extension of the process consisting of mixing water gas with an olefin, ketones and aldehydes were synthesized. The synthesis of a mixture of acetone (30 per cent) and propionaldehyde (70 per cent) by the interaction of water gas and ethylene in presence of cobalt oxide has been referred to on p. 113.

Raw materials for the Fischer–Tropsch process

Different methods of obtaining the mixture of carbon monoxide and hydrogen are now in use in different countries.

(1) By the action of steam on white hot coke:

$$C + H_2O \longrightarrow CO + H_2$$

The proportion of hydrogen in the mixture can be increased by passing it with steam over heated iron(III) oxide (and promoters):

$$\underbrace{CO + H_2}_{\text{water gas}} + H_2O \xrightarrow{Fe_2O_3} CO_2 + 2H_2$$

The carbon dioxide is removed by compressing and scrubbing with water.

(2) By the partial oxidation of natural gas in oil wells and in coal mines, e.g.

$$2CH_4 + O_2 \longrightarrow 2CO + 4H_2$$

To obtain a product of a lower carbon content, methane and steam may be passed over heated nickel (p. 13):

$$CH_4 + H_2O \longrightarrow CO + 3H_2$$

(3) *Coal gas* consisting chiefly of hydrogen, methane and carbon monoxide may be used as the raw material after treatment as above. Of great interest in this connection, are the experiments on underground gasification which are being conducted in many countries.

Catalysts

In the early experiments of Fischer and Tropsch the catalysts used were iron and cobalt, but nickel was soon added to the list and more recently, ruthenium. In general, a higher temperature is needed for iron than when cobalt and nickel are used. Cobalt gives a higher yield of olefins in the product than is obtained with nickel. Ruthenium gives a greater yield of solid hydrocarbons and furthermore lasts longer, not being so rapidly ' poisoned '.

In all cases, the catalyst is mixed with another substance which is itself a catalyst, and this is known as a *promoter*. There may be various reasons for this: the promoter may supply some catalytic effect not possessed by the actual catalyst; it may prohibit poisoning; or it may provide a suitable supporting material. Promoters (some of which are specific to certain catalysts) in common use are the oxides of aluminium, magnesium, silicon (as kieselguhr), thorium, cobalt, copper, etc.

The physical condition of the catalyst is varied. Very recently (and with some success), finely divided particles of the catalyst have been suspended in the gas stream.

The nature of the catalyst action is not fully understood. It is suggested that the metal first reacts with carbon monoxide to form a carbide. At higher temperatures, this reacts with hydrogen to give a hydrocarbon, an alkane or an alkene. These may subsequently ' crack ' to form hydrocarbons or hydrocarbon radicals, e.g. methyl or methylene, which then join up to form carbon chains.

EXAMINATION QUESTIONS

(1) Outline how the following changes may be effected. Give equations for every reaction discussed:

(*a*) acetic acid → methane;
(*b*) acetylene → acetic acid;
(*c*) acetic acid → methyl cyanide;
(*d*) ethylamine → ethyl bromide;
(*e*) toluene → sodium benzoate.

(C. A.)

(2) Give an account of the synthesis of some substances of industrial importance starting with (*a*) acetylene, (*b*) formalin. (O. S.)

(3) By means of formulae and brief explanatory notes, indicate how the following transformations could be effected:

(*a*) $CH_3 \cdot CH_2 \cdot CH_2 \cdot OH$ → $CH_3 \cdot CO \cdot CH_3$
(*b*) $(CH_3)_2 CH \cdot CO \cdot OH$ → $(CH_3)_2 CH \cdot CH(CH_3)_2$
(*c*) C_6H_6 (benzene) → $C_6H_5 \cdot CH_2 \cdot CO \cdot OH$

(Liverpool B.Sc., Part I)

(4) Outline possible syntheses of the following compounds, starting in each case with an aliphatic carboxylic acid $R \cdot CO \cdot OH$ and any other necessary reagents and materials:

$$R \cdot COO \cdot CH_2 \cdot R$$
$$R \cdot CH_2 \cdot CH{=}CH \cdot CH_2 \cdot R \ (cis \text{ isomer})$$
$$R \cdot CO \cdot NH \cdot R$$
$$R \cdot CH_2 \cdot N{\Big\langle}{}^{C_6H_5}_{CH_3}$$
$$R \cdot CH_2 D$$

(Liverpool B.Sc., Part I)

(5) Alkyl halides have many applications in synthesis, either directly or in the form of their Grignard derivatives.

Construct a chart which illustrates this statement, representing the alkyl halide as RX.

(6) State the reagents and conditions required to perform the following syntheses, using the letters over the arrows to show the reactions to which you are referring. (Further notes are not required.)

(i) $CaC_2 \xrightarrow{a} C_2H_2 \xrightarrow{b} CH_3CHO \xrightarrow{c} CHI_3 \xrightarrow{d} H \cdot COONa$

(ii) $CH_3COOH \xrightarrow{e} CH_2ClCOOH \xrightarrow{f} CH_2NH_2COONH_4$
$\xrightarrow{g} CH_3CO \cdot NH \cdot CH_2COONH_4$

(iii) $C_6H_6 \xrightarrow{h} C_6H_5SO_3H \xrightarrow{i} C_6H_5ONa \xrightarrow{j} C_6H_2(NO_2)_3OH.$

Write the full equations for the reactions (c) and (d). Reaction (c) is used also as a test. Name two other substances which respond to the same test. (S. A.)

(7) Give full equations accompanied by concise explanatory comments, for a series of reaction which would accomplish the conversion of the acid $CH_3 \cdot CH_2 \cdot CO \cdot OH$ into its next lower homologue, viz. acetic acid. Name the class of compound produced at each intermediate stage in the conversion. (W. S.)

(8) How is acetic acid obtained industrially?

Suggest in outline, how each of the following compounds could be synthesised from acetic acid: (i) methyl alcohol, (ii) hydroxyacetic acid, (iii) propylamine. (L. A.)

17

AROMATIC HYDROCARBONS

TOWARDS the end of the eighteenth century an illuminating gas was manufactured in this country by the destructive distillation of certain oils, e.g. whale oil. The gas was compressed in cylinders for easier distribution. It was found that after standing under pressure for some time, a colourless inflammable liquid separated out. In 1820 some of this liquid was given to Faraday who, by distillation and subsequent crystallization, isolated a new hydrocarbon to which, later, the name of *benzene* was given.

The chemists of this period had obtained some organic compounds from gums, resins and some of the strongly smelling naturally occurring oils, e.g. oil of cloves, cinnamon, etc. These compounds possessed certain characteristics distinguishing them from other organic compounds already isolated. They were, for example, richer in carbon, they were more stable and not so easily oxidized to carbon dioxide and water. One of these substances had been obtained by the destructive distillation of *gum benzoin*. It was an acid and hence called *benzoic acid*. In 1833, Mitscherlich heated this compound with lime and obtained the same hydrocarbon as Faraday had isolated a few years previously. He gave it the name of benzin which has since been changed to *benzene*. In 1845, this same hydrocarbon was found by Hofmann to be a constituent of coal tar (which is still a useful source of this important compound to-day). It was soon realized that those more stable organic compounds which were rich in carbon were derivatives of benzene, as the aliphatic compounds studied in previous chapters are derivatives of methane. Because of the pleasant smell of some of the derivatives of benzene first obtained, they were called *aromatic compounds*.

The molecular formula of benzene is C_6H_6, a molecule which by comparison with aliphatic hydrocarbons would seem to have a high degree of unsaturation. It is possible to write straight chain structural formulae for the molecule C_6H_6 in which all the carbon atoms are tetravalent, e.g. $CH{\equiv}C{\cdot}CH_2{\cdot}CH_2{\cdot}C{\equiv}CH$. A compound possessing this constitution is actually known* and, as would be expected, it shows the strongly additive properties of acetylene. Although benzene does show some unsaturation—for under certain conditions it adds on hydrogen, chlorine, bromine, and ozone—

* It is called dipropargyl.

it does not form addition compounds with the halogen hydracids and sulphuric acid, nor does it add on hydroxyl groups on treatment with potassium permanganate, as do compounds containing a double or triple bond. It does in fact, rather resemble an alkane, for it is quite stable and forms substitution products more readily than additive ones.

Further, when substitution occurs, only one monoderivative of formula C_6H_5X is known, indicating that all the six hydrogen atoms are similarly situated. Three isomeric disubstituted derivatives have been obtained.

To meet these chemical characteristics, Kekulé, in 1865, proposed a closed ring symmetrical formula as below (I). Because of the quadrivalency of carbon and also the power (even though weak) of benzene to form addition compounds similar to those obtained from ethylene, double bonds were inserted between alternate pairs of carbon atoms, to give formula II.

While this latter formula satisfies the power of benzene to add on hydrogen, chlorine, bromine and ozone, it did not explain why it undergoes no additive reaction with hydrogen chloride, sulphuric acid and potassium permanganate. Again, while it accounted satisfactorily for the formation of only one monoderivative, it did not explain why there are only three isomers of formula $C_6H_4X_2$. According to formula II, four isomers should be possible, in which the two substituent atoms or groups X take up respective positions on the following carbon atoms, viz. 1:2, 1:3, 1:4, and 1:6. The groupings 1:2 and 1:6 are not identical for between 1 and 2 there is a double bond, and between 1 and 6 there is a single one. Kekulé was led to conclude that the alternate double bonds occupied no fixed positions but oscillated round the molecule.

Thus, according to this theory, benzene could be regarded as an equilibrium mixture of the two forms (usually referred to as Kekulé forms):

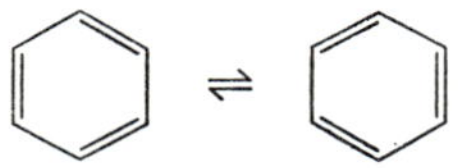

(N.B. The hydrogen atoms are usually omitted)

so that no C—C bond is wholly single or double.

A modern variant of Kekulé's view, taking account of the electronic nature of chemical bonds, is that the Kekulé forms are in fact extreme or *limiting forms* and that the true state of the molecule is between the two (i.e. the extra electrons of the ' double bonds ' are *delocalized*):

As in the case of the carboxylate anion discussed previously (p. 124), *delocalization is associated with stabilization*, so that benzene does not have the properties one would expect of a ' triene ' (*one* of the Kekulé structures) but is much more stable. It is the presence of this particular type of stabilized system which distinguishes an aromatic from an aliphatic compound.

Kekulé introduced the terms ' ring ' or ' nucleus ' for the benzene system and also the useful term ' side chain ' for an attached alkyl group (e.g. in the compound $C_6H_5 \cdot CH_2 \cdot CH_3$, the ethyl group is the side chain).

Saturated ring compounds are also known, e.g. cyclohexane, of structural formula:

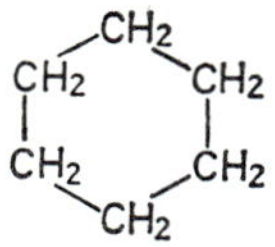

Here, however, all the carbon bonds are used to their maximum combining capacity—there is no fourth bond on each carbon atom not easily accounted for—and cyclohexane behaves in many respects like an alkane and not like benzene.

There are several ways of representing the benzene ring, some of which attempt to show that all the six C—C bonds are equivalent, e.g. (III); a monoderivative where X is the substituting atom or group is shown in (IV).

III a benzene molecule   IV a monoderivative of benzene   V

In formula (IV) it is immaterial at which corner of the hexagon, the atom or group X is placed. However, for convenience and following the recommendation of The Chemical Society, the formula (V) is used subsequently.

The monovalent group C_6H_5— is called the *phenyl* group and the compound of formula C_6H_5Cl could be called *phenyl chloride*, but is more commonly known as *chlorobenzene*. The phenyl group is the typical *aryl* group—a term corresponding to the *alkyl* group of aliphatic chemistry.

The three isomeric types of disubstitution product of benzene and their names are given below:

ortho   meta   para

The relative positions of the substituting atoms or groups determine which isomer the compound is and not the actual pair of corners chosen. Thus:

and

are both the same *para*-derivative. If X = Br, for example, both structures represent *para*-dibromobenzene.

The prefixes *ortho-*, *meta-* and *para-* are usually represented by the initial letters *o-*, *m-* and *p-*.

Benzene is the first member of a homologous series of hydrocarbons. The second member of the series, the formula of which is obtained by replacing a hydrogen atom by the —CH_3 group, is *toluene* $C_6H_5 \cdot CH_3$. If (in theory) we now replace a hydrogen atom in toluene by a methyl group, we can do this in four different ways, leading to four isomeric compounds.

The respective names and formulae of these are:

CH_3 ... CH_3 ... CH_3 ... $CH_2 \cdot CH_3$

ethylbenzene

o-   *m-*   *p-*

xylenes

These simple members of the benzene series are important raw materials in chemical industry and are obtained from coal, and also from petroleum.

Coal results from the slow decomposition of wood under pressure and in absence of air, the complicated chemical compounds in the wood gradually breaking up into simple ones and finally, in very old coal, into carbon. The decomposition can be speeded up by heating, when the simple volatile products distil over. The process is called the *destructive distillation* or *carbonization of coal*. In relatively recent years, much research work has been done on this subject and three processes of carbonization are in use.

(1) Coal is heated in retorts in *gas works* to a temperature in the region of 1000°C, when coal gas, ammonia, and coal tar distil off and coke is left in the retorts. Although this process is carried on primarily for the production of coke (for the extraction of iron) and coal gas (for use as a fuel) the ammonia is of value in the fertilizer industry and the coal tar is an extremely important source of raw materials for chemical industry. Among other substances, it contains the aromatic hydrocarbons benzene, toluene, the xylenes, and also carbolic acid (phenol).

(2) In another *lower temperature carbonization* process, coal is heated to about 600°C under reduced pressure. A smokeless fuel which can be burned in an open grate is obtained instead of coke. There is a smaller volume of coal gas but twice the amount of coal tar is obtained. The coal tar is of a different composition for it consists of the aliphatic hydrocarbons (paraffins and naphthenes) from which motor spirit is made.

(3) During World War I, a process invented by Bergius, and hence known as the *berginization* of coal, was used extensively in Germany for the production of petrol. There are two stages to the process. Powdered coal is first heated with heavy oil (p. 42) to a temperature of 450°C and under a pressure of 250 atmospheres in the presence of iron(III) oxide as a catalyst. The products are passed along with

hydrogen over further catalysts. In the presence of the additional hydrogen, the coal is almost completely liquefied to give a product consisting of alkanes, alkenes and aromatic hydrocarbons which are most suitable for making motor spirit.

Distillation of coal tar from high temperature carbonization

The tar is subjected to fractional distillation and the following four fractions usually collected.

(1) *Light oil*, collected up to 170°C, containing benzene, toluene and the xylenes.

(2) *Middle oil*, collected between 170° and 230°C, containing naphthalene and phenol.

(3) *Heavy oil* (or creosote oil), collected between 230° and 270°C, containing naphthalene and naphthol.

(4) *Anthracene oil*, distilling between 270° and 400°C, containing anthracene.

Pitch (used as road binding material, the manufacture of roofing felt, etc.) is left.

A further supply of light oil is obtained by scrubbing the coal gas with creosote oil, when the benzene and toluene, present as vapour in the coal gas, are dissolved out. They are recovered from the creosote oil by distillation.

The light oil (so called because of its low density which allows it to float on water) is agitated with sulphuric acid. The oily layer is washed with water and then with caustic soda and subjected to further fractionation during which benzene (b.p. 80°C), toluene (b.p. 110°C) and the xylenes (boiling at around 140°C) are obtained. The benzene is further purified by the method used by Faraday, namely crystallization (m.p. 5·5°C).

Until about 1940, light oil was the sole source of benzene. The increasing demand for this compound has made its manufacture from certain kinds of petroleum economically possible and now the greater part of the world's supply comes from petroleum (p. 44).

Benzene, C_6H_6

Soon after benzene was discovered, Berthelot showed that it was a polymer of acetylene. If this gas is heated to a dull-red heat (about 500°C) benzene is obtained, mixed with other hydrocarbons:

$$3C_2H_2 \longrightarrow C_6H_6$$

Preparation of benzene

Benzene can be more conveniently prepared in the laboratory by the following methods:

(1) By heating benzoic acid or its salts of the alkali or alkaline earth metals with soda-lime. This method was used by Mitscherlich:

$$C_6H_5 \cdot COONa + NaOH \longrightarrow Na_2CO_3 + C_6H_6$$
sodium benzoate

It is also prepared by heating the acid in a basic solvent (quinoline) in presence of copper:

$$C_6H_5 \cdot COOH \xrightarrow{Cu} C_6H_6 + CO_2$$

The latter is a rather general method of decarboxylation.

(2) By the reduction of *phenol*, C_6H_5OH, by passing its vapour over heated zinc dust:

$$C_6H_5OH + Zn \rightarrow C_6H_6 + ZnO$$

There is no reaction comparable to this in aliphatic chemistry.

(3) By the reduction of the *diazonium salt* (p. 266).

Properties of benzene

Benzene is a colourless inflammable liquid of boiling point 80°C and melting point 5·5°C. Because of its high carbon content, it burns with a luminous smoky flame. It is a good solvent, readily dissolving fats, many resins, iodine, etc.

Apart from its inflammability, it is remarkably stable. It is not affected by alkalis, hydrochloric acid, or oxidizing agents such as aqueous potassium permanganate and chromic acid (potassium dichromate and sulphuric acid). As mentioned above, it forms both addition and substitution products. The formation of addition products is a characteristic of unsaturated compounds, but it must be emphasized that benzene does not undergo addition reactions easily, and *its stability towards oxidizing agents is in striking contrast to the behaviour of alkenes and dienes.*

Addition reactions

(1) Benzene adds on hydrogen atoms in pairs, giving finally *cyclo-hexane* (or hexahydrobenzene) of formula C_6H_{12}. The reaction is brought about in the presence of nickel at high temperatures, using hydrogen under considerable pressure, or in the presence of very finely-divided platinum (platinum black) under milder conditions.

$$C_6H_6 + 3H_2 \xrightarrow[200°C]{Ni} C_6H_{12}$$
cyclohexane

(2) In the absence of any catalyst, but in presence of sunlight, *addition* of chlorine occurs, giving finally the hexahalogen compound:

$$C_6H_6 + 3Cl_2 \xrightarrow{\text{sunlight}} C_6H_6Cl_6$$
$$\text{benzene hexachloride*}$$

(3) Ozone adds rather slowly to benzene giving a 'triozonide' (see p. 20) which yields the expected dialdehyde, glyoxal, on hydrolysis:

$$\bigcirc \xrightarrow{O_3} \text{triozonide} \xrightarrow{H_2O} \begin{array}{c} \text{CHO} \\ | \\ \text{CHO} \end{array}$$
$$\text{glyoxal}$$

Substitution reactions

(1) When benzene is shaken or stirred with hot concentrated sulphuric acid, *sulphonation* occurs, leading to benzene sulphonic acid:

$$\bigcirc \xrightarrow{H_2SO_4} \bigcirc - \underset{\underset{O}{\|}}{\overset{\overset{O}{\|}}{S}} - OH$$

Sulphonic acids are 'strong' acids, like sulphuric acid. Note that the sulphur atom is directly linked to the benzene ring.

(2) When benzene is shaken with a mixture of concentrated nitric and sulphuric acids, an exothermic reaction leading to *nitrobenzene* occurs; this is an example of a *nitration reaction*:

$$\bigcirc + HO\cdot NO_2 \xrightarrow[\text{H}_2\text{SO}_4]{\text{conc.}} \bigcirc -NO_2 + H_2O$$

*In 1825, Faraday attempted to find whether chlorine had any reaction on benzene. He found there was none until he raised the blind of his laboratory and let the sunlight on to the vessel, when he obtained the hexachloride above.

This substance has been found to possess powerful insecticidal properties and is manufactured in large quantities, by the action of chlorine on benzene under the influence of ultra-violet light.

The compound shows stereoisomerism (p. 338). Only one isomer, the γ-isomer and hence called *gammexane*, possesses the insecticidal power. It is most effective against cockroaches, cabbage root fly, red mite in poultry houses, mosquito larvae, sheep scab (and hence used in sheep and cattle dips), etc.

If the temperature is not kept low, a second nitro-group enters the ring. This second group takes up a *meta* position; the nitro-group is therefore said to be *meta-orientating*:

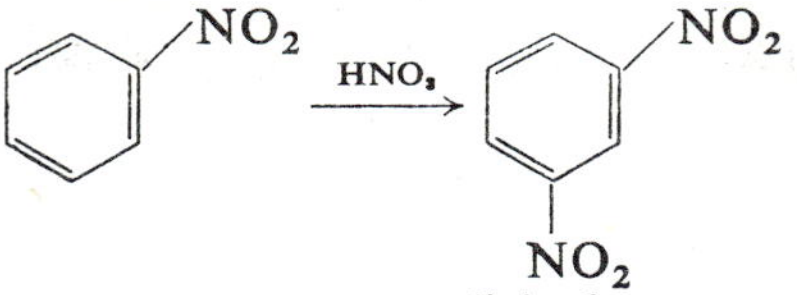

m-dinitrobenzene

i.e. the presence of the first nitro-group dictates the position of entry of the second group.

(3) In the presence of certain catalysts [iron, iron(III) chloride, aluminium chloride, iodine etc.] but in the absence of sunlight, chlorine and bromine react with benzene to give the substitution products chlorobenzene, C_6H_5Cl and bromobenzene, C_6H_5Br:

$$C_6H_6 + Cl_2 \longrightarrow C_6H_5Cl + HCl$$

(4) In the presence of aluminium chloride as a catalyst, benzene reacts with aliphatic halogen compounds. Dry methyl chloride, for example, is passed into a mixture of dry benzene and aluminium chloride to yield toluene, $C_6H_5 \cdot CH_3$. (Unless all reagents are thoroughly dry the yield of product is very poor for aluminium chloride is readily hydrolysed by water.) This is the *Friedel–Crafts reaction*, most useful in organic syntheses.

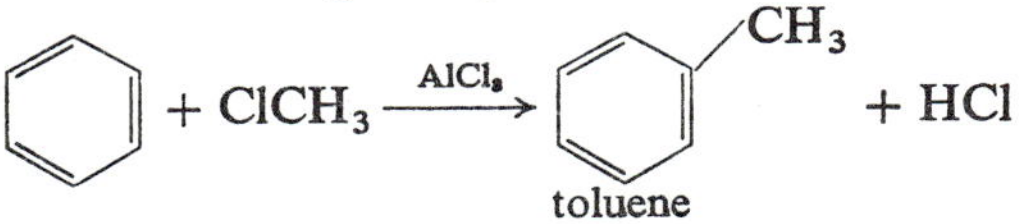

toluene

The reaction of benzene and acetyl chloride (in the presence of aluminium chloride) is another example:

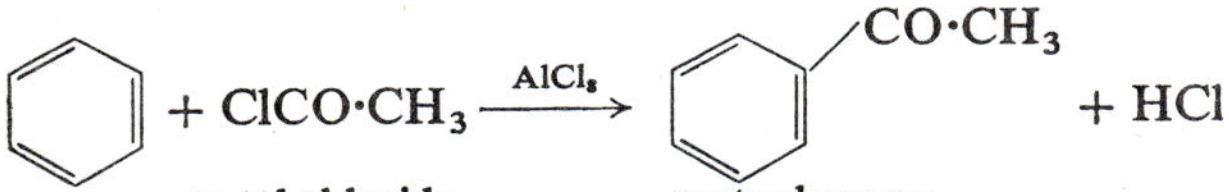

acetyl chloride   acetophenone

The yield of the product in the reaction with an acyl halide is better than with an alkyl halide, for in this latter case, further substitution occurs to give some *ortho-* and *para*-xylenes,

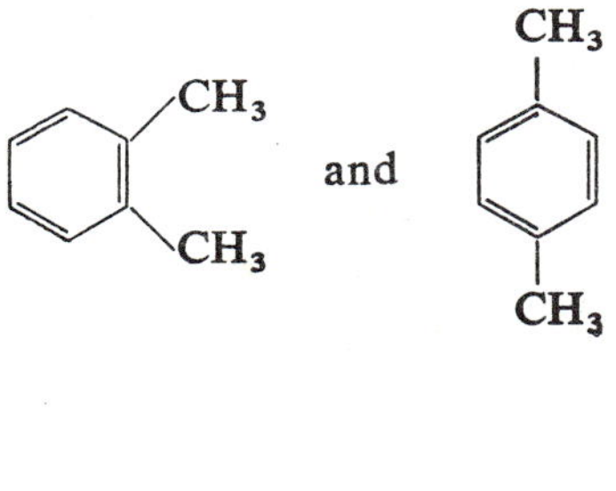

and

Mechanism of benzene substitution reactions

In all the substitution reactions of benzene set out above, the benzene molecule is believed to act as a source of electrons (i.e. as a *nucleophile**) and the attacking reagents are all electrophilic* in character.

It is instructive to compare the *addition* of bromine to an alkene with the *substitution* of bromine in benzene (which requires a catalyst):

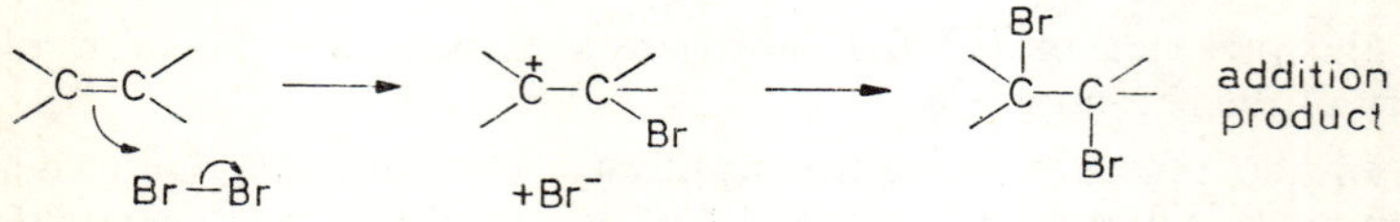

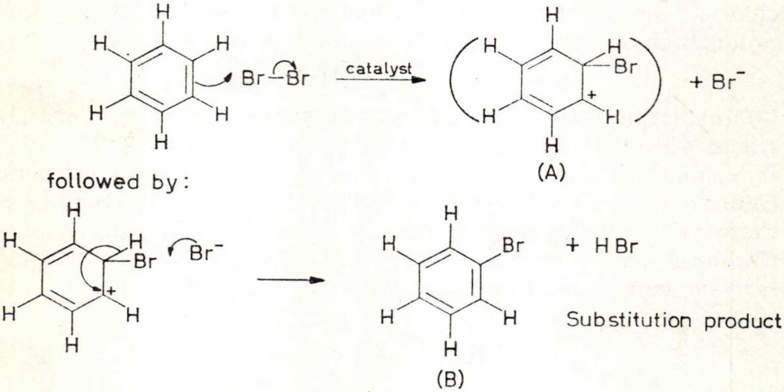

followed by:

In the alkene reaction, the last step is the formation of a second C—Br bond, leading to a stable addition product. In the benzene reaction, the positively charged intermediate (A) ' prefers ' not to form a second C—Br bond, but instead reverts to the stabilized ' aromatic ' state (B) by loss of a proton to the bromide anion. The overall result is then substitution rather than addition.

Three further points should be noted:

(*i*) Benzene is less reactive than alkenes, and its reaction with bromine requires a catalyst. The function of the catalyst is to assist in the ' ionic ' fission of the bromine molecule—in effect making it a more reactive electrophile, e.g.

$$\overset{\delta+}{Br} \longrightarrow \overset{\delta-}{Br} \quad FeBr_3$$

* The reader should refer to the discussion of the ' ionic ' addition reactions of alkenes on p. 21 and also to p. 97.

(*ii*) The substitution of benzene by bromine (or chlorine) is quite different *in mechanism* from the halogenation of methane, although both reactions give rise to substitution products. The benzene reaction is ' ionic ', the methane reaction ' free radical ' (p. 10). In fact, under conditions favouring the formation of halogen atoms (radicals), benzene undergoes the addition reaction previously mentioned (p. 206).

(*iii*) The intermediate (A) above should strictly be represented as a system with delocalized electrons:

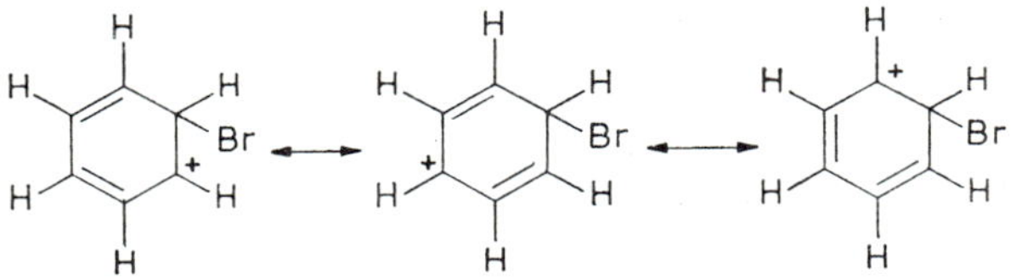

The three structures differ only in the distribution of the double bond electrons. In the actual ion these electrons will be ' smeared out ' over five carbon atoms, a situation which we can attempt to represent by the following single structure:

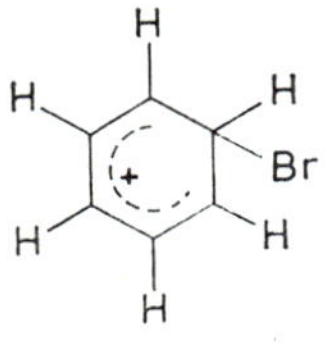

The nitration of benzene is analogous in mechanism to the bromination reaction. There is good evidence that the steps involved are:

$$H_2SO_4 + HO \cdot NO_2 \rightleftharpoons HSO_4^- + H_2\overset{+}{O} \cdot NO_2$$
$$H_2\overset{+}{O} \cdot NO_2 \rightleftharpoons H_2O + \overset{+}{N}O_2$$
$$H_2O + H_2SO_4 \rightleftharpoons H_3O^+ + HSO_4^-$$

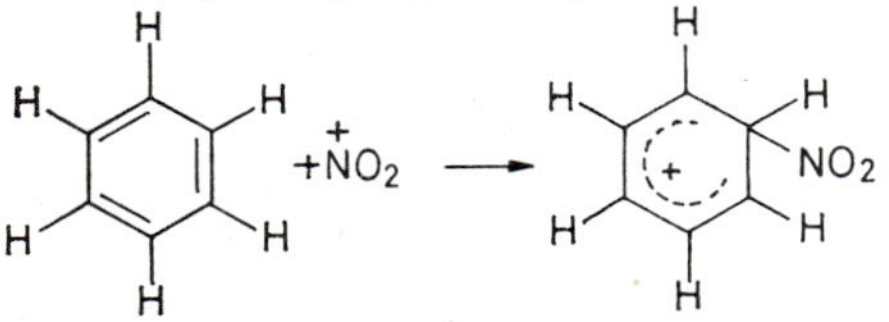

(analogous to the intermediate A in bromination)

(A′)

nitrobenzene

Again, the benzene ring provides a pair of electrons to form a new bond with the electrophilic *nitronium* ion, $\overset{+}{N}O_2$. In a subsequent step, the intermediate (A') loses a proton, regaining the aromatic structure in the product, nitrobenzene.

Similarly, in Friedel–Crafts reactions, the function of the aluminium chloride is to aid the formation of very reactive electrophilic species from the alkyl halide or the acyl chloride, e.g.

$$RCOCl + AlCl_3 \rightleftharpoons \overset{+}{R}CO + AlCl_4^-$$

followed by:

' Sandwich ' compounds of benzene

One further reaction of benzene is worthy of mention. When benzene, chromium(III) chloride, aluminium chloride and aluminium are heated together under pressure at 180°C, a product is obtained which on reduction forms *dibenzene chromium*, $(C_6H_6)_2Cr$. Here, the benzene rings are not attached through any one carbon atom to the chromium, for the structure is a ' sandwich ', thus:

the chromium being attached to each ring as a whole.

Uses of benzene

Benzene is used extensively as a solvent and also for blending with petrol. Being the ' parent ' of aromatic chemistry, it is the raw material from which many different products are manufactured. These include explosives, dyes and pharmaceutical products.

Toluene, $C_6H_5 \cdot CH_3$

Toluene was first obtained by the distillation of *balsam of tolu*, a fragrant oil found in certain plants. The commercial product is isolated from the light oil fraction from the distillation of coal tar and is also obtained from petroleum, by cracking with the use of a suitable catalyst.

Preparation of toluene

It may be prepared in the laboratory by any of the following methods:

(1) By the distillation of the appropriate acid or its salts of the alkali or alkaline earth metals with soda-lime, or by heating the acid in quinoline in presence of copper. Any one of the following acids may be used:

$$CH_3 \cdot C_6H_4 \cdot COOH \xrightarrow[\text{or copper}]{\text{soda-lime}} C_6H_5 \cdot CH_3 + CO_2$$

(2) By the *Friedel–Crafts reaction*:

$$C_6H_6 + CH_3Cl \xrightarrow{AlCl_3} C_6H_5 \cdot CH_3 + HCl$$

N.B. Other homologues of benzene can be obtained by a modification of the Friedel–Crafts reaction, e.g. ethylene adds on to benzene in presence of aluminium chloride to give *ethylbenzene*:

$$C_6H_6 + CH_2{=}CH_2 \xrightarrow{AlCl_3} C_6H_5 \cdot CH_2 \cdot CH_3$$
$$\text{ethylbenzene}$$

This is the method of manufacturing this compound from which, by dehydrogenation, the compound *styrene* $C_6H_5 \cdot CH = CH_2$ (used to prepare the important plastic *polystyrene*) is obtained (p. 26).

(3) By the action of sodium on a mixture of the monohalogen derivatives of benzene and methane in dry ether. Here, as in method (2) the reagents must be thoroughly dry to obtain a good yield:

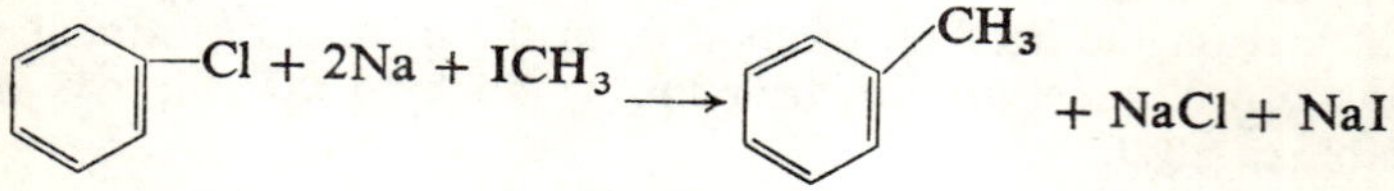

This was the method first used by *Fittig* in 1863 to synthesize the homologues of benzene. It compares with the similar *Wurtz reaction* (p. 51) of the aliphatic series.

(4) By the reduction of the hydroxy derivative of toluene with zinc dust.

$$C_6H_4(CH_3)OH + Zn \xrightarrow{\text{heat}} C_6H_5 \cdot CH_3 + ZnO$$
$$\text{a cresol}$$

The reaction is given by all three isomers, *ortho-*, *meta-* and *para-* cresols.

Properties of toluene

Toluene is a colourless liquid boiling at 110°C. It is immiscible with water and volatile in steam.

While the predominant properties of toluene would be expected to be those of the benzene nucleus, the presence of the alkyl group in the side chain will exert some influence on the chemical behaviour also. In fact, the methyl group in toluene makes the ring somewhat more reactive (than it is in benzene) towards electrophiles, i.e. substitution reactions occur more easily. A further, very interesting effect of the presence of the methyl group is that these substitution reactions give almost entirely *ortho* and *para* derivatives. Special methods are needed to make the *meta* derivatives. Further, the hydrogen atoms of the methyl group, like those of alkanes, are replaceable by the halogens, and again, the methyl group—modified by its attachment to the aromatic ring—is capable of oxidation.

(1) Toluene adds on six hydrogen atoms per molecule as does benzene, to give *hexahydrotoluene* or *methylcyclohexane*, $C_6H_{11} \cdot CH_3$:

$$C_6H_5 \cdot CH_3 + 3H_2 \longrightarrow C_6H_{11} \cdot CH_3$$
$$\text{methylcyclohexane}$$

(2) Toluene reacts with chlorine as follows:

(a) In the presence of a halogen carrier (e.g. iron or iodine), substitution in the ring occurs to give a mixture of *ortho-* and *para*-chlorotoluenes:

(b) In the absence of a catalyst, substitution occurs in the side chain, to form finally the trichlor derivative. The reaction takes place with boiling toluene in the dark or with cold toluene in sunlight (i.e. free radical conditions) (see also p. 227):

(3) With concentrated sulphuric acid, toluene gives a mixture of *o*- and *p*-toluene sulphonic acids:

(4) Toluene is nitrated by a mixture of concentrated nitric and sulphuric acids, to give both *o*- and *p*-nitro derivatives:

* The monovalent group $C_6H_5 \cdot CH_2$—is called the *benzyl* group, cf. the mono-valent group C_6H_5—which is the *phenyl* group.

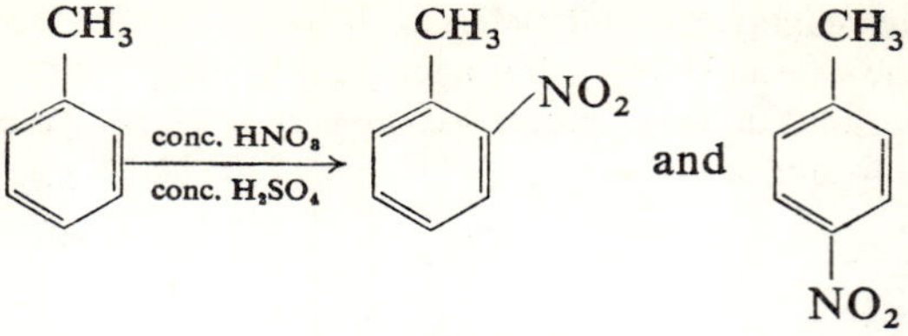

$o-$ ‿‿‿‿‿‿‿‿‿‿‿‿‿‿‿‿ $p-$
nitrotoluenes

(5) Toluene is oxidized by chromic acid (sodium dichromate and concentrated sulphuric acid), the side chain being attacked and converted ultimately to a carboxyl group.

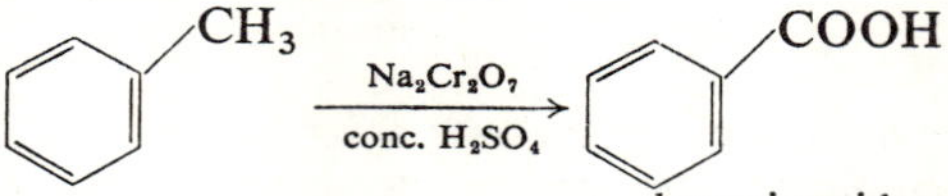

(Oxidizing agents have little or no action on benzene.)

Uses of toluene

Toluene is used along with benzene for blending in motor spirit. Its most important use, however, is as a raw material for the manufacture of explosives, dyes and other aromatic derivatives. Saccharin,

of formula

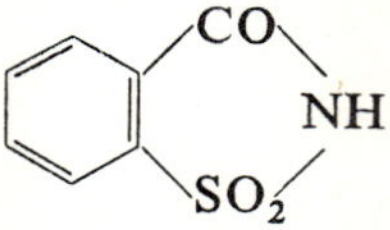

NH and with a sweetening power three

hundred times that of sucrose, is one such interesting derivative of toluene.

SUMMARY OF THE CHEMISTRY OF THE AROMATIC HYDRO-CARBONS BENZENE AND TOLUENE

The first member is C_6H_6, benzene, conventionally written

The monovalent radical C_6H_5— is the phenyl radical.
The second member is $C_6H_5 \cdot CH_3$, toluene.
The monovalent radical $C_6H_5 \cdot CH_2$— is the benzyl radical.
The xylenes of formula $C_6H_4(CH_3)_2$ exist as the following 3 isomers:

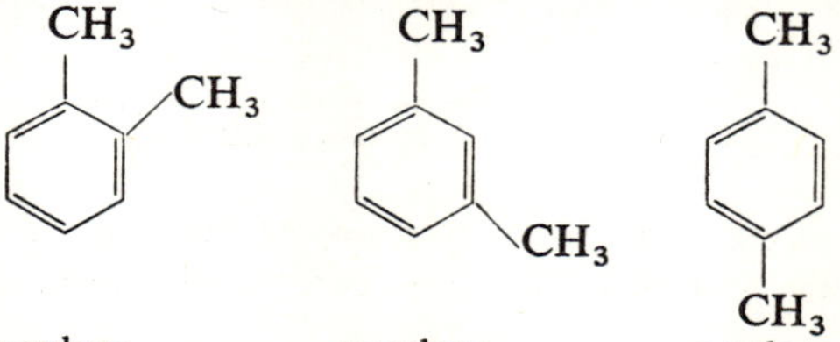

o-xylene   *m*-xylene   *p*-xylene

These are isomeric with $C_6H_5 \cdot CH_2 \cdot CH_3$, ethyl benzene.

Benzene, C_6H_6

Sources

(1) Coal tar from coal carbonization (high and low temperature and berginization).

(2) Petroleum (by cracking and then hydrodealkylation, p. 45).

Synthesis

Polymerization of acetylene:

$$3C_2H_2 \xrightarrow{500°C} C_6H_6$$

(1) *Preparation*

(a) Decarboxylation of benzoic acid, $C_6H_5\cdot COOH$:

$$\text{or} \quad \begin{matrix} C_6H_5\cdot COONa \\ C_6H_5\ COOH \end{matrix} \xrightarrow{\text{NaOH}} C_6H_6 \text{ (cf. methane preparation)}$$

$$C_6H_5\cdot COOH \xrightarrow[\text{basic solvent}]{\text{Cu}} C_6H_6$$

(b) Reduction of phenol, C_6H_5OH:

$$C_6H_5OH + Zn \xrightarrow{\text{heat}} C_6H_6 + ZnO$$

(c) Reduction of diazonium salt (p. 266):

$$[C_6H_5\cdot N_2]^+Cl^- + C_2H_5OH \longrightarrow C_6H_6 + N_2 + CH_3\cdot CHO + HCl$$

(2) *Addition reactions*

(a) Adds on hydrogen in presence of Ni or Pt black:

$$C_6H_6 \xrightarrow[\text{3H}_2,\text{ catalyst}]{} \underset{\text{cyclohexane}}{C_6H_{12}}$$

(b) Reacts with chlorine in absence of catalyst, but in presence of sunlight:

$$C_6H_6 \xrightarrow{\text{Cl}_2} \underset{\text{benzene hexachloride}}{C_6H_6Cl_6}$$

(c) Reacts slowly with ozone, forming triozonide.

(3) *Substitution reactions*

(a) Reacts with halogens (Cl_2 or Br_2) in presence of a catalyst (e.g. Fe, I_2):

$$C_6H_6 + Cl_2 \longrightarrow \underset{\text{chlorobenzene}}{C_6H_5Cl} + HCl$$

(b) With hot conc. sulphuric acid, forms a sulphonic acid:

$$C_6H_6 + HO\cdot SO_2\cdot OH \longrightarrow \underset{\text{benzene sulphonic acid}}{C_6H_5\cdot SO_3H} + H_2O$$

(c) With a mixture of conc. nitric and sulphuric acids, forms a nitro-derivative:

$$C_6H_6 + HNO_3 \xrightarrow{\underset{H_2SO_4}{conc.}} C_6H_5NO_2 + H_2O$$
$$\text{nitrobenzene}$$

(d) Reacts with halogen derivatives, under the influence of aluminium chloride as a catalyst (the *Friedel–Crafts reaction*):

$$C_6H_6 + ClCH_3 \xrightarrow{AlCl_3} C_6H_5 \cdot CH_3 + HCl$$
$$\text{toluene}$$

$$C_6H_6 + ClCOCH_3 \xrightarrow{AlCl_3} C_6H_5 \cdot CO \cdot CH_3 + HCl$$
$$\text{acetophenone}$$
$$\text{(a ketone)}$$

Toluene, $C_6H_5 \cdot CH_3$

Sources

Coal tar, petroleum.

(1) *Preparation*

(a) Decarboxylation of any of the three toluic acids or phenyl acetic acid (an interesting example of isomerism):

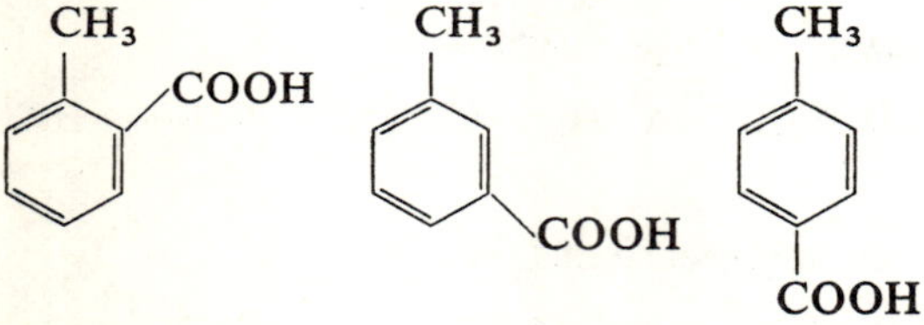

$$C_6H_5 \cdot CH_2 \cdot COOH$$

(b) By the *Friedel–Crafts reaction* [see (d) above]:

$$C_6H_5H + ClCH_3 \xrightarrow{AlCl_3} C_6H_5 \cdot CH_3 + HCl$$

(c) By *Fittig's synthesis:* the action of sodium on the mono-halogen derivatives of benzene and methane:

$$C_6H_5Cl + 2Na + ClCH_3 \longrightarrow C_6H_5 \cdot CH_3 + 2NaCl$$

(d) Reduction of any ring-substituted hydroxy-derivative of toluene by zinc dust:

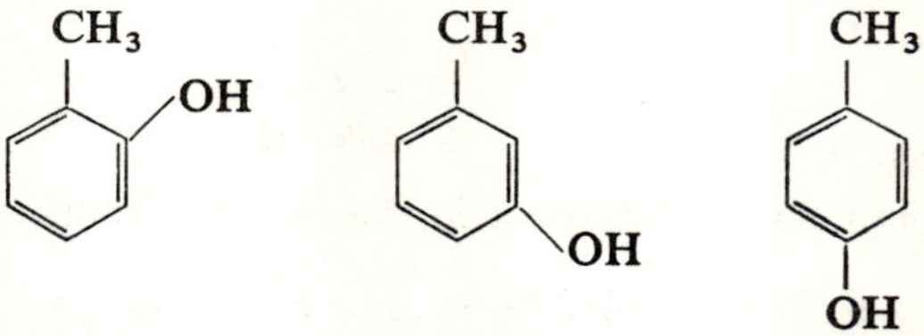

o-cresol   m-cresol   p-cresol

(2) Properties

(a) Reacts with chlorine:

(i) in presence of a catalyst (which favours 'ionic' substitution reaction):

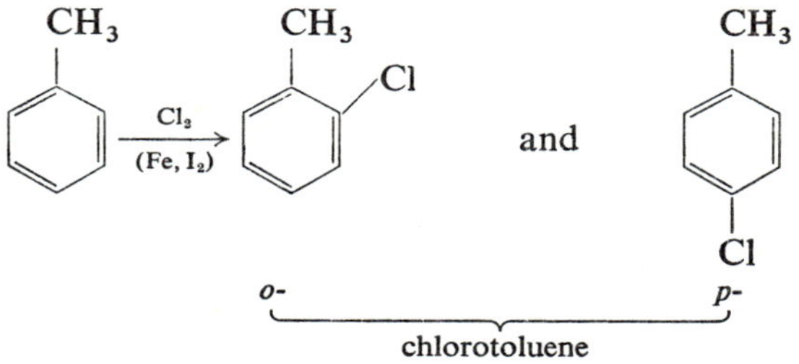

o- and p- chlorotoluene

(ii) in sunlight, no catalyst (favouring free radical reaction):

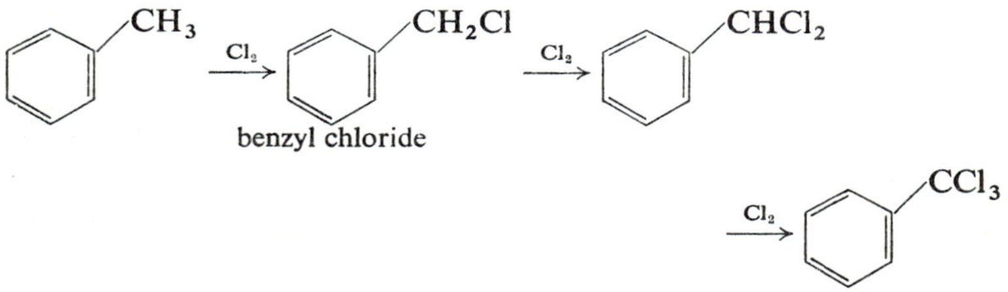

benzyl chloride

(b) With conc. sulphuric acid:

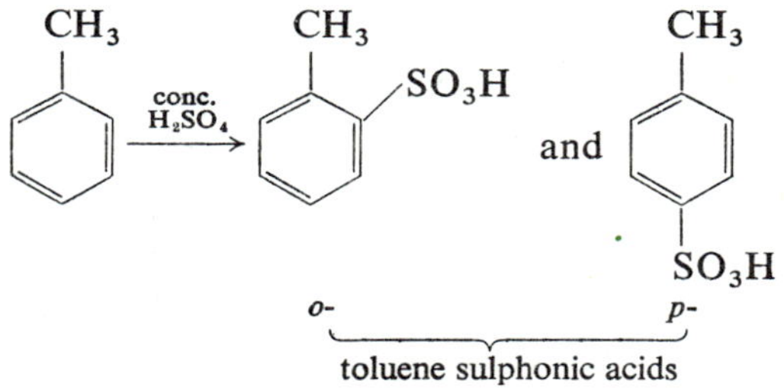

o- and p- toluene sulphonic acids

(c) Forms nitro compounds:

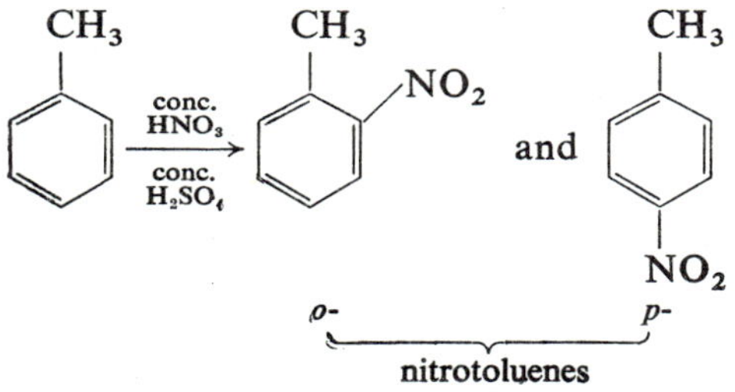

o- and p- nitrotoluenes

218 AROMATIC HYDROCARBONS

(*d*) Is oxidized by chromic acid:

$$CH_3 \xrightarrow[\text{conc. } H_2SO_4]{Na_2Cr_2O_7} COOH$$

benzoic acid

EXAMINATION QUESTIONS

(1) State how, and under what conditions, benzene reacts with (*a*) chlorine, (*b*) sulphuric acid, (*c*) nitric acid. Name the organic compounds formed, and describe how you would isolate the sodium salt of the organic product in (*b*).

Give the names and structural formulae of the dichlorobenzenes. (O. A.)

(2) Describe briefly how you would convert benzene to (*a*) chlorobenzene, (*b*) nitrobenzene. How would you show that the former compound contains chlorine and the latter nitrogen? (C. A.)

(3) How and under what conditions does toluene react with the following substances: (*a*) chlorine, (*b*) sulphuric acid?

Starting from toluene, how could you obtain (*i*) benzaldehyde, (*ii*) benzene? Write the equations and give the reagents and conditions for each step.

Indicate briefly how you could obtain specimens of each of the three constituents from a mixture of toluene, phenol and aniline, using chemical methods of separation. (J. M. B. A.)

(4) Define the terms 'saturated hydrocarbon' and 'unsaturated hydrocarbon'. Distinguish between ethane, ethylene and acetylene, and compare their chemical properties with those of benzene. (Liverpool, B.Sc. Inter.)

(5) Explain the difference between saturated and unsaturated compounds. Give two examples of each category, other than hydrocarbons.

State, with reasons, whether you regard benzene as saturated or unsaturated.

67·2 cc of a mixture of ethane and ethylene (measured at N.T.P.) were passed over red-hot copper oxide in a combustion apparatus. The issuing gases were led through a calcium chloride tube which was afterwards found to have increased in weight by 0·126 gm. What was the composition by volume of the hydrocarbon mixture? (J. M. B. A.)

(6) What products would be obtained from the reaction of benzene with the following: (*a*) bromine, (*b*) acetyl chloride, (*c*) a mixture of concentrated nitric and sulphuric acids?

Indicate the conditions necessary for the reactions you describe. Illustrate the meaning of the term electrophilic substitution by reference to these reactions, indicating the probable functions of any catalysts used.

18

AROMATIC NITRODERIVATIVES

Functional Groups —NO₂

Nitrobenzene, $C_6H_5 \cdot NO_2$

NITROBENZENE is a pale yellow liquid boiling at 211°C. It has a strong smell of almonds.

Laboratory preparation

Fifty ml of concentrated sulphuric acid are added in small quantities to an equal volume of concentrated nitric acid, contained in a 1-litre flask. During the mixing, the temperature should be kept low by immersing the flask into a trough of cold water. An equal volume (50 ml) of benzene is now added, as before in small quantities, with vigorous shaking between each addition, for benzene does not mix with the acids:

$$\text{(forming } \overset{+}{N}O_2) \qquad \xrightarrow[\text{H}_2\text{SO}_4]{\text{conc.}} \qquad + H_2O$$

The action is exothermic and the temperature of the reaction mixture must be kept below 60°C, otherwise a second nitro group will enter the benzene ring. When the addition of benzene is finished (an operation which should take about half an hour), the flask is warmed on a water bath for an hour or so, shaking at frequent intervals.

The contents of the flask are now transferred to a separating funnel and the lower layer of unused acids is drawn off. The remaining yellow oily layer of nitrobenzene (of density about 1·2) is shaken up in a separating funnel with water and then with aqueous sodium carbonate until, in the latter case, there is no further effervescence. This treatment removes any acids which may have dissolved in the nitrobenzene. The lower layer of nitrobenzene is withdrawn. It contains as impurity unchanged benzene and a little water—the latter being the cause of the milky appearance. A few pieces of anhydrous calcium chloride are now added and the mixture allowed to stand in a stoppered vessel for a few days. It is then distilled using an air condenser (see *Figure 11.2*). Unchanged benzene comes over first at about 80°C and the temperature then rises quickly. The fraction of nitrobenzene distilling between 204° and 211°C is collected in a separate receiver. The residue in the distillation flask is chiefly dinitrobenzene.

In the absence of sulphuric acid, the yield of nitrobenzene is poor. It is believed that in a mixture of concentrated nitric and sulphuric acids, the following equilibrium exists:

$$HNO_3 + 2H_2SO_4 \rightleftharpoons NO_2^+ + H_3O^+ + 2HSO_4^-$$
$$\text{(nitronium ion)}$$

Support for this view comes from (a) freezing point depression measurements of such mixtures which tend to indicate the presence of four ions and (b) spectroscopic analysis which shows the presence of the nitronium ion.

The nitronium ion is the actual nitrating agent (see discussion in previous chapter, pp. 209-210).

Nitration

The introduction of the nitro group, —NO₂, into a molecule is known as *nitration*.

Further nitration of nitrobenzene occurs at temperatures above 60°C (i.e. nitration of nitrobenzene is more difficult than nitration of benzene) and the product formed is mainly the *meta*-isomer:

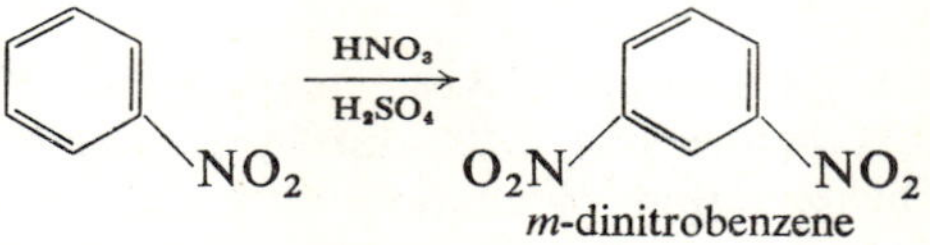

m-dinitrobenzene

Toluene is more easily nitrated than is benzene and gives a mixture of the *ortho*- and *para*-isomers:

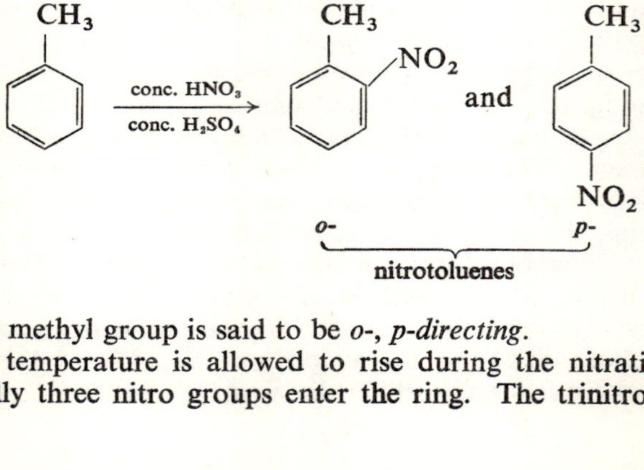

nitrotoluenes

Thus the methyl group is said to be *o-, p-directing*.

If the temperature is allowed to rise during the nitration, two and finally three nitro groups enter the ring. The trinitrotoluene, commonly referred to as T.N.T., has the formula

the groups entering the *ortho* and *para* positions.

Phenol, $C_6H_5 \cdot OH$ (p. 234), is very easily nitrated, for the influence of the hydroxy group is very great. Like the methyl group, this is *ortho*- and *para*-directing. In the preparation of the monoderivatives, the phenol is added slowly, with constant shaking, to fairly dilute nitric acid* (1 part of the concentrated acid to 4 parts of water), the temperature being kept low, not above 20°C. A black resinous mass is formed which contains the *ortho*- and *para*-isomers. It is subjected to steam distillation (p. 310) when the *ortho*-isomer distils over and separates as a yellow crystalline solid. The residue in the distillation flask contains the *para*-isomer, which is fairly soluble in hot water. This residue is boiled with water to which a little decolorizing charcoal has been added, and then filtered. As the filtrate cools, needle-shaped crystals of colourless *p*-nitrophenol separate.

If phenol is treated with a mixture of concentrated nitric and sulphuric acids and the temperature is allowed to rise, the final

product is a trinitrophenol of formula   which is

explosive and a dye stuff commonly known as *picric acid*.

As mentioned on p. 11, nitroalkanes cannot be prepared as above, but by treatment with nitric acid at 400°C and under 10 atmospheres pressure (' vapour phase nitration ') these compounds are now being manufactured.

Properties of nitroderivatives

Although aromatic nitro compounds undergo a number of reactions, only one of these is of sufficient importance to justify inclusion in an introductory text. This reaction, reduction to aromatic primary amines:

---

* The nitration of phenol is catalysed by nitrous acid (HNO_2). It is suggested that *ortho*- and *para*-nitrosophenols ($HO \cdot C_6H_4 \cdot NO$) are first formed and that the dilute nitric acid oxidizes these to the corresponding nitro compounds.

e.g.

aniline

is very important indeed for, as we shall see later, aromatic primary amines can be converted into many other types of aromatic compound. The reduction can be brought about by mixtures such as iron and hydrochloric acid, tin and hydrochloric acid, tin(II) chloride and hydrochloric acid, zinc and acetic acid, or most conveniently and cleanly, by catalytic hydrogenation using Raney nickel.

If there are two nitro groups in the molecule as in *m*-dinitrobenzene, for example, it is possible, by regulating the amount of reducing agent used, to reduce one group only. This is sometimes of value in organic synthesis:

m-dinitrobenzene   *m*-nitroaniline

Aliphatic nitro compounds are reduced similarly:

$$C_2H_5 \cdot NO_2 \longrightarrow C_2H_5 \cdot NH_2$$
ethylamine

The introduction of a nitro group into an organic compound tends to make the product unstable, for there is always the possibility of an intramolecular oxidation occurring to give carbon dioxide and steam and release nitrogen. If this should occur, the relatively enormous volumes of these gases set free cause an explosion. The more nitro groups substituted in the benzene ring, the greater is the chance of this taking place. Thus while nitrobenzene is quite stable, trinitrotoluene and picric acid are dangerously explosive. For the same reason, the nitrate esters of cellulose (p. 282) and glycerol (p. 79) (often wrongly referred to as nitro compounds) are also explosive.

Uses

(1) Nitrobenzene is the starting point in the manufacture of aniline dyes. It is also the raw material for the manufacture of many aromatic compounds, including, for example, pharmaceuticals.

(2) Poly-nitroderivatives are used as explosives.

SUMMARY OF THE CHEMISTRY OF NITROBENZENE AND NITRO-TOLUENES

Nitrobenzene, $C_6H_5 \cdot NO_2$

Preparation

Action of a mixture of conc. nitric and sulphuric acids on benzene at temperatures below 60°C:

$$C_6H_6 + HNO_3 \xrightarrow{\text{conc. } H_2SO_4} C_6H_5 \cdot NO_2 + H_2O$$

The nitrating agent is probably the *nitronium* ion, NO_2^+, formed:

$$HNO_3 + 2H_2SO_4 \rightleftharpoons NO_2^+ + H_3O^+ + 2HSO_4^-$$

Nitrotoluenes

Preparation

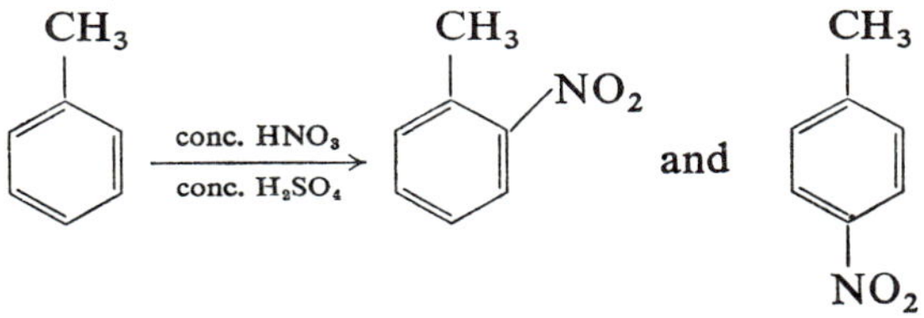

o-   p-
nitrotoluenes

(The CH_3— group is *ortho*- and *para*-directing.)

At high temperature:

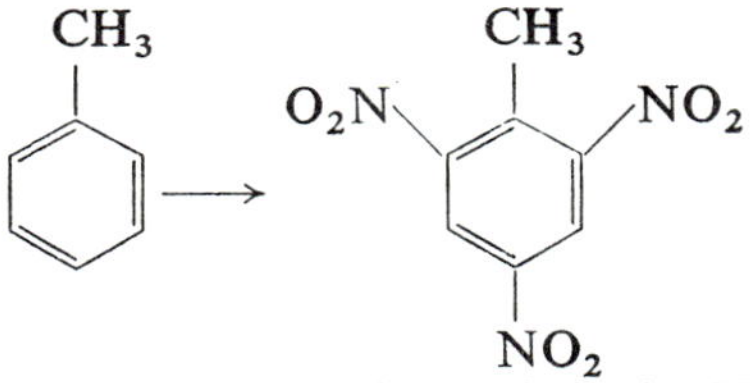

trinitrotoluene (T.N.T.)

Owing to the presence of the methyl group, it is somewhat easier to introduce nitro groups into the toluene molecule than into that of benzene. It is very much easier to nitrate phenol:

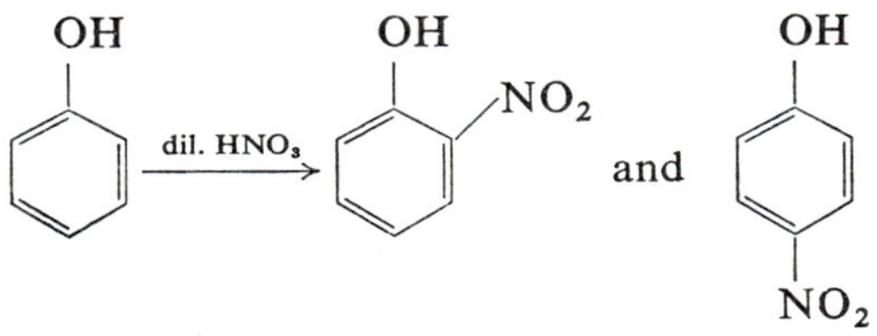

o-   p-
nitrophenols

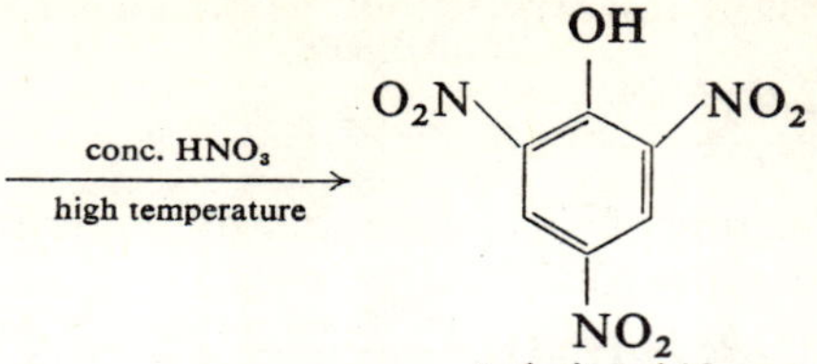

$$\xrightarrow[\text{high temperature}]{\text{conc. HNO}_3}$$

OH

O_2N \ | / NO_2

NO_2

' picric acid '

EXAMINATION QUESTION

(1) Describe, with practical details, the preparation of a pure sample of nitrobenzene from benzene. How may nitrobenzene be converted into (a) *meta*dinitrobenzene, and (b) phenol?

By what tests would you establish the purity of the *meta*dinitrobenzene?

(J. M. B. A.)

19

AROMATIC HALOGEN COMPOUNDS

Functional Groups —Cl, —Br, —I

MONOHALOGEN DERIVATIVES OF BENZENE

THE chlorine and bromine derivatives can be prepared by the direct action of the halogen on benzene in the presence of a carrier (p. 207):

$$C_6H_6 + Cl_2 \longrightarrow C_6H_5Cl + HCl$$
chlorobenzene or
phenyl chloride

The reaction must be carried out in absence of sunlight which causes addition (p. 206).

Under these conditions, iodine has no action on benzene. Iodobenzene and also the above chlorine and bromine derivatives are readily obtained from the benzene diazonium salt as described on p. 267.

Chlorobenzene is manufactured by direct substitution, chlorine being passed through cold benzene in the presence of iron as a catalyst.

It is also manufactured by passing benzene vapour, hydrogen chloride and air over a copper/iron catalyst at a temperature of about 230°C. This is *Raschig's Process*:

$$2C_6H_6 + 2HCl + O_2 \longrightarrow 2C_6H_5Cl + 2H_2O$$

(The process is, in fact, a chlorination of benzene by chlorine produced by the oxidation of the hydrogen chloride by air. It is a reminder of the Deacon Process in which chlorine is manufactured by passing air and hydrogen chloride over heated copper(II) chloride as a catalyst.)

Properties of the monohalogenobenzenes

The phenyl halides (typical examples of *aryl halides*) are unreactive for the halogen atom is closely united to the benzene ring. They therefore contrast strongly with alkyl halides which, because the halogen atom is so readily replaced, are so valuable in organic syntheses. Aromatic halogen derivatives where the halogen is directly united to the benzene ring are of little value in this way*.

Aryl halides are not hydrolysed by boiling with water or aqueous alkali (under normal conditions). They do not react with potassium

---

* Alkaline hydrolysis of chlorobenzene under violent conditions is one industrial method for the manufacture of phenol. The recently-discovered reactions of aryl halides with copper derivatives of alkynes may have considerable uses.

225

cyanide, silver nitrite or alcoholic ammonia. They show their aromatic character in undergoing nitration and sulphonation, mixtures of o- and p-isomers being obtained:

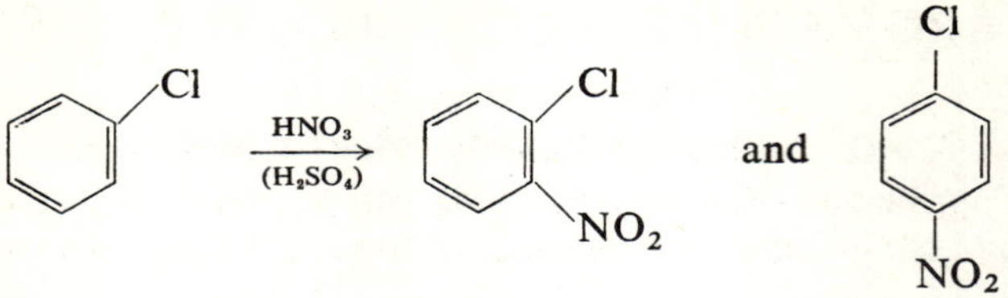

nitrochlorobenzenes

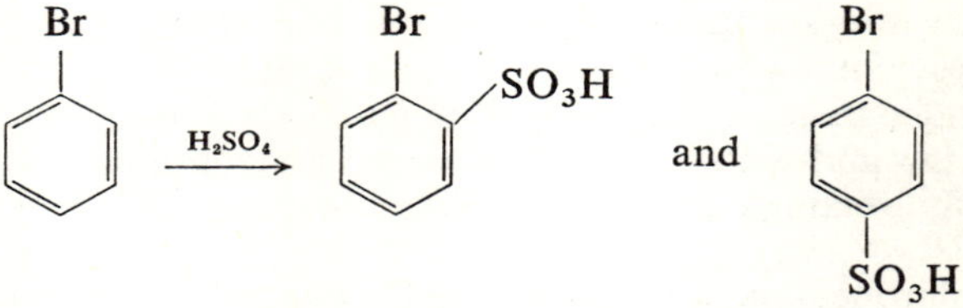

bromobenzenesulphonic acids

Aryl halides do, however, resemble alkyl halides in the following respects:

(1) They are reduced by sodium amalgam or lithium aluminium hydride to the hydrocarbon:

$$C_6H_5Br + (2H) \longrightarrow C_6H_6 + HBr$$

(2) They react with an alkyl halide in presence of metallic sodium (Fittig's reaction, p. 212):

$$C_6H_5Br + 2Na + IC_2H_5 \longrightarrow NaI + NaBr +$$

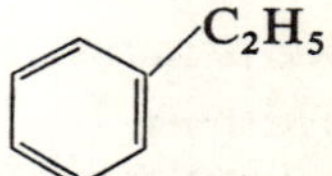

ethylbenzene

(3) The bromo- and iodo-compounds readily form Grignard compounds (p. 51) on treatment with magnesium in ether; these Grignard derivatives are extremely useful in organic synthesis.

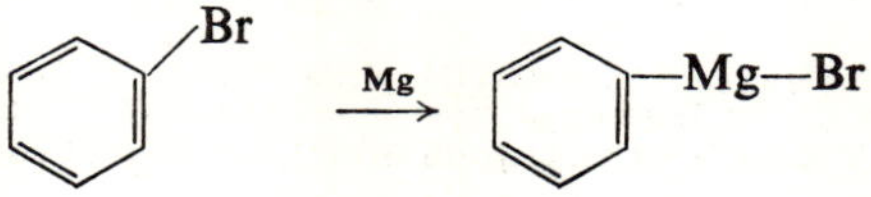

Uses of chlorobenzene

Chlorobenzene is a valuable intermediate product in the manufacture of phenol (p. 235), aniline (p. 256) and also *p*-dichlorobenzene. This last compound is an extremely efficient clothes moth larvicide. Chlorobenzene is also an intermediate in the manufacture of D.D.T., which is a product of its condensation with chloral, in the presence of sulphuric acid:

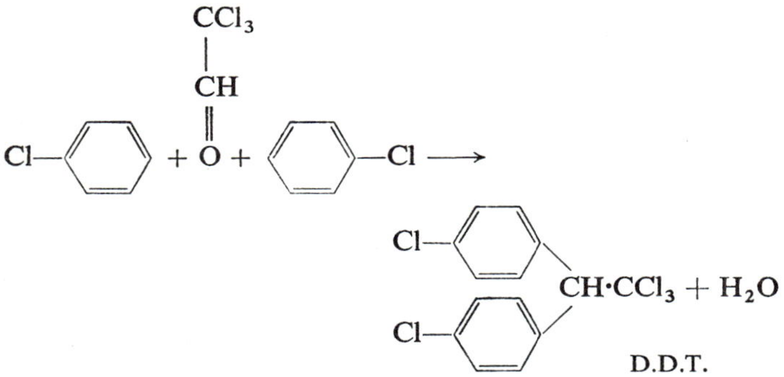

HALOGEN DERIVATIVES OF TOLUENE

Four isomeric monoderivatives are known. Taking chlorine as the halogen, their names and formulae are:

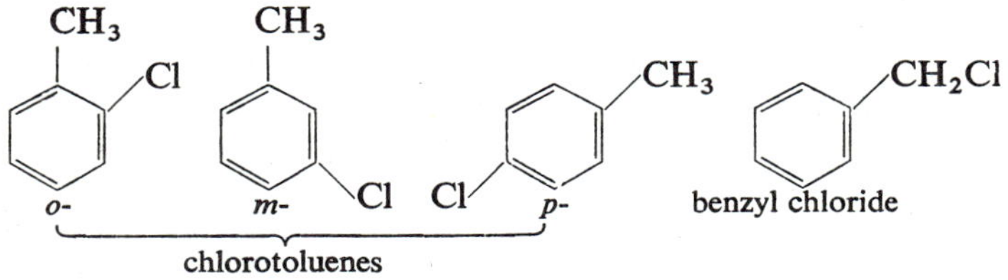

The *o*- and *p*-chlorotoluenes and the corresponding bromine compounds can be prepared by direct substitution, working at room temperature, in the dark and in presence of a carrier. Neither the *meta* derivatives nor any of the iodine isomers are obtainable by this method.

In chemical behaviour, all these compounds, *except benzyl chloride*, resemble the corresponding monohalogen derivatives of benzene.

Benzyl chloride, $C_6H_5 \cdot CH_2Cl$

Benzyl chloride (or the corresponding bromide) is prepared by passing chlorine (or bromine) through boiling toluene in absence of a catalyst, when successive substitution of the hydrogen atoms in the methyl side chain takes place. These successive reactions are:

benzyl chloride

benzylidene chloride
(benzal chloride)

benzylidyne chloride
(benzotrichloride)

As the halogenation proceeds, the weight of the product increases. By calculating the theoretical weight of the desired product and weighing at intervals, it is possible to stop the halogenation at the required stage. (A similar ' weight ' method can be used to prepare the separate chloracetic acids, p. 127.)

The mechanism of this ' side-chain ' halogenation process is thought to be very similar to the mechanism of methane chlorination, involving ' free radicals ', whereas nuclear halogenation is an ' ionic ' reaction.

With the halogen in the side chain, the benzyl halides might be expected to resemble the alkyl halides. The properties given below show this to be so.

Alternative methods of preparing benzyl chloride, similar to those used for alkyl halides, are:

(1) The action of phosphorus pentachloride or thionyl chloride on benzyl alcohol:

Here, as in aliphatic chemistry, a hydroxyl group is being replaced by a chlorine atom.

(2) The action of sodium chloride and concentrated sulphuric acid on benzyl alcohol:

$$NaCl \xrightarrow{H_2SO_4} HCl$$

$$+ HCl \xrightarrow{H_2SO_4} + H_2O$$

Properties of benzyl halides

By contrast with the monohalogen derivatives of benzene, the halogen atom in the benzyl halides is easily replaced. Benzyl halides thus resemble alkyl halides and are equally useful in organic syntheses. The following properties show this:

(1) Benzyl halides are hydrolysed by boiling water, aqueous sodium carbonate or any alkali. (See also p. 244 for hydrolysis followed by oxidation which occurs when benzyl chloride is boiled with an aqueous solution of copper(II) nitrate or lead nitrate.)

$$C_6H_5 \cdot CH_2Cl \xrightarrow{H_2O} C_6H_5 \cdot CH_2OH$$

(2) With potassium cyanide, benzyl halides form a nitrile or cyanide:

$$C_6H_5 \cdot CH_2Cl + KCN \xrightarrow[\text{aq. alc.}]{\text{heat in}} C_6H_5 \cdot CH_2 \cdot CN + KCl$$
$$\text{benzyl cyanide}$$

(3) With alcoholic ammonia, benzyl halides give an amine, e.g.:

$$C_6H_5 \cdot CH_2Cl + 2NH_3 \longrightarrow C_6H_5 \cdot CH_2 \cdot NH_2 + NH_4Cl$$
$$\text{benzylamine}$$

but further reactions may occur.

Benzyl halides differ from alkyl halides in their ability to be nitrated and sulphonated, showing aromatic character.

Benzylidene chloride, $C_6H_5 \cdot CHCl_2$

This compound, also called benzal chloride, is manufactured by the chlorination of toluene. It may be prepared by the action of phosphorus pentachloride on benzaldehyde (p. 244). This is an application of a general method of preparing dichloro-compounds where the two chlorine atoms are attached to the same carbon atom:

$$C_6H_5 \cdot C \overset{H}{\underset{O}{\diagup}} + PCl_5 \longrightarrow C_6H_5 \cdot CHCl_2 + POCl_3$$
$$\text{benzaldehyde}$$

On hydrolysis with a mild alkali, benzaldehyde is regenerated (two hydroxyl groups attached to one carbon atom is not, in general, a stable arrangement).

$$C_6H_5 \cdot C \underset{Cl}{\overset{H}{\diagdown}} Cl \longrightarrow C_6H_5 \cdot C \underset{O}{\overset{H}{\diagdown}}$$

Benzylidyne chloride, $C_6H_5 \cdot CCl_3$

This compound is the final product of the chlorination of toluene, and is an intermediate in the manufacture of benzoic acid, which is obtained from it by hydrolysis, by boiling with water or an alkali, e.g. milk of lime. If an alkali is used, a salt of benzoic acid is obtained and to liberate the free benzoic acid, a mineral acid, e.g. hydrochloric acid, must be added:

$$C_6H_5 \cdot C \underset{Cl}{\overset{Cl}{\diagdown}} Cl \xrightarrow{H_2O} C_6H_5 \cdot COOH + 3HCl$$
benzoic acid

In *Table 19.1* the properties of ethyl chloride, chlorobenzene and benzyl chloride are compared. The name of the reagent is given in the first column and the products, if any, with the different chlorides in the remaining columns.

Table 19.1.—A comparison of Ethyl Chloride, Chlorobenzene and Benzyl Chloride

Reagent	Ethyl chloride	Chlorobenzene	Benzyl chloride
1. Aq. alkali	Ethyl alcohol	No action except at high temperatures (300°C)	Benzyl alcohol
2. Reducing agent	Ethane	Benzene	Toluene
3. Potassium cyanide	Propionitrile	No action	Benzyl cyanide
4. Alcoholic ammonia	Ethylamine (and other amines)	No action	Benzylamine (and other amines)
5. Silver nitrite	Nitromethane	No action	Phenylnitro-methane
6. Magnesium	Grignard compound	Grignard compound*	Grignard compound
7. Nitric acid	No action	Nitro compound	Nitro compound and oxidation products
8. Sulphuric acid	No action	Sulphonic acid	Sulphonic acid

* Not formed in diethyl ether as solvent (a higher-boiling ether is required), but bromobenzene reacts normally with magnesium in diethyl ether.

SUMMARY OF THE CHEMISTRY OF THE MONOHALOGEN DERIV-
ATIVES OF BENZENE, C_6H_5X

The compound C_6H_5Cl is chlorobenzene (the name phenyl chloride is not commonly used).

(1) *Preparation*

Chloride and bromide (but not iodide) by direct substitution in presence of a carrier:

$$C_6H_6 + Cl_2 \xrightarrow{\text{Fe}} C_6H_5Cl + HCl$$

(2) *Manufacture*

Benzene vapour, hydrogen chloride and air over copper/iron catalyst at 230°C (*Raschig's process*):

$$2C_6H_6 + 2HCl + \underset{\text{(air)}}{O_2} \longrightarrow 2C_6H_5Cl + 2H_2O$$

(3) *Properties*

Not very reactive—no reaction with water, potassium cyanide, silver nitrite or alcoholic ammonia (cf. alkyl halide); reacts with sodium hydroxide only at high temperatures (300°C).

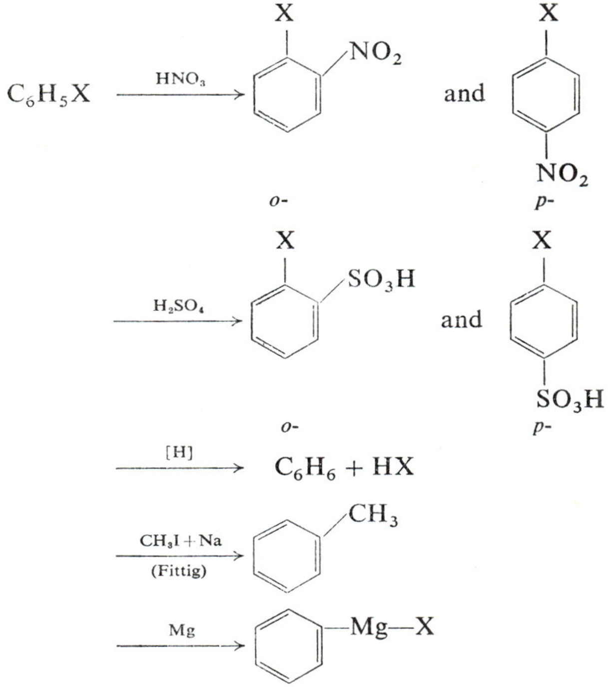

a Grignard compound useful in synthesis

SUMMARY OF THE CHEMISTRY OF THE HALOGEN DERIVATIVES OF TOLUENE

Substitution may be in the *nucleus* or *side chain* (different conditions, different mechanisms):

There are four isomeric monoderivatives, e.g.:

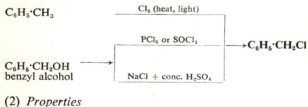

o-   m-   p-

chlorotoluenes   benzyl chloride

Further substitution products in the side chain are:

benzylidene chloride   benzylidyne chloride

The chlorotoluenes resemble chlorobenzene.

Benzyl chloride, $C_6H_5 \cdot CH_2Cl$

Apart from its ability to be *nitrated* and *sulphonated*, it resembles an alkyl halide.

(1) *Preparation*

$C_6H_5 \cdot CH_3$

Cl$_2$ (heat, light)

PCl$_5$ or SOCl$_2$ ——→ $C_6H_5 \cdot CH_2Cl$

$C_6H_5 \cdot CH_2OH$
benzyl alcohol ——→   NaCl + conc. H$_2$SO$_4$

(2) *Properties*

$C_6H_5 \cdot CH_2Cl$ ——→

hydrolysis ——→ $C_6H_5 \cdot CH_2OH$

reduction ——→ $C_6H_5 \cdot CH_3$

KCN ——→ $C_6H_5 \cdot CH_2 \cdot CN$

alc. NH$_3$ ——→ $C_6H_5 \cdot CH_2 \cdot NH_2$

Benzylidene chloride, $C_6H_5 \cdot CHCl_2$

$C_6H_5 \cdot CHO$ $\xrightarrow[\text{(preparation)}]{PCl_5}$

$C_6H_5 \cdot CH_3$ $\xrightarrow[\text{(manufacture)}]{Cl_2}$

$\longrightarrow C_6H_5 \cdot CHCl_2 \xrightarrow[\text{(alkali)}]{\text{hydrolysis}} C_6H_5 \cdot CHO$ benzaldehyde

Benzylidyne chloride, $C_6H_5 \cdot CCl_3$

Final product in the chlorination of toluene in absence of a catalyst. This is the method by which it is manufactured.

Hydrolysed by acid or alkali to benzoic acid:

$$C_6H_5C\underset{Cl}{\overset{Cl}{\underset{\displaystyle}{\diagup}}} \kern-1.5em Cl \xrightarrow{H_2O} C_6H_5 \cdot COOH + 3HCl$$

For comparison of properties of ethyl, phenyl and benzyl chlorides, see *Table 19.1* (p. 230).

EXAMINATION QUESTION

(1) Explain, with the aid of suitable examples, the differences between organic free radical reactions and organic ionic reactions.

20

PHENOLS AND AROMATIC ALCOHOLS

Functional Group —OH

BOTH these classes are hydroxy compounds. In the phenols, the hydroxyl group is attached directly to the benzene ring and in the aromatic alcohols it is present in the side chain. Thus, the simplest phenol will have the formula of C_6H_5OH. This is *phenol* proper (old name, *carbolic acid*). The corresponding compounds of toluene are called *cresols* of formula $C_6H_4(OH)\cdot CH_3$. There are three isomers, viz. the *ortho*, *meta* and *para* compounds. The simplest aromatic alcohol is *benzyl alcohol* of formula $C_6H_5\cdot CH_2OH$ and isomeric with the three cresols.

Phenol, C_6H_5OH

Phenol is a colourless hygroscopic crystalline substance of melting point 41°C and boiling point 181°C. It gradually turns pink in air, probably owing to oxidation, is partly miscible with water and readily soluble in many organic solvents, e.g. ether. It is dangerously toxic and if left on the skin for any length of time, causes severe burns. It is famous in preventive medicine as being the first antiseptic used in surgery by the great British surgeon Joseph Lister in 1867.

Laboratory preparation of phenol

Two methods are in common use.

(1) By the fusion of sodium benzene sulphonate with caustic soda at a temperature of about 250°C in a nickel crucible.

The melt is stirred frequently. Sodium phenate and sodium sulphite are formed:

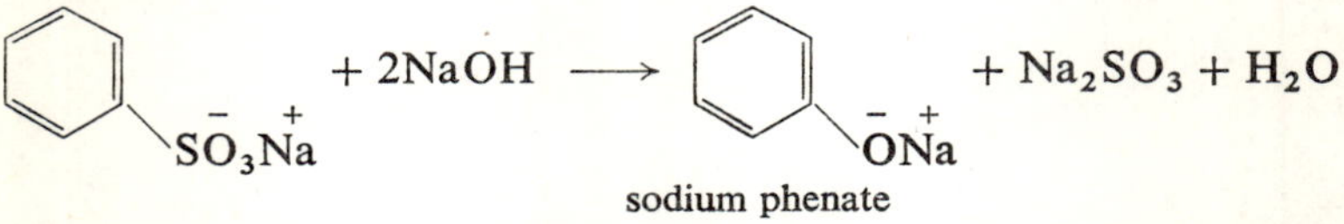

The residue is extracted with water and acidified with either hydrochloric or sulphuric acids, when the phenol is liberated, some of it collecting on the surface as an oil:

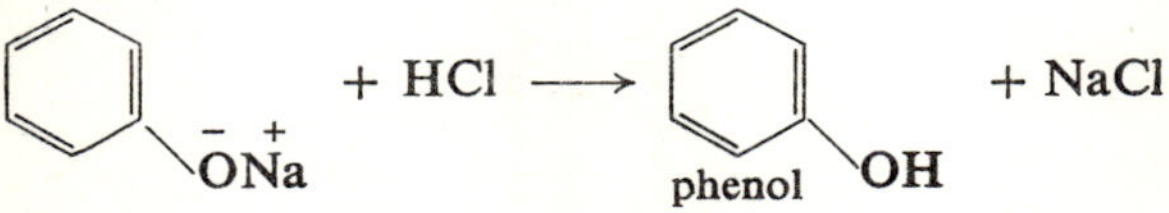

The phenol is extracted with ether* and the ether distilled off.

(2) By warming a solution of the phenyl diazonium salt (p. 266) until effervescence of nitrogen stops:

$$[C_6H_5N_2]^+Cl^- + HOH \longrightarrow C_6H_5OH + HCl + N_2$$

The residue is steam distilled (p. 310) and the phenol extracted from the distillate with ether.

Manufacture of phenol

Phenol is one of the most important aromatic compounds in industrial chemistry. Various methods are used to obtain it.

(1) By extraction from coal tar.

Phenol is present along with the hydrocarbon naphthalene in the middle oil fraction. The oil is agitated with aqueous caustic soda which dissolves the phenol as sodium phenate:

$$C_6H_5OH + NaOH \longrightarrow C_6H_5\overset{-}{O}\overset{+}{Na} + H_2O$$

The aqueous layer is then withdrawn and carbon dioxide passed through to liberate the phenol:

$$2C_6H_5\overset{-}{O}\overset{+}{Na} + CO_2 + H_2O \longrightarrow 2C_6H_5OH + Na_2CO_3$$

(2) By the fusion of sodium benzene sulphonate with caustic soda as in the laboratory preparation (Method 1).

(3) Via chlorobenzene (p. 225):

$$2C_6H_6 + 2HCl + O_2 \xrightarrow[\substack{(air)}]{\substack{Cu/Fe \\ 230°C}} 2C_6H_5Cl + H_2O$$

Several methods have been developed for the (difficult) hydrolysis of chlorobenzene, e.g. treatment with steam at a temperature of over 400°C under pressure in presence of a catalyst, e.g. a copper salt:

$$C_6H_5Cl + H_2O \xrightarrow[\text{catalyst}]{425°C} C_6H_5OH + HCl$$

(4) An increasingly important method involves the following stages: reaction of benzene with propene under pressure at 250°; the benzene is alkylated to give *cumene* (the reaction is related to Friedel–Crafts alkylation):

$$C_6H_6 + CH_2{=}CH{\cdot}CH_3 \xrightarrow[\substack{\text{phosphoric} \\ \text{acid} \\ \text{catalyst}}]{} \underset{\substack{\text{cumene or} \\ \text{isopropyl benzene}}}{C_6H_5{\cdot}CH(CH_3)_2}$$

* See p. 311.

The cumene is then oxidized by air and treated with dilute sulphuric acid to give phenol and acetone (the reactions include formation of a hydroperoxide and its rearrangement).

Properties of phenol

The following are among the more important properties of phenol. In general, these properties are also shown by the cresols.

(1) Phenol is a *weak acid* (p. 291) dissociating slightly in aqueous solution—hence the justification for its older name of carbolic acid. In this respect, it is in strong contrast with aliphatic and aromatic alcohols:

$$C_6H_5OH \rightleftharpoons C_6H_5O^- + H^+$$

It will therefore affect some acid/alkali indicators and form a salt with caustic soda:

$$C_6H_5OH + NaOH \longrightarrow C_6H_5\overset{-}{O}\overset{+}{Na} + H_2O$$

Phenol is set free from the phenate by means of carbon dioxide:

$$2C_6H_5\overset{-}{O}\overset{+}{Na} + CO_2 + H_2O \longrightarrow 2C_6H_5OH + Na_2CO_3$$

(or, of course, by stronger acids).

Phenol is thus incapable of evolving carbon dioxide from a carbonate and therefore is unaffected by addition of the carbonate or bicarbonate of sodium, cf. benzoic acid, p. 250.

Note that a phenol which is insoluble in water will dissolve in aqueous sodium hydroxide *because it forms a salt*, whereas a water-insoluble alcohol will not dissolve in aqueous alkali.

(2) Phenol forms esters, but not by the direct action of an acid, either mineral or carboxylic. Phenol has no reaction with hydrogen halides; there is some reaction with phosphorus pentachloride, yielding hydrogen chloride, but the main product is *not* chlorobenzene.

Esters are formed by the interaction of phenol and acid chlorides or acid anhydrides, thus:

$$CH_3 \cdot COOC_6H_5 + HCl$$
phenyl acetate

acetyl chloride

$$C_6H_5 \cdot COOC_6H_5 + HCl$$
phenyl benzoate

benzoyl chloride

These reactions are usually carried out in the presence of a basic catalyst, e.g. pyridine, which has a structure similar to that of benzene:

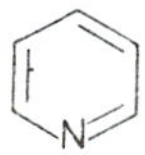
N

Benzoyl chloride is considerably less reactive than acetyl chloride (p. 137) and can therefore be used in the presence of aqueous sodium hydroxide. *Benzoylation* of a phenol with alkali as the basic catalyst is known as the *Schotten–Baumann reaction* [also possible with aromatic amines (p. 257)].

Anhydrides also react with phenol:

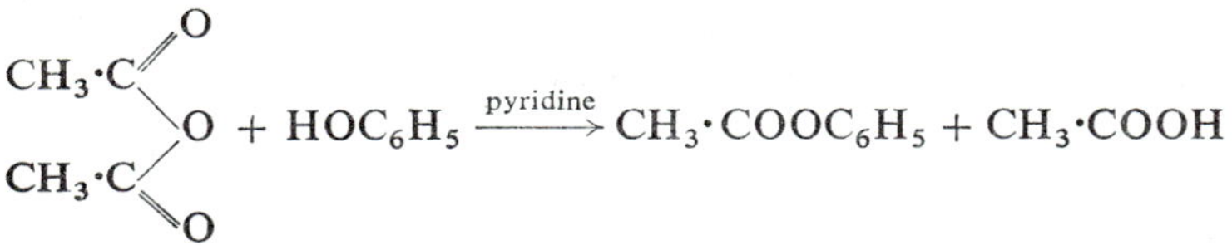
O
CH3·C
O + HOC6H5
pyridine
CH3·COOC6H5 + CH3·COOH
CH3·C
O

(3) Sodium (or potassium) phenate readily forms ethers by *Williamson's method* (p. 85):

$$C_6H_5\overset{-}{O}\overset{+}{Na} + ICH_3 \longrightarrow \underset{\text{phenyl methyl ether}}{C_6H_5OCH_3} + NaI$$

Better methods of preparing such ethers consist in refluxing phenol with methyl iodide in acetone in the presence of either anhydrous potassium carbonate or silver oxide:

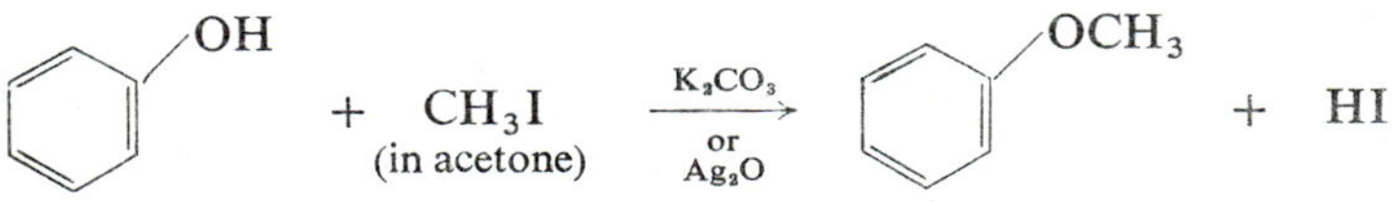
OH
+ CH3I
(in acetone)
K2CO3
or
Ag2O
OCH3
+ HI

(4) Phenol is reduced when heated strongly with zinc dust:

$$C_6H_5OH + Zn \longrightarrow C_6H_6 + ZnO$$

(5) Phenol gives a characteristic *blue or purple colour* with aqueous iron(III) chloride*. (Most phenols give a characteristic colour with this reagent.)

(6) The *ortho*- and *para*-orientating hydroxyl group has a powerful influence on the carbon atoms of the benzene ring, making substitution in the nucleus relatively easy. The following reactions illustrate this:

* Properties which provide convenient tests to identify phenol. See also the reaction on p. 269 with a diazonium salt.

(a) On adding bromine water to phenol, a precipitate of tribromo-phenol is formed:

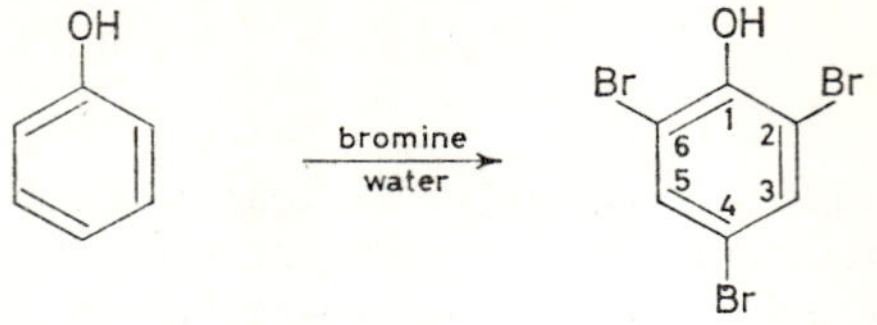

This tribromophenol is strictly named 2,4,6-tribromophenol, a system of nomenclature based on the numbering round the ring of the carbon atoms, starting from the hydroxyl group. The ease with which this reaction occurs illustrates the *activating* effect which the hydroxyl group has on the benzene nucleus (i.e. the ring is activated towards electrophiles).

(b) With dilute nitric acid, phenol gives a mixture of *ortho*- and *para*-nitrophenol (p. 221):

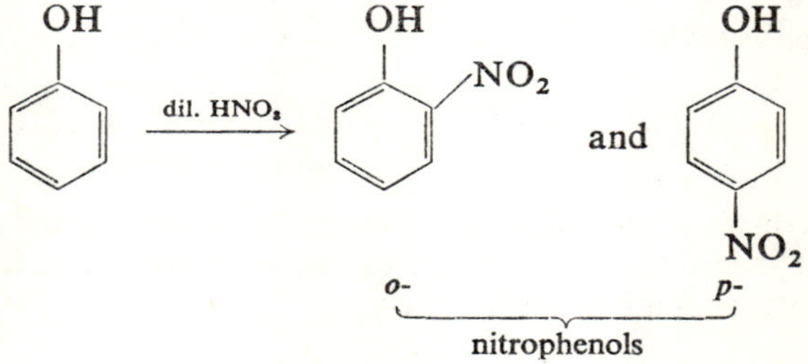

Using a mixture of concentrated nitric and sulphuric acids, three nitro groups enter the ring to give picric acid (p. 221):

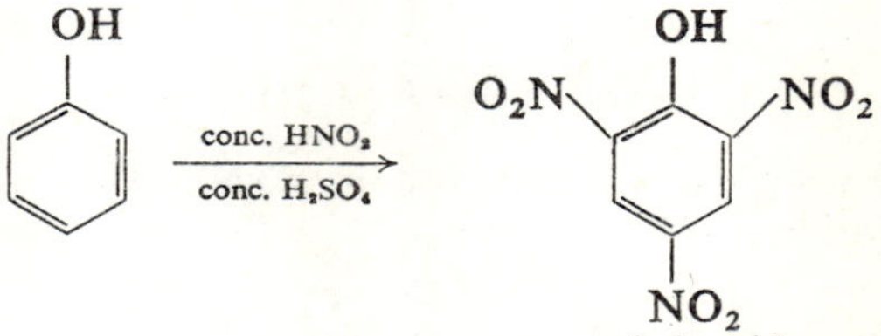

picric acid

(c) Phenol undergoes other substitution reactions, e.g. it is sulphonated by concentrated sulphuric acid. The preparation of phenolphthalein, used as an indicator in alkali/weak acid titrations, is a more complicated example of substitution.

Uses of phenol

In addition to its uses as an antiseptic (now largely superseded by better, less dangerous substances), phenol is extensively used in the manufacture of plastics (p. 106), picric acid (the explosive lyddite), salicylic acid and many pharmaceutical products.

The thermo-setting phenol-formaldehyde plastics (the first of which was bakelite, are made by heating together phenol and formaldehyde in the presence of either an alkali (ammonia or caustic soda) or an acid (hydrochloric acid). In both cases, big molecules are made by condensation processes, by which chains of benzene rings, cross-linked by methylene groups derived from the formaldehyde, are built up. Fillers of different kinds are added to alter such physical properties as, for example, mechanical strength and electrical conductivity. Dyes are also added to give colour.

Salicylic acid, the *ortho*-carboxylic derivative of phenol, is made by heating sodium phenate to about 150°C in an atmosphere of carbon dioxide under pressure:

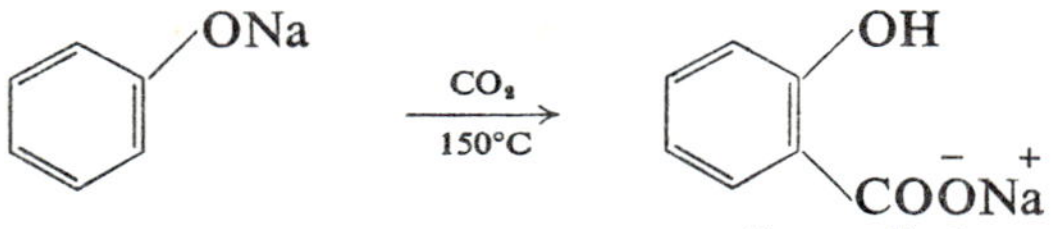

sodium salicylate

The methyl ester—the strong-smelling *oil of wintergreen*—is used in treating rheumatism:

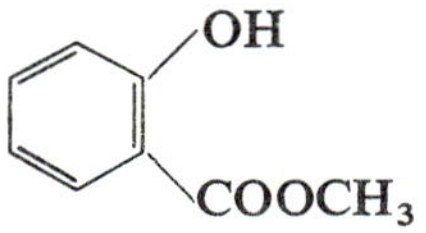

Reaction of acetyl chloride or acetic anhydride with salicylic acid, gives the well-known compound *aspirin*:

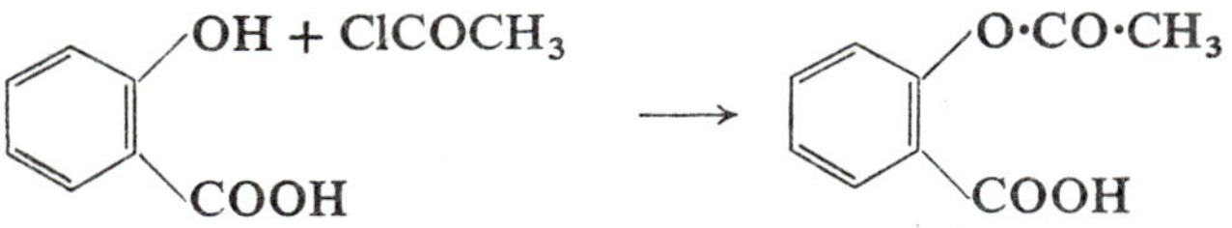

acetyl salicylic acid
(aspirin)

Benzyl alcohol, $C_6H_5 \cdot CH_2OH$

Benzyl alcohol is a colourless liquid of b.p. 206°C which closely resembles an aliphatic primary alcohol. Like ethyl alcohol, for example, it gives hydrogen and an ionic derivative, $C_6H_5 \cdot CH_2\overset{-}{O}\overset{+}{Na}$, with sodium, benzyl chloride with phosphorus pentachloride, esters with acids, first benzaldehyde and then benzoic acid with an oxidizing agent, etc.

Preparation of benzyl alcohol

The following methods of preparation (with the exception of the third) resemble methods applicable to aliphatic alcohols.

(1) By hydrolysis of benzyl chloride, on boiling with water or more quickly and completely by alkalis or sodium carbonate:

$$C_6H_5 \cdot CH_2Cl + H_2O \xrightarrow[\substack{\text{or} \\ Na_2CO_3}]{NaOH} C_6H_5 \cdot CH_2OH + HCl \\ \downarrow \\ NaCl$$

This is a manufacturing method using either milk of lime or sodium carbonate as the hydrolysing agent.

(2) By the reduction of benzaldehyde (p. 244) e.g. by lithium aluminium hydride:

$$C_6H_5 \cdot CHO \xrightarrow[\substack{\text{or} \\ LiAlH_4}]{Na/Hg} C_6H_5 \cdot CH_2OH$$

(3) By the action of hot caustic potash on benzaldehyde.

The process is a disproportionation, discovered by Cannizzaro and hence known as *Cannizzaro's reaction* (p. 105):

$$2C_6H_5 \cdot CHO \xrightarrow{KOH} C_6H_5 \cdot CH_2OH + C_6H_5 \cdot COOH \\ \downarrow \\ C_6H_5 \cdot COOK$$

SUMMARY OF THE CHEMISTRY OF THE AROMATIC HYDROXY-DERIVATIVES

In the *phenols* the hydroxyl group is attached directly to the benzene ring, e.g.:

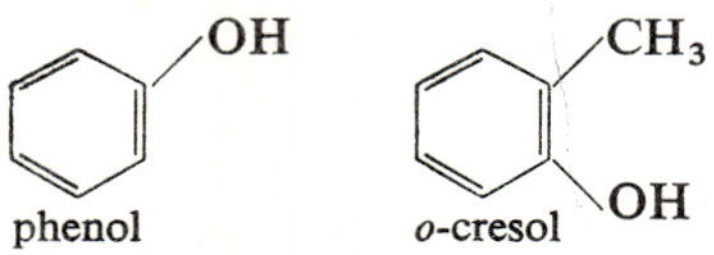

phenol   *o*-cresol

In the *aromatic alcohols* the hydroxyl group is in the side chain, e.g.:

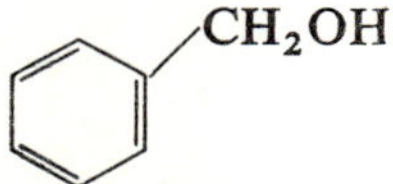

benzyl alcohol

Phenol, C_6H_5OH

(1) *Preparation*

(*a*) Fusion of sodium benzene sulphonate with alkali and then acid treatment:

$$C_6H_5 \cdot \overset{-}{S}O_3\overset{+}{Na} + 2NaOH \longrightarrow C_6H_5 \cdot \overset{-}{O}\overset{+}{Na} + Na_2SO_3 + H_2O$$
$$\text{sodium phenate}$$

$$C_6H_5 \cdot \overset{-}{O}\overset{+}{Na} + HCl \longrightarrow C_6H_5OH + NaCl$$

(*b*) Warming phenyl diazonium salt solution (p. 266):

$$C_6H_5\overset{+}{N}_2\overset{-}{Cl} + HOH \longrightarrow C_6H_5OH + \underset{.}{N}_2 + HCl$$

(2) *Manufacture*

(*a*) Extraction from coal tar.

(*b*) From sodium benzene sulphonate [see above, *Preparation (a)*].

(*c*) Hydrolysis of chlorobenzene:

$$C_6H_5Cl + H_2O \xrightarrow[\text{catalyst}]{425°C} C_6H_5OH + HCl$$

(*d*) From cumene (p. 235).

(3) *Properties*

(*a*) *Weak acid:* reacts with alkalis but does not evolve carbon dioxide from carbonates:

$$C_6H_5OH \rightleftharpoons C_6H_5O^- + H^+$$

(*b*) Forms esters, but not by direct treatment with acid:

$$CH_3 \cdot COCl + HOC_6H_5 \xrightarrow{\text{pyridine}} CH_3 \cdot COOC_6H_5 + HCl$$
$$\text{phenyl acetate}$$

$$C_6H_5 \cdot COCl + HOC_6H_5 \xrightarrow{\text{NaOH}} C_6H_5 \cdot COOC_6H_5 + H\underset{.}{C}l$$
$$\text{phenyl benzoate}$$

(The last reaction is an example of the *Schotten–Baumann reaction*.)

(*c*) Phenates form ethers by *Williamson's reaction*, p. 85:

$$C_6H_5 \cdot \overset{-}{O}\overset{+}{Na} + CH_3I \longrightarrow C_6H_5OCH_3 + NaI$$
$$\text{phenyl methyl ether}$$

(*d*) Is reduced by zinc, i.e. hydroxyl group removed:

$$C_6H_5OH + Z\underset{.}{n} \xrightarrow{\text{heat}} C_6H_6 + ZnO$$

(*e*) Gives purple colour with aqueous iron(III) chloride.

(*f*) Gives precipitate of tribromophenol with bromine water, illustrating activating influence of hydroxyl group on nucleus (no catalyst required):

$$C_6H_5OH + 3Br_2 \longrightarrow C_6H_2Br_3{\cdot}(OH) + 3HBr$$
2, 4, 6-tribromophenol

(*g*) With dilute nitric acid, gives mixture of *o*- and *p*-nitrophenols:

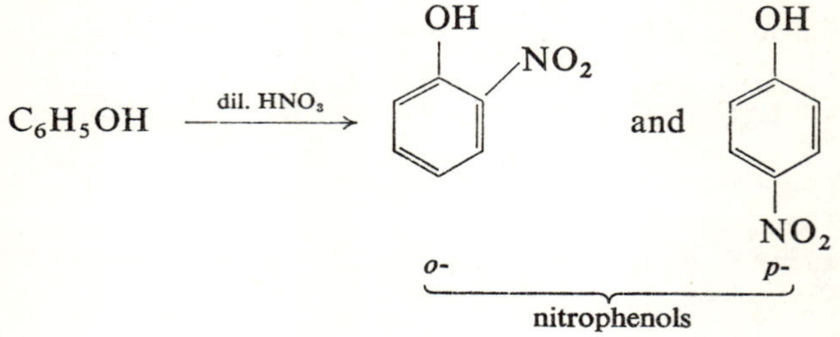

With concentrated nitric acid, gives the trinitro-derivative, 2,4,6-trinitrophenol or *picric acid*.

(*h*) On fusion with phthalic anhydride, forms *phenolphthalein*.

(4) *Uses*

Antiseptic; manufacture of plastics, salicylic acid (for aspirin) and other pharmaceutical products.

Benzyl alcohol, $C_6H_5{\cdot}CH_2OH$

(1) *Preparation*

(*a*) Hydrolysis of benzyl chloride:

$$C_6H_5{\cdot}CH_2Cl + H_2O \xrightarrow{NaOH} C_6H_5{\cdot}CH_2OH + HCl$$

(*b*) Reduction of benzaldehyde:

$$C_6H_5{\cdot}CHO \xrightarrow[\substack{or \\ Na/Hg}]{LiAlH_4} C_6H_5{\cdot}CH_2OH$$

(*c*) Disproportionation of benzaldehyde in presence of alkali (Cannizzaro's reaction):

$$2C_6H_5{\cdot}CHO \xrightarrow{KOH} C_6H_5{\cdot}COOH + C_6H_5{\cdot}CH_2OH$$

(2) *Properties*

Closely resembles an aliphatic primary alcohol.

EXAMINATION QUESTIONS

(1) Give **three** substitution reactions of benzene, indicating clearly the conditions under which they are carried out.

Starting from benzene, by what series of reactions could you prepare (*a*) phenol, (*b*) benzyl alcohol? Give the necessary reagents and conditions.

Give **one** definite chemical reaction which is given by phenol but not by benzyl alcohol, and **one** reaction which is given by benzyl alcohol but not by phenol. (J. M. B. A.)

(2) Give **two** methods by which phenol may be prepared from benzene. Compare the behaviour of the —OH group in phenol and in ethyl alcohol.
(O. and C. S.)

(3) Give *three* examples of organic compounds which show acidic properties. Compare and contrast the chemical properties of ethyl alcohol with those of phenol. (Liverpool, B.Sc. Inter., 1st. M.B., Ch.B., B.V.Sc.)

21

AROMATIC ALDEHYDES AND KETONES

Functional Groups $-C{\overset{H}{\underset{O}{\diagdown}}}$, $\diagup C{=}O$

THE simplest member of this class of compounds is benzaldehyde of formula $C_6H_5 \cdot CHO$. It is a colourless oily liquid of boiling point 179°C, which is not miscible with water but dissolves readily in most organic liquids. It has a strong smell of almonds. It occurs naturally in a complex compound with glucose and prussic acid in almonds and in peach and cherry seeds.

Laboratory preparation of benzaldehyde

This consists of a hydrolysis of benzyl chloride followed by an oxidation of the alcohol to benzaldehyde:

$$C_6H_5 \cdot CH_2Cl \xrightarrow{\text{H}_2\text{O}} C_6H_5 \cdot CH_2OH \xrightarrow{\text{oxidation}} C_6H_5 \cdot CHO$$

Since benzaldehyde is so readily oxidized to benzoic acid, a mild oxidizing agent, e.g. copper(II) nitrate or lead nitrate, must be used.

Benzyl chloride (50 g) is added to a solution of the same weight of copper(II) nitrate in half a litre of water. The mixture is heated in a 2-litre flask fitted with a reflux condenser, for about 5 or 6 hours. As oxidation proceeds, oxides of nitrogen are formed. These are removed from the reaction mixture by passing a stream of carbon dioxide (or some other gas inert to the mixture) through the flask, thus preventing further oxidation to benzoic acid.

The resulting benzaldehyde is extracted by ether and the solution is shaken with some freshly prepared sodium bisulphite, when the solid benzaldehyde bisulphite compound separates out:

$$C_6H_5 \cdot C{\overset{H}{\underset{O}{\diagdown}}} + Na\overset{+}{H}\overset{-}{S}O_3 \rightarrow C_6H_5 \cdot C{\overset{H}{\underset{\underset{\overset{-}{S}O_3\overset{+}{Na}}{\diagdown}}{\diagup}}}OH$$

This is filtered off and the free aldehyde obtained by distillation with dilute sulphuric acid. It is recovered from the distillate by ether extraction.

Other methods of preparation of benzaldehyde

(1) By the hydrolysis of benzylidene chloride with an aqueous alkali:

$$C_6H_5 \cdot CH \overset{Cl}{\underset{Cl}{\diagdown}} + 2OH^- \longrightarrow C_6H_5 \cdot CHO + 2Cl^- + H_2O$$

benzylidene chloride

The hydrolysis can be brought about by boiling water to which some iron filings have been added.

(2) By the oxidation of toluene by chromyl chloride or a mixture of manganese dioxide and fairly concentrated sulphuric acid:

$$C_6H_5 \cdot CH_3 \xrightarrow{\text{oxidation}} C_6H_5 \cdot CHO$$

*Properties of benzaldehyde**

Benzaldehyde shows most of the properties of a typical aldehyde, but being an aromatic compound, it can, in addition, be nitrated and sulphonated.

(1) On exposure to air or on boiling in water for some time, benzaldehyde is oxidized to benzoic acid. It is a *reducing agent*, reducing:

 (*a*) Acidified permanganate and dichromate.

 (*b*) Ammoniacal silver oxide.

Note that benzaldehyde has little effect on Fehling's solution.

(2) Benzaldehyde is reduced to the primary alcohol, benzyl alcohol $C_6H_5 \cdot CH_2 \cdot OH$, by lithium aluminium hydride and by sodium borohydride.

(3) Benzaldehyde forms typical addition products with hydrogen cyanide and sodium bisulphite:

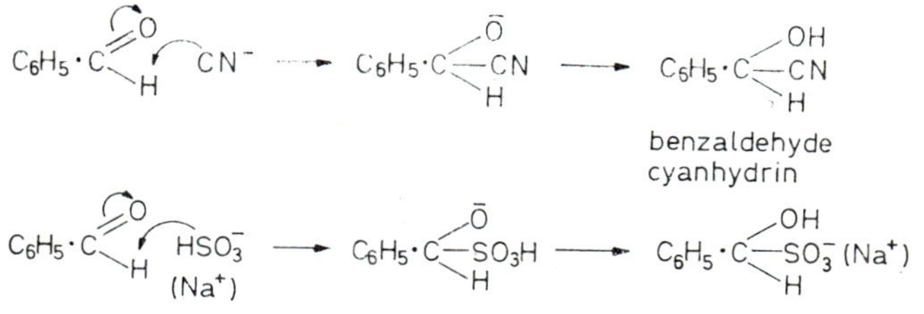

benzaldehyde cyanhydrin

* These should be compared with the properties of acetaldehyde (p. 93).

(4) Benzaldehyde undergoes addition-elimination reactions (p. 99) with hydroxylamine, hydrazine, phenylhydrazine and similar reagents:

$$C_6H_5 \cdot CHO + NH_2 \cdot OH \longrightarrow C_6H_5 \cdot CH = NOH + H_2O$$

benzaldehyde oxime

$$C_6H_5 \cdot CHO + NH_2 \cdot NH \cdot C_6H_5 \longrightarrow C_6H_5 \cdot CH = N \cdot NH \cdot C_6H_5 + H_2O$$

benzaldehyde phenylhydrazone

The products are of value as crystalline derivatives (i.e. they can be used to identify, or characterize, aromatic aldehydes).

(5) Benzaldehyde obviously differs from acetaldehyde in that it contains an aromatic ring which can be attacked in electrophilic substitution reactions such as nitration. However, these reactions require more vigorous conditions with benzaldehyde than with benzene itself and it may be inferred that the aldehyde group attached to the ring has a deactivating effect. When substitution does take place, it occurs in the *meta*-position (giving, for example, *m*-nitrobenzaldehyde). Thus the —CHO group, like the —NO$_2$ group, is *meta*-orientating; both are deactivating.

Benzaldehyde also differs from acetaldehyde (but resembles formaldehyde) in its behaviour towards alkalis. With fairly concentrated aqueous alkalis, benzaldehyde undergoes a *disproportionation* (Cannizzaro's reaction):

$$2C_6H_5 \cdot CHO \xrightarrow{OH^-} C_6H_5 \cdot CO_2H + C_6H_5 \cdot CH_2OH$$

benzoic acid   benzyl alcohol
(as a salt)

whereas, under mild alkaline conditions, acetaldehyde is converted to aldol:

$$2CH_3 \cdot CHO \xrightarrow{OH^-} CH_3 \cdot CH(OH) \cdot CH_2 \cdot CHO$$

With stronger alkalis, acetaldehyde is converted to a resin (probably of much higher molecular weight).

Aromatic ketones

The most representative member is the diphenyl ketone of formula $C_6H_5 \cdot CO \cdot C_6H_5$, known as *benzophenone*. A well known 'mixed' aromatic/aliphatic ketone is phenyl methyl ketone of formula $C_6H_5 \cdot CO \cdot CH_3$ commonly called *acetophenone*.

Both these ketones can be prepared by the Friedel–Crafts reaction (p. 207) as below:

$$C_6H_6 + ClCO \cdot C_6H_5 \xrightarrow{AlCl_3} C_6H_5 \cdot CO \cdot C_6H_5 + HCl$$
benzoyl chloride   benzophenone

$$C_6H_6 + ClCO \cdot CH_3 \xrightarrow{AlCl_3} C_6H_5 \cdot CO \cdot CH_3 + HCl$$
acetyl chloride   acetophenone

Acetic anhydride is commonly used in the latter reaction in place of acetyl chloride.

Benzophenone is also obtainable by heating calcium benzoate (a general method of preparing a ketone):

$$(C_6H_5 \cdot COO)_2Ca \longrightarrow C_6H_5 \cdot CO \cdot C_6H_5 + CaCO_3$$

Aromatic ketones show similarities to the corresponding aliphatic compounds, but reactions at the carbonyl group occur rather less readily.

SUMMARY OF THE CHEMISTRY OF AROMATIC ALDEHYDES AND KETONES

Benzaldehyde, $C_6H_5 \cdot CHO$

(1) *Preparation*

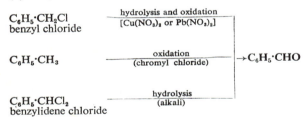

(2) *Properties*

Resembles aliphatic aldehydes in the following respects:

(*a*) Is easily oxidized and acts as a reducing agent (but has little effect on Fehling's solution).

(*b*) Can be reduced, e.g. by using complex hydrides, to a primary alcohol (benzyl alcohol).

(*c*) Gives addition products (cyanhydrin, sodium bisulphite compound).

(*d*) Undergoes typical addition-elimination reactions with NH_2OH, $NH_2 \cdot NH_2$, $NH_2 \cdot NH \cdot C_6H_5$ etc., forming crystalline derivatives.

Differs from acetaldehyde in the following respects:
(a) Reacts with alkalis thus:

$$2C_6H_5 \cdot CHO + KOH \longrightarrow C_6H_5 \cdot CO_2^- K^+ + C_6H_5 \cdot CH_2OH$$

(b) Undergoes ring-substitution reactions (e.g. nitration).

Benzophenone, $C_6H_5 \cdot CO \cdot C_6H_5$

Benzophenone is an aromatic ketone.

Preparation

C_6H_6 + ClCO·C_6H_5 $\xrightarrow{\quad AlCl_3 \quad}$
benzoyl chloride

$\rightarrow C_6H_5 \cdot CO \cdot C_6H_5$

$(C_6H_5 \cdot COO)_2Ca$ $\xrightarrow{\quad heat \quad}$

Acetophenone, $CH_3 \cdot CO \cdot C_6H_5$

Acetophenone is a 'mixed' aliphatic aromatic ketone.
It is prepared by the action of acetyl chloride or acetic anhydride on benzene:

$$C_6H_6 + ClCOCH_3 \xrightarrow{\quad AlCl_3 \quad} CH_3 \cdot CO \cdot C_6H_5$$

(or $CH_3 \cdot CO \cdot O \cdot CO \cdot CH_3$)

Both benzophenone and acetophenone show some resemblances to acetone ($CH_3 \cdot CO \cdot CH_3$) in chemical character.

EXAMINATION QUESTIONS

(1) (a) A white solid A, formula $C_7H_7SO_4Na$, reacts with a dilute acid giving sulphur dioxide and a colourless liquid B, formula C_7H_6O. B is readily oxidised to C, formula $C_7H_6O_2$, and on heating C with soda-lime, benzene is formed. Identify A, B and C, and explain these reactions.

(b) A colourless liquid D, formula $C_3H_6O_2$, reacts with sodium carbonate giving carbon dioxide. Identify D.
Give the names and structural formulae of the isomers of D, and state briefly how you could distinguish between them chemically. (O. A.)

(2) Using formulae and brief explanatory notes, illustrate the more important reactions of acetaldehyde and benzaldehyde.
To what extent can the mechanisms of these reactions be generalised?
(Liverpool, B.Sc., Part I)

22

AROMATIC CARBOXYLIC ACIDS

Functional Group $-C\diagdown{}^O_{OH}$

Benzoic acid, $C_6H_5 \cdot COOH$

THIS, the simplest of the aromatic carboxylic acids, occurs free in many resins. It is a white solid of m.p. 122°C, which readily sublimes when heated. It is thus obtained when resins in which it is present are heated, and was, in fact, first prepared in the sixteenth century by heating *gum benzoin* which is a resin secreted by certain species of plants. Its composition was not established until the nineteenth century when quantitative methods of analysis had been improved.

Benzoic acid also occurs in the urine of the horse and also most other herbivorous animals, as a compound $C_6H_5CO \cdot NH \cdot CH_2 \cdot COOH$ (derived from aminoacetic acid, $NH_2 \cdot CH_2 \cdot COOH$) (p. 283). The compound was thus named *hippuric acid* (Greek, *hippos* = horse). It was prepared from urine by Scheele in 1785.

Preparation of benzoic acid

(1) By the oxidation of toluene and, in fact, almost any aromatic compound with a single side chain, e.g. benzyl alcohol, benzyl chloride, benzaldehyde, etc.:

$$C_6H_5 \cdot CH_3 \xrightarrow{\text{oxidation}} C_6H_5 \cdot COOH$$

A variety of oxidizing agents can be used, e.g. potassium permanganate, or air in the presence of a catalyst.

(2) By the hydrolysis of benzylidyne chloride (p. 230).

Benzoic acid is manufactured by this method using milk of lime as the hydrolysing agent. The free acid is obtained as a white precipitate from the solution of calcium benzoate by the addition of hydrochloric acid.

(3) By the hydrolysis of benzonitrile by acids or alkalis (cf. the preparation of acetic acid from acetonitrile):

$$C_6H_5 \cdot CN \xrightarrow{H_2O} C_6H_5 \cdot COOH$$
benzonitrile
or
phenyl cyanide

Properties of benzoic acid

Benzoic acid is only slightly soluble in cold water but appreciably soluble in hot water. It is readily soluble in benzene in which it is associated as double molecules $(C_6H_5 \cdot COOH)_2$; in water solution

249

it occurs as single molecules which undergo partial dissociation.

It is a slightly stronger acid than acetic acid, but, in general, resembles it in its chemical behaviour*.

(1) On heating the acid (or better, its salts of the alkali and alkaline earth metals) with soda-lime, decarboxylation occurs:

$$C_6H_5 \cdot COOH \xrightarrow{\text{soda-lime}} C_6H_6$$

(2) Benzoic acid forms esters with alcohols, the esterification being catalysed by sulphuric acid or by hydrogen chloride:

$$C_6H_5 \cdot COOH + HOC_2H_5 \rightleftharpoons C_6H_5 \cdot COOC_2H_5 + H_2O$$
ethyl benzoate

(3) Benzoic acid forms an acid chloride on treatment with either thionyl chloride or phosphorus pentachloride:

$$C_6H_5 \cdot COOH + SOCl_2 \longrightarrow C_6H_5 \cdot COCl + SO_2 + HCl$$
benzoyl chloride

Although benzoyl chloride is not so chemically active as acetyl chloride, it possesses similar properties. It is hydrolysed by acids and alkalis or merely by boiling with water to give benzoic acid:

$$C_6H_5 \cdot COCl \xrightarrow{H_2O} C_6H_5 \cdot COOH$$

It reacts with alcohols to give esters:

$$C_6H_5 \cdot COCl + C_2H_5OH \longrightarrow C_6H_5 \cdot COOC_2H_5 + HCl$$

Benzoyl chloride also reacts with phenols and amines, the benzoyl group $C_6H_5 \cdot CO$— being introduced. This process of *benzoylation* is helped by the presence of an alkali. This is the Schotten–Baumann reaction (p. 237):

$$C_6H_5OH + ClCOC_6H_5 \longrightarrow C_6H_5O \cdot CO \cdot C_6H_5 + HCl$$
phenyl benzoate

$$C_6H_5 \cdot NH_2 + ClCOC_6H_5 \longrightarrow C_6H_5NH \cdot CO \cdot C_6H_5 + HCl$$
aniline
benzanilide
(the phenyl substituted amide of benzoic acid)

---

* Benzoic acid (like acetic acid) liberates carbon dioxide from sodium bicarbonate on which phenol has no action. This fact enables phenol and benzoic acid to be separated as follows:

$$C_6H_5OH + C_6H_5 \cdot COOH \xrightarrow{\text{NaHCO}_3} C_6H_5OH + C_6H_5 \cdot \overset{-}{C}OO\overset{+}{Na}$$
phenol   benzoic acid   soln.   sodium benzoate

$$\downarrow \text{extraction with ether}$$

$$C_6H_5OH + C_6H_5 \cdot \overset{-}{C}OO\overset{+}{Na}$$
(in ether)   (in aq. phase)
$$\downarrow H^+$$
$$C_6H_5 \cdot COOH$$

(4) Benzoic acid forms an acid anhydride by the interaction of benzoyl chloride and sodium benzoate:

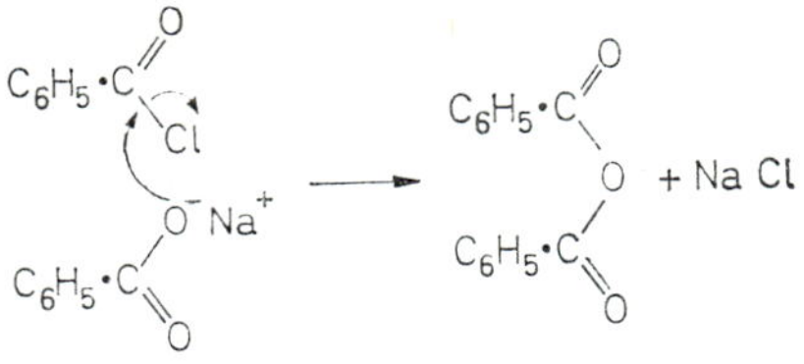

Benzoic anhydride

(5) Benzoic acid shows *aromatic character* by forming substitution products in the nucleus. The carboxyl group, like the aldehyde and nitro groups, is *meta*-directing. Benzoic acid can be nitrated, sulphonated and chlorinated:

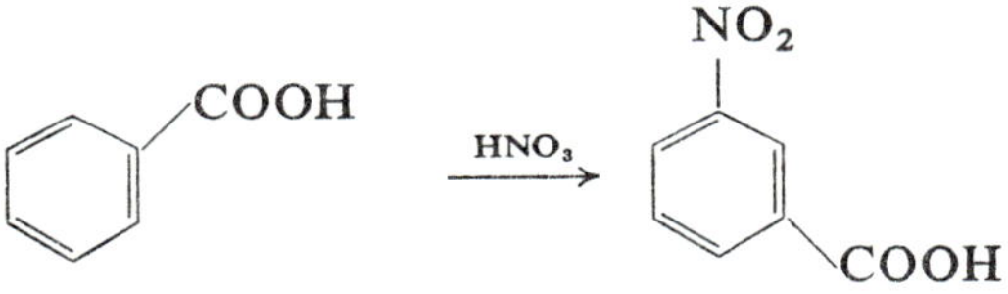

m-nitrobenzoic acid

Toluic acids, $CH_3 \cdot C_6H_4 \cdot COOH$

The three methyl-benzoic acids are usually known as toluic acids; these are isomeric with phenylacetic acid:

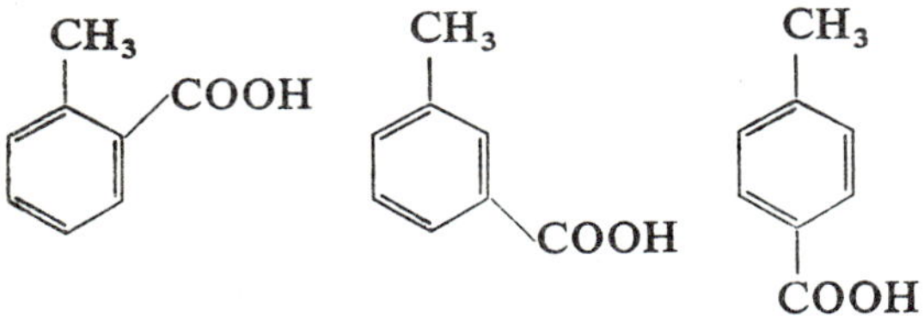

cf.
$C_6H_5 \cdot CH_2 \cdot COOH$
phenyl acetic acid

All these isomers possess properties characteristic of the carboxyl group and also show aromatic character by their ability to undergo substitution in the benzene ring.

Oxidation of the methyl group in a toluic acid gives a dicarboxylic acid $C_6H_4(COOH)_2$, a *phthalic acid*. One of these acids is terephthalic acid, usually prepared by direct oxidation of *p*-xylene:

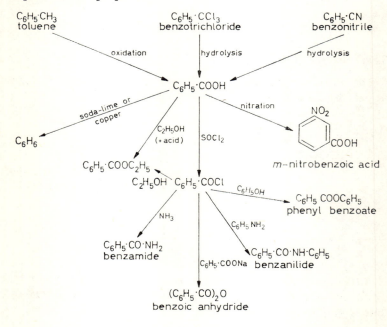

The methyl ester of this acid, dimethyl terephthalate, when condensed with ethylene glycol, yields the important polymer ' terylene '.

The *ortho*-isomer, known simply as phthalic acid, gives rise to a number of important derivatives, e.g. phenolphthalein and fluorescein.

SUMMARY

Benzoic acid, $C_6H_5 \cdot COOH$

Benzoic acid is the simplest aromatic carboxylic acid.

Preparation and properties

$C_6H_5 \cdot CH_3$
toluene

$C_6H_5 \cdot CCl_3$
benzotrichloride

$C_6H_5 \cdot CN$
benzonitrile

oxidation   hydrolysis   hydrolysis

$C_6H_5 \cdot COOH$

soda-lime or copper

C_2H_5OH (+ acid)

nitration

$SOCl_2$

C_6H_6

NO_2

$COOH$

m−nitrobenzoic acid

$C_6H_5 \cdot COOC_2H_5$

C_2H_5OH $C_6H_5 \cdot COCl$ C_6H_5OH $C_6H_5 \, COOC_6H_5$
phenyl benzoate

NH_3

$C_6H_5 \, NH_2$

$C_6H_5 \cdot CO \cdot NH_2$
benzamide

$C_6H_5 \cdot CO \cdot NH \cdot C_6H_5$
benzanilide

$C_6H_5 \cdot COONa$

$(C_6H_5 \cdot CO)_2O$
benzoic anhydride

Toluic acid, $CH_3 \cdot C_6H_4 \cdot COOH$

There are three isomers:

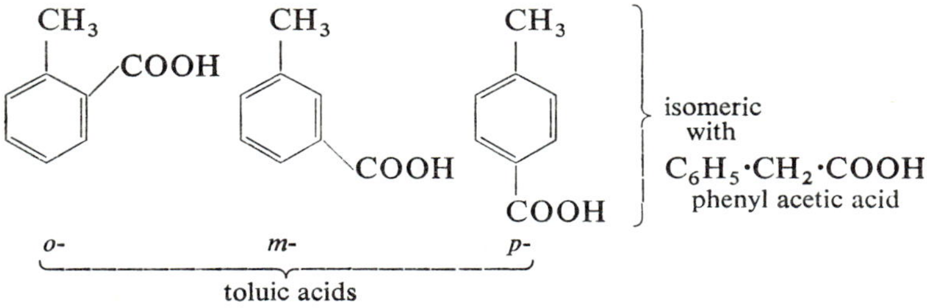

$o-$ $m-$ $p-$

toluic acids

isomeric with

$C_6H_5 \cdot CH_2 \cdot COOH$
phenyl acetic acid

EXAMINATION QUESTIONS

(1) Outline two general methods for the preparation of aliphatic carboxylic acids, mentioning the necessary reagents and conditions.

Give two reactions, of different types, which are given by formic acid but not by other simple carboxylic acids.

How does benzoic acid differ from acetic acid in physical properties?

Describe in each case one simple test-tube experiment by which you would differentiate between (a) benzoic acid and phenol, (b) aqueous solutions of sodium acetate and sodium benzoate. State clearly what you do and what you would see. (J. M. B. A.)

(2) Starting from toluene, describe in detail how you would prepare pure specimens of (a) benzaldehyde, (b) benzoic acid.

Compare and contrast the properties of acetaldehyde and benzaldehyde. (O. S.)

(3) Outline two methods by which benzoic acid may be prepared from benzene, giving the essential reagents and conditions. Describe briefly the reactions by which benzoic acid may be converted into (a) benzene, (b) benzoic anhydride, (c) phenyl cyanide (benzonitrile). (S. A.)

23

AROMATIC AMINES

Functional Group —NH$_2$, —NHR etc.
(attached to benzene nucleus)

PRIMARY, secondary and tertiary aromatic amines similar to those in the aliphatic series are well known and some are extremely important. Other types are secondary and tertiary amines with both alkyl and aryl radicals. Examples of some of the important amines are given below.

Primary amines

Aniline: $C_6H_5 \cdot NH_2$ Benzylamine: $C_6H_5 \cdot CH_2 \cdot NH_2$*

These two primary amines differ in that in aniline the amino group is attached directly to the benzene ring, whereas in benzylamine it is in the side chain. As we shall see later, these two amines differ markedly. Benzylamine, which resembles aliphatic primary amines in many respects, is isomeric with the three amino derivatives of toluene, viz.:

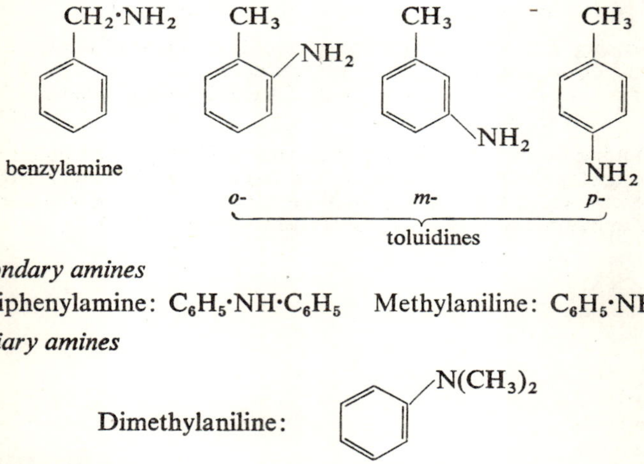

benzylamine   *o-*   *m-*   *p-*
                                                 toluidines

Secondary amines

Diphenylamine: $C_6H_5 \cdot NH \cdot C_6H_5$ Methylaniline: $C_6H_5 \cdot NH \cdot CH_3$

Tertiary amines

Dimethylaniline:

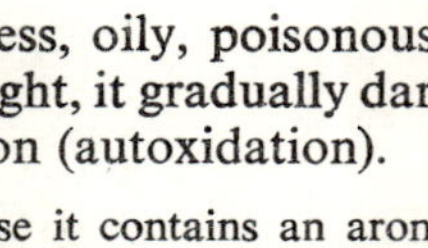

Aniline, $C_6H_5 \cdot NH_2$

Aniline when pure is a colourless, oily, poisonous liquid boiling at 182°C. On standing in air and light, it gradually darkens in colour, as a result of atmospheric oxidation (autoxidation).

---

* Benzylamine is included here because it contains an aromatic ring, and is thus an aromatic amine. However, it should be noted that the term *aromatic amine* is usually taken to mean a compound in which the amino group is directly linked to the aromatic nucleus, as in aniline and the toluidines.

Although only slightly soluble in water, aniline mixes readily with most organic solvents.

Aniline was first obtained in 1826 by the destructive distillation of indigo, which is one of the oldest of the natural dyes (the word *anil* is the Spanish for *indigo*). Eight years later, aniline was shown to be present in small amounts in coal tar. In 1841, it was obtained by the reduction of nitrobenzene. Hofmann showed that this was the same compound as was obtained from indigo and from coal tar.

Preparation of aniline

Under the conditions of temperature and pressure obtainable in an ordinary laboratory, ammonia has no action on monochlor-benzene (aromatic halogen compounds are much less reactive than aliphatic halides, p. 225).

Aniline is the final product of the reduction of most nitrogen deriv-atives of benzene where the nitrogen is attached directly to the ben-zene nucleus. The commonest of such compounds is nitrobenzene, readily prepared from benzene by direct nitration (p. 219). A variety of reducing agents can be used, e.g. tin and hydrochloric acid, tin(II) chloride and hydrochloric acid, zinc and acetic acid, iron and hydrochloric acid, and even ammonium sulphide (this was the reducing agent actually used in 1841 when aniline was first obtained from nitrobenzene). · In general, these methods of reduction are less convenient than the catalytic method using hydrogen and Raney nickel.

The following method may be used in the school laboratory:

To 100 g of iron filings in a 1-litre flask, about 200 ml of water are added and then 20 ml of concentrated hydrochloric acid are added. The flask is fitted with a reflux condenser and when hydrogen starts to be evolved, 50 ml of nitrobenzene are added, being poured down the condenser tube, a few ml at a time at intervals. The flask is shaken vigorously between additions of the nitrobenzene. When all the nitrobenzene has been added, the flask is heated over a water bath until the almond-like smell of nitrobenzene disappears.

The condenser is then removed and a paste of slaked lime and water added to the flask until the contents are alkaline. The aniline now set free from its hydrochloride is recovered from the mixture by steam distillation (see *Figure 27.5*). The distillate contains water and aniline, a small amount of the aniline being dissolved in the water. Common salt is added to throw some of this dissolved aniline out of solution. After standing for some time, the lower layer of aniline is drawn off. To recover the small amount of aniline in the water layer, it may be shaken up three times successively using 20 ml of ether each time (see p. 311).

These ether extracts are added to the main volume of aniline separated above and the mixture is freed from water with caustic potash pellets or anhydrous sodium sulphate. After standing for two or three days, the ether is distilled off over a water bath. The aniline is then distilled over a gauze, using an air condenser, that fraction coming over between 180° and 183°C being collected:

$$C_6H_5NO_2 + (6H) \longrightarrow C_6H_5NH_2 + 2H_2O$$

Manufacture of aniline

Two methods are in use.

(1) By the reduction of nitrobenzene (as above) by iron borings in the presence of water to which is added hydrochloric acid (a smaller proportion even than that used above) or iron(III) chloride. After addition of lime, the free aniline is recovered by steam distillation. Some tri-iron tetroxide (Fe_3O_4) is present in the residue.

(2) By heating a mixture of chlorobenzene and aqueous ammonia to a temperature of 200°C and under a pressure of 60 atmospheres, in presence of copper(I) chloride as a catalyst:

$$C_6H_5Cl + 2NH_3 \xrightarrow[\text{CuCl}]{200°C} C_6H_5NH_2 + NH_4C$$

Properties of aniline

(1) The amino group attached directly to a benzene ring is much less basic than when linked to an alkyl group. Aniline is only slightly soluble in water and is neutral to litmus (cf. ethylamine). It shows basic character by dissolving in dilute hydrochloric and and dilute sulphuric acids to form respectively the hydrochloride $[C_6H_5\cdot NH_3]^+Cl^-$ and the sulphate $[C_6H_5\cdot NH^+_3]_2SO_4^{2-}$, the latter being only sparingly soluble. The free aniline can be obtained from these salts by the addition of alkalis or sodium carbonate.

Both salts are freely dissociated in aqueous solution.

(2) Aniline reacts with an alkyl halide to give first a secondary and then a tertiary amine; the aniline molecule acts as a nucleophile:

$$C_6H_5 \cdot NH_2 \quad CH_3 \overset{\frown}{-} I \longrightarrow C_6H_5 \cdot \overset{+}{N} \overset{H}{\underset{CH_3}{\overset{\nearrow H}{\diagdown}}} \quad I^-$$

$$C_6H_5 \cdot \overset{+}{\underset{CH_3}{\overset{|}{N}H_2}} \ I^- + C_6H_5 \cdot NH_2 \rightleftharpoons C_6H_5 \cdot NH \cdot CH_3$$
$$+ C_6H_5 \cdot \overset{+}{N}H_3$$

$$C_6H_5 \cdot N\overset{\frown}{H} \cdot CH_3 \quad CH_3 \overset{\frown}{-} I \longrightarrow C_6H_5 \cdot \overset{+}{N} \overset{H}{\underset{CH_3}{\overset{\nearrow H}{\diagdown} CH_3}} \quad I^-$$

On treatment with an alkali, methylaniline and dimethylaniline are obtained.

(3) Aniline is acetylated (a) by warming it with acetic anhydride, (b) allowing it to react at room temperature with the much more reactive acetyl chloride, or (c) by prolonged heating with acetic acid:

$$C_6H_5 \cdot NH_2 + ClCOCH_3 \longrightarrow \underset{\text{acetanilide}}{CH_3 \cdot CO \cdot NH \cdot C_6H_5} + HCl$$

In the commonly used preparation using acetic anhydride, the mixture is poured into water, when the acetanilide separates out as a crystalline solid (white after purification by re-crystallization).

Alternatively, acetanilide can be prepared by suspending aniline in 2N NaOH and shaking with an equimolar proportion of acetic anhydride; after a few minutes acetanilide separates out.

(4) Aniline is similarly *benzoylated* by shaking with benzoyl chloride in the presence of caustic soda. This is the Schotten–Baumann reaction (p. 237):

$$C_6H_5 \cdot NH_2 + ClCO \cdot C_6H_5 \xrightarrow{\text{NaOH}} \underset{\text{benzanilide}}{C_6H_5 \cdot CO \cdot NH \cdot C_6H_5} + HCl \downarrow$$
$$NaCl$$

(5) Aniline gives an isocyanide (carbylamine) on treatment with chloroform (or iodoform) and alcoholic potash:

$$C_6H_5 \cdot NH_2 + Cl_3HC \xrightarrow{\text{KOH}} \underset{\text{phenyl isocyanide}}{C_6H_5 \cdot NC} + 3HCl \downarrow$$
$$3KCl$$

Like the corresponding compounds of aliphatic chemistry, phenyl isocyanide possesses an objectionable smell (the reaction is a convenient *test for an aromatic primary amine* as well as for the aliphatic compound).

(6) Aniline reacts with nitrous acid as follows:

(a) If an aqueous solution of sodium or potassium nitrite is added to a solution of aniline in excess dilute hydrochloric acid at a temperature between 0° and 5°C, the compound *phenyl diazonium chloride* or *benzene diazonium chloride* is obtained:

$$NaNO_2 + HCl \longrightarrow HNO_2 + NaCl$$

$$C_6H_5 \cdot NH_2 + HCl + HO \cdot NO \longrightarrow [C_6H_5 \cdot N_2]^+ Cl^- + 2H_2O$$
phenyldiazonium chloride
or
benzene diazonium chloride

Aliphatic primary amines give similar compounds, but *these are so unstable that they decompose immediately*, even at low temperatures, to yield alcohols and other products:

$$R{-}NH_2 \xrightarrow[\text{(HCl)}]{\text{HO} \cdot \text{NO}} [R{-}\overset{+}{N}_2Cl^-] \xrightarrow{\text{H}_2\text{O}} R \cdot OH + N_2$$

aliphatic
primary
amine

Aromatic diazonium compounds are extremely important both in synthetic organic chemistry and in chemical industry. They are studied more fully in Chapter 24.

(*b*) If the reaction with nitrous acid is carried out at room temperature, and particularly at higher temperatures, nitrogen and the hydroxyl derivative are formed as with aliphatic primary amines. The diazonium salt which is no doubt first formed is hydrolysed:

$$[C_6H_5 \cdot N_2]^+ Cl^- + HOH \longrightarrow C_6H_5OH + N_2 + HCl$$
phenol

This constitutes an important general method for the preparation of phenols. The diazonium salt is prepared at 0°C and the solution is then heated.

(7) On oxidation, aniline undergoes complex reactions giving mixtures of compounds, many of which are highly coloured. The reactions provide useful *tests for aniline*.

(*a*) With aqueous bleaching powder or sodium hypochlorite, an intense *purple colour* appears.

(*b*) With a mixture of sulphuric acid and potassium dichromate a black deposit is obtained.

In 1856, William Perkin (later Sir William Perkin), the young research student of Hofmann in London, added potassium dichromate to a sulphuric acid solution of crude aniline, in what now seems an extremely optimistic attempt to prepare quinine (the only then known cure for malaria). He obtained a black mass from which he extracted a mauve coloured substance. This product proved to be an excellent dye which was not affected by light; its manufacture was started a year later. The dye was used to colour the penny mauve stamp of the reign of Queen Victoria.

(8) The following properties of aniline are due to the benzene ring in its molecule and are therefore characteristic of an aromatic as distinct from an aliphatic compound.

(a) Chlorine and bromine readily substitute in the ring. On adding bromine water to aniline, there is a white precipitate of tribromoaniline:

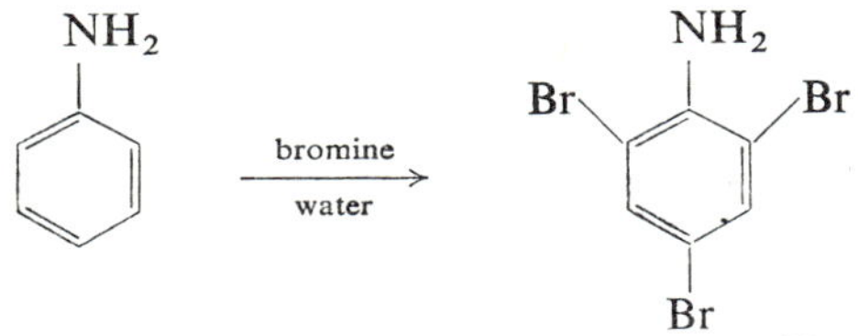

2,4,6-tribromoaniline

The amino group is *ortho*- and *para*-directing and has a marked influence on the reactivity of the ring in these positions.

(b) With sulphuric acid, sulphonation occurs giving the *para*-sulphonic acid, an important compound in industry and commonly called *sulphanilic acid*:

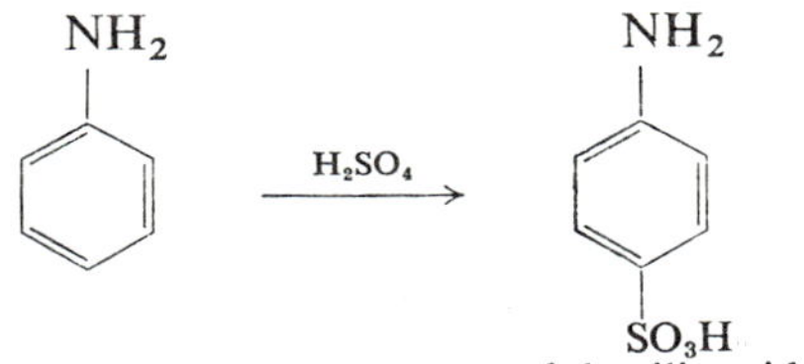

sulphanilic acid

(c) Nitric acid tends to cause oxidation, but if the amino group is first 'protected' by acetylating it (p. 165), then nitro groups enter the *ortho* and *para* positions:

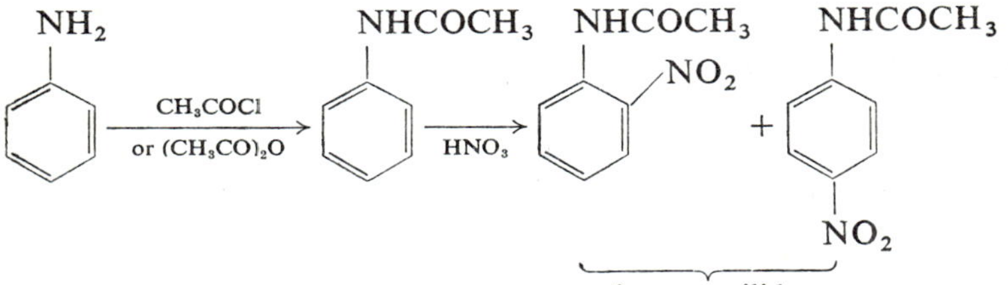

nitroacetanilides

If these nitroacetanilides are warmed with alkali, the acetyl group is removed to leave the free nitroaniline.

m-Nitroaniline cannot be prepared by direct nitration. If, however, *m*-dinitrobenzene is treated with ammonium sulphide, reduction can be stopped at a stage when one nitro group only is attacked:

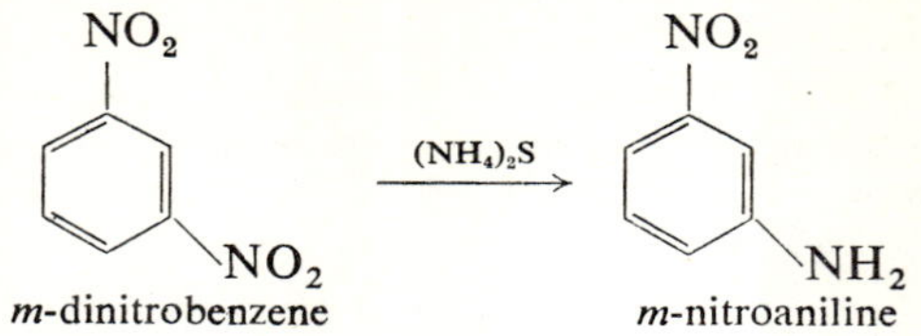

m-dinitrobenzene m-nitroaniline

Uses of Aniline

(1) In the manufacture of the *aniline dyes* (p. 269).

(2) In the manufacture of a variety of pharmaceutical products, and other ' fine chemicals '.

The toluidines and benzylamine

The toluidines strongly resemble aniline. They are prepared by the reduction of the corresponding nitrotoluene. In addition to showing the properties of all primary amines, they form diazonium salts on treatment with nitrous acid at low temperatures. These compounds, like the corresponding ones from aniline, are used to manufacture dyes. On oxidation, the toluidines give mixtures of complex coloured substances.

Benzylamine, on the other hand, rather resembles an aliphatic primary amine.

Preparation of benzylamine

(1) By the reduction of the corresponding nitrile:

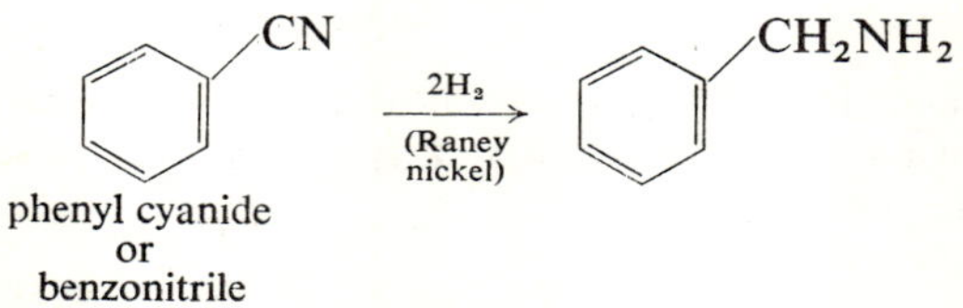

phenyl cyanide
or
benzonitrile

(2) By the action of alcoholic ammonia on a benzyl halide, e.g.:

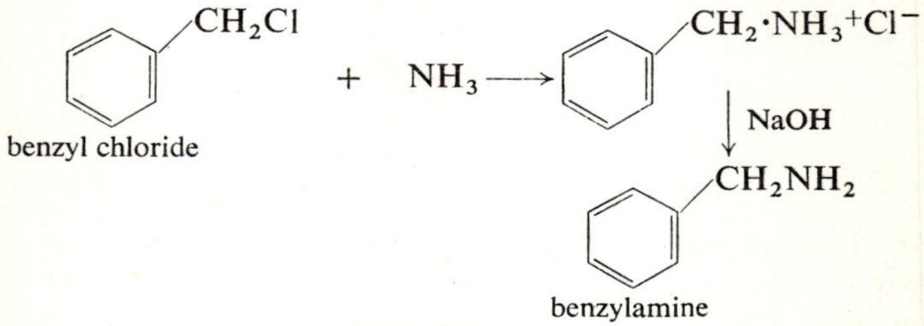

benzyl chloride

benzylamine

Other products may be formed (p. 162).

Like an aliphatic primary amine and in contrast to aniline, benzyl-amine is strongly basic, being readily soluble in water and alkaline to litmus. On treatment with nitrous acid, even at low temperatures, nitrogen is evolved and the alcohol formed:

$$C_6H_5 \cdot CH_2 \cdot NH_2 \xrightarrow[\substack{\text{and} \\ \text{HCl}}]{\text{NaNO}_2} [C_6H_5 \cdot CH_2 \cdot N_2^+ Cl^-] \longrightarrow$$
$$C_6H_5 \cdot CH_2OH + N_2$$

SUMMARY OF THE CHEMISTRY OF AROMATIC AMINES

Primary amines

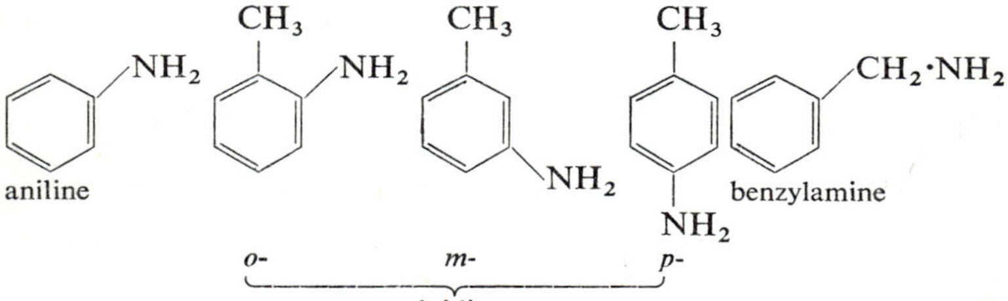

aniline   *o-*   *m-*   *p-*   benzylamine

toluidines

Secondary amines

Diphenylamine: $C_6H_5 \cdot NH \cdot C_6H_5$ Methylaniline: $C_6H_5 \cdot NH \cdot CH_3$

Tertiary amines

Triphenylamine: $(C_6H_5)_3N$ Dimethylaniline: $C_6H_5 \cdot N(CH_3)_2$

Aniline, $C_6H_5 \cdot NH_2$

(1) *Preparation*

Reduction of nitrobenzene in acid solution or with H_2/catalyst:

$$C_6H_5 \cdot NO_2 \longrightarrow C_6H_5 \cdot NH_2$$

Reducing agents may be Fe, Sn or $SnCl_2$ and HCl, Zn and $CH_3 \cdot COOH$. In hydrochloric acid, the final product is $C_6H_5 \cdot NH_3^+Cl^-$—aniline hydrochloride—from which the free aniline is liberated by alkali.

(2) *Manufacture*

(*a*) Reduction above using Fe and HCl.

(*b*) Ammonia on chlorobenzene at 200°C and 60 atm in presence of copper(I) chloride as catalyst. The conditions are necessary on account of inactivity of chlorobenzene:

$$C_6H_5Cl + 2NH_3 \xrightarrow[\text{CuCl}]{\text{200°C 60 atm}} C_6H_5 \cdot NH_2 + NH_4Cl$$

(3) *Properties*

$$C_6H_5{\cdot}NH_2 \xrightarrow{HCl} C_6H_5{\cdot}NH_3{}^+Cl^-$$

$$\xrightarrow{\text{dil. } H_2SO_4} [C_6H_5{\cdot}NH_3{}^+]_2SO_4{}^{2-}$$

$$\xrightarrow{CH_3I} [C_6H_5{\cdot}NH_2{\cdot}CH_3]^+I^- \text{ and then } [C_6H_5{\cdot}NH(CH_3)_2]^+I^-$$

$$\xrightarrow[\text{or } (CH_3CO)_2O]{CH_3{\cdot}COCl} CH_3{\cdot}CO{\cdot}NH{\cdot}C_6H_5$$

acetanilide

$$\xrightarrow{CHCl_3 + KOH} C_6H_5{\cdot}NC$$

phenyl isocyanide

$$\xrightarrow[(NaNO_2 + HCl)]{HNO_2} \begin{cases} \xrightarrow[\text{temp.}]{\text{low}} [C_6H_5{\cdot}N_2]^+Cl^- \text{ phenyldiazonium chloride} \\ \xrightarrow[\text{temp.}]{\text{ordinary}} C_6H_5OH \end{cases}$$

$[C_6H_5{\cdot}N_2]^+Cl^-$ —heat solution→ C_6H_5OH (*via* diazonium salt)

(cf. primary aliphatic amine

$$\xrightarrow{Br_2}$$

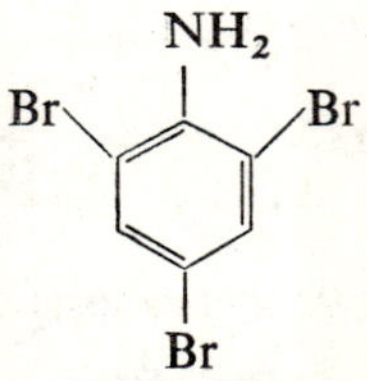

2,4,6-tribromoaniline

$$\xrightarrow{\text{conc. } H_2SO_4}$$

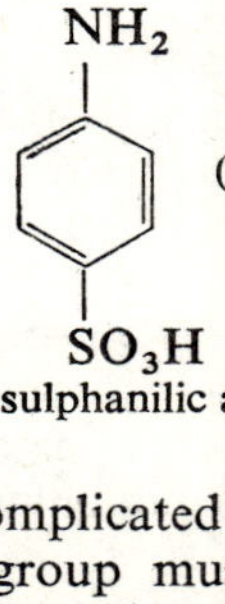

(very little *ortho*-isomer formed)

sulphanilic acid

Aniline is readily oxidized, yielding complicated products. Nitric acid oxidizes aniline, and the amino group must be 'protected' before a nitro group can be introduced into the nucleus:

N.B. *m*-Nitroaniline is prepared by partial reduction of *m*-dinitrobenzene:

m-nitroaniline

EXAMINATION QUESTIONS

(1) Name and give the structural formulae, one in each case, of compounds in which the —NH₂ group is attached to (*a*) an alkyl group, (*b*) a phenyl group, (*c*) a carbonyl group.

How do the compounds you have named differ in physical properties? Give two reactions typical of each compound. (C. A.)

(2) Describe the usual laboratory preparation of aniline from nitrobenzene.

Compare the chemical properties of aniline with those of methylamine. How would you detect the presence of nitrogen in aniline? (A. A.)

(3) Given a mixture of aniline and nitrobenzene, describe how you would obtain a pure sample of each compound.

State how aniline reacts with (a) acetyl chloride, (b) chloroform and potassium hydroxide, naming the organic products.

Briefly describe how aniline may be converted into phenol (extraction and purification not required). (O. A.)

(4) Review the methods available for the preparation of aliphatic and aromatic amines. How would you attempt to distinguish between compounds whose structures are represented by the formulae $R \cdot CH_2 \cdot NH_2$ and $R \cdot CO \cdot NH_2$? (Liverpool, B.Sc. Part I)

(5) Indicate the essential meaning of the term ' aromatic character '.

If benzene, benzaldehyde and aniline were available, how would you attempt to prepare the compounds shown below?

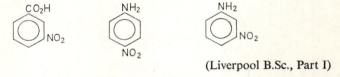

(Liverpool B.Sc., Part I)

(6) Show how you could distinguish, using chemical reactions, between the following pairs of isomers:

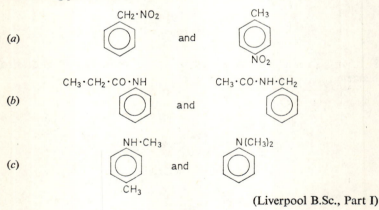

(Liverpool B.Sc., Part I)

24

DIAZONIUM SALTS

THE value of diazonium salts in synthetic aromatic chemistry and in chemical industry is due to their reactivity. Chlorobenzene, C_6H_5Cl, is of little use in laboratory work because the chlorine atom directly attached to the benzene ring is not easily replaced by other groups. In great contrast, phenyldiazonium chloride, $[C_6H_5 \cdot N_2]^+Cl^-$, is extremely reactive.

Similar derivatives of the toluidines, and in fact of many other aromatic amines are known.

Diazonium compounds (first prepared in 1858 by the German-born Peter Griess, working as a chemist in an English brewery) are true salts, being completely dissociated into ions in aqueous solution. In synthetic work, diazonium salts are normally used in the solution in which they are prepared. They are not easily obtained in the pure state on account of their instability. Further, when dry, some of them are very explosive.

Preparation of phenyldiazonium chloride (and sulphate)

Phenyldiazonium chloride is prepared from aniline by first dissolving it in dilute hydrochloric acid, the amounts being in the ratio of one mole of the amine to two and a half of acid. This solution is cooled to a temperature of about 5° by surrounding it with a freezing mixture. An aqueous solution of sodium or potassium nitrite is then run in drop by drop during constant stirring. The amount of nitrite used is in the ratio of one mole.

The nitrous acid required for the *diazotization* (as the making of a diazonium salt is called) is generated *in situ*, one mole of the acid being used for the reaction.

$$NaNO_2 + HCl \longrightarrow NaCl + HNO_2$$

Then: $C_6H_5NH_2 + HCl + HONO \longrightarrow [C_6H_5 \cdot N_2]^+Cl^- + 2H_2O$

The remaining acid (0·5 mole) ensures the absence of any free aniline, which would react with the phenyldiazonium chloride as soon as some was formed.

The temperature range must be strictly maintained during the addition of the nitrite. If the temperature is allowed to exceed 10°C, the diazonium salt tends to decompose. The reaction by

265

which the diazonium salt is actually formed is exothermic, hence constant stirring to prevent local rise of temperature is most essential.

Below 5°C the reaction is slow. It is in fact, advisable to wait some time after all the nitrite has been added for the reaction to finish, before the solution of the diazonium salt is used.

If the hydrogen sulphate, $[C_6H_5 \cdot N_2]^+HSO_4^-$ is required, dilute sulphuric acid is used instead of hydrochloric in the same molar proportions. With certain reagents it is to be preferred to the chloride.

Properties of diazonium salts

The diazonium cation, e.g. $C_6H_5 \cdot \overset{+}{N_2}$ can be represented as:

$$C_6H_5 \cdot \overset{+}{N}\!\!\equiv\!\!N\!: \longleftrightarrow C_6H_5 \cdot \overset{..}{N}\!\!=\!\!\overset{+}{N}\!:$$

where the double-headed arrow has the significance previously discussed. In the first structure, both nitrogen atoms have octets of electrons, and the atom attached to the benzene ring is tetravalent and positively charged, as is the nitrogen atom in quarternary ammonium salts, $R_4N^+X^-$. In the second structure, the terminal nitrogen atom is not only positively charged but is *electron-deficient* (only six electrons around the nucleus). It is thus not surprising that, in some reactions, diazonium cations behave as electrophiles (e.g. the coupling reaction discussed below).

The reactions of diazonium salts can conveniently be divided into two groups, viz (a) those in which the nitrogen is evolved and (b) those where it remains in the molecule of the product. Some of the more important examples of each type are given below.

The following reactions involve elimination of nitrogen:

(1) *Replacement by a hydroxyl group:*—If the temperature of the diazonium salt solution is allowed to rise or if it is warmed cautiously, hydrolysis occurs to give a phenol:

$$[C_6H_5 \cdot N_2]Cl + H_2O \longrightarrow C_6H_5OH + N_2 + HCl$$

This is an alternative method of preparing phenol, which can be obtained from the reaction mixture by steam distillation, and subsequent ether extraction.

(2) *Replacement by a hydrogen atom:*—Various reagents, e.g. hypophosphorous acid, alkaline sodium stannite, ethyl alcohol, convert the benzene diazonium salt to benzene, the process being a reduction:

$$[C_6H_5 \cdot N_2]Cl + H_3PO_2 + H_2O \longrightarrow C_6H_6 + N_2 + HCl + H_3PO_3$$
orthophosphorous
acid

$$[C_6H_5 \cdot N_2]HSO_4 + C_2H_5OH \longrightarrow C_6H_6 + H_2SO_4$$
$$+ CH_3 \cdot CHO + N_2$$

This reaction is important in organic syntheses for it makes possible the complete *removal of an amino group and also a nitro group* (after previous reduction) *from a benzene ring*. The following reaction sequence illustrates its use:

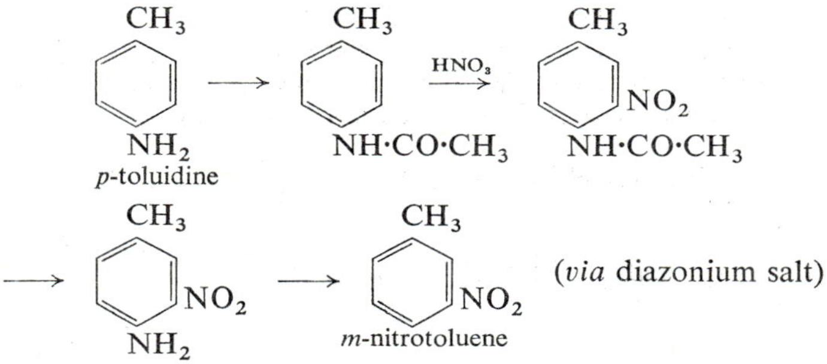

(*via* diazonium salt)

m-Nitrotoluene cannot be made by direct nitration of toluene; *p*-nitrotoluene, on the other hand, is easily available (from toluene) and is reduced to give the starting material above, *p*-toluidine.

(3) *Sandmeyer reactions:*—The following reactions of replacement by chlorine, bromine, and the cyanogen radical were discovered by Sandmeyer in 1884.

(*a*) Replacement by *chlorine:*—
By the action of a solution of copper(I) chloride in hydrochloric acid:

$$[C_6H_5 \cdot N_2]^+ Cl^- \xrightarrow[\text{HCl}]{\text{CuCl}} C_6H_5Cl + N_2$$

Gatterman found later that this conversion could be brought about by the addition of powdered copper:

$$[C_6H_5 \cdot N_2]^+ Cl^- \xrightarrow{\text{Cu}} C_6H_5Cl + N_2$$

(*b*) Replacement by *bromine:*—
By the action of a solution of copper(I) bromide in hydrobromic acid:

$$[C_6H_5 \cdot N_2]^+ HSO_4^- \xrightarrow[\text{HBr}]{\text{CuBr}} C_6H_5Br + N_2$$

(c) Replacement by the *cyanogen radical:*—

By the action of a solution of copper(I) cyanide in potassium cyanide, i.e. $[K_3Cu(CN)_4]$:

$$[C_6H_5 \cdot N_2]^+HSO_4^- \xrightarrow{\text{K}_3\text{Cu(CN)}_4} C_6H_5 \cdot CN + N_2$$
benzonitrile or
phenyl cyanide

(The hydrogen sulphate is used in (b) and (c) to prevent formation of chlorobenzene.)

Reactions (a) and (b) above are most useful in view of the fact that introduction of halogen atoms into the benzene ring is not always easy by direct methods; (c) is of particular value because of the ease of conversion of the —CN group to carboxyl —COOH by hydrolysis and to the —$CH_2 \cdot NH_2$ by reduction. Benzene may thus be converted into benzoic acid or into benzylamine as follows:

$$C_6H_6 \xrightarrow[\text{H}_2\text{SO}_4]{\text{HNO}_3} C_6H_5 \cdot NO_2 \longrightarrow C_6H_5 \cdot NH_2 \xrightarrow[\text{H}_2\text{SO}_4]{\text{NaNO}_2} [C_6H_5 \cdot N_2]^+HSO_4^-$$

$$[C_6H_5 \cdot N_2]^+HSO_4^- \xrightarrow{\text{K}_3[\text{Cu(CN)}_4]} C_6H_5 \cdot CN \overbrace{\begin{array}{l} \longrightarrow C_6H_5 \cdot COOH \\ \\ \longrightarrow C_6H_5 \cdot CH_2 \cdot NH_2 \end{array}}$$

(4) *Replacement by iodine:*—By the addition of the diazonium solution to potassium iodide solution. Again, the diazonium sulphate gives a better yield than does the chloride:

$$[C_6H_5 \cdot N_2]^+HSO_4^- + KI \longrightarrow C_6H_5I + N_2 + KHSO_4$$

This is extremely useful in synthetic work, since iodine is not easily introduced directly into a benzene ring.

In the following reactions the nitrogen is not removed:

(1) *Reduction to phenylhydrazine:*—Any of the following mild reducing agents will bring about this reaction:—tin(II) chloride in hydrochloric acid solution, sulphur dioxide, sodium bisulphite:

$$[C_6H_5 \cdot N_2]^+Cl^- \xrightarrow{\text{4H}} [C_6H_5 \cdot NH \cdot NH_3]^+Cl^-$$
phenylhydrazine hydrochloride

The free hydrazine can be obtained by warming with an alkali. Phenylhydrazine forms *derivatives* with aldehydes and ketones and has been particularly useful in studies of *carbohydrates*. With the simpler sugars, phenylhydrazine yields crystalline products (known as osazones) which help to characterize and relate them.

(2) *Coupling:*—Diazonium salts readily react with phenols and tertiary aromatic amines. Here, the diazonium cation behaves as

an electrophile and the nucleus of the aromatic compound acts as a nucleophile (p. 208):

$$C_6H_5-N_2^+Cl^- + C_6H_5-OH \longrightarrow$$

$$C_6H_5-N{=}N-C_6H_4-OH + HCl$$

The —N=N— group is known as the *azo* group and the above compound is thus called p-*hydroxyazobenzene*. (This compound possesses a bright orange colour and the reaction provides another *test* for *aniline* or for *phenol*.) The aniline is first diazotized, then an alkaline solution of phenol added; a crystalline orange precipitate is obtained. (If β-naphthol,

[naphthalene ring with OH]

, is added instead

of phenol, a brilliant scarlet precipitate is obtained. This reaction provides an even more striking test for aniline than does phenol.)

An example of coupling with a tertiary aromatic amine is given by:

$$C_6H_5-N_2^+Cl^- + C_6H_5-N(CH_3)_2 \longrightarrow$$

$$C_6H_5-N{=}N-C_6H_4-N(CH_3)_2 + HCl$$

p–dimethylaminoazobenzene

If in the coupling compound, the *para* position is already occupied, the reaction occurs in the *ortho* position. If both *ortho* and *para* positions are occupied, no coupling takes place.

This coupling process is the basis of the aniline dye industry. A dye is a coloured substance not affected by light, which can attach itself to the fibres of material so tightly that it is not removed by washing. Groups which impart colour to a compound are known as *chromophore* groups. Both azo compounds above are thus dyes. (They are not used in industry for other better compounds are known.)

The methyl orange used as an acid/alkali indicator is an azo dye which can be prepared as follows:—

The amine p-aminosulphonic acid, commonly called *sulphanilic acid* (p. 259), is first diazotized:

$$HO_3S-C_6H_4-NH_2 \xrightarrow[\text{HCl}]{\text{NaNO}_2} HO_3S-C_6H_4-N_2^+Cl^-$$

This is then coupled with dimethylaniline:

$$HO_3S-\!\!\!\langle\ \rangle\!\!\!-N_2^+Cl^- + \langle\ \rangle\!\!\!-N(CH_3)_2 \longrightarrow$$

$$HCl + HO_3S-\!\!\!\langle\ \rangle\!\!\!-N\!\!=\!\!N-\!\!\langle\ \rangle\!\!\!-N(CH_3)_2$$

The sodium salt of this compound is methyl orange.

SUMMARY OF THE CHEMISTRY OF DIAZONIUM SALTS

Phenyl diazonium chloride (or benzene diazonium chloride),

$$C_6H_5 \cdot N_2^+Cl^-, \text{ i.e. } C_6H_5 \cdot \overset{+}{N}\!\!=\!\!N \longleftrightarrow C_6H_5 \cdot N\!\!=\!\!\overset{+}{N}$$

(1) *Preparation*

By the slow addition (with constant stirring) of an aqueous solution of sodium or potassium nitrite to a solution of aniline in dilute hydrochloric acid, cooled to a temperature of 5°C:

$$C_6H_5 \cdot NH_2 + HCl + HNO_2 \longrightarrow [C_6H_5N_2]Cl + 2H_2O$$

(2) *Properties*

(*a*) Reactions in which *nitrogen is evolved* when the temperature rises:

$$[C_6H_5N_2]^+Cl^- \xrightarrow{\;\;H_2O\;\;} C_6H_5OH + N_2 + HCl$$

$$\xrightarrow[\text{(C}_2\text{H}_5\text{OH or H}_3\text{PO}_2)]{\text{reduction}} C_6H_6 + N_2$$

$$\xrightarrow{\;\;CuCl\;\;} C_6H_5Cl + N_2 \;\;\Big\rangle$$

$$[C_6H_5N_2]^+HSO_4^- \xrightarrow{\;\;CuBr\;\;} C_6H_5Br + N_2 \;\;\Big\rangle \begin{array}{l}\text{Sandmeyer}\\\text{reactions}\end{array}$$

$$[C_6H_5N_2]^+HSO_4^- \xrightarrow[\text{(and KCN)}]{\;\;CuCN\;\;} C_6H_5 \cdot CN + N_2 \;\;\Big\rangle$$

$$[C_6H_5N_2]^+HSO_4^- \xrightarrow{\;\;KI\;\;} C_6H_5I + N_2$$

(*b*) Reactions in which *nitrogen is retained*:

$$(i)\ [C_6H_5N_2]^+Cl^- \xrightarrow[\text{(SnCl}_2\text{, NaHSO}_3\text{, etc.)}]{\text{reduction}} C_6H_5NH \cdot NH_3^+Cl^-$$

phenylhydrazine hydrochloride

(*ii*) Condenses with phenol:

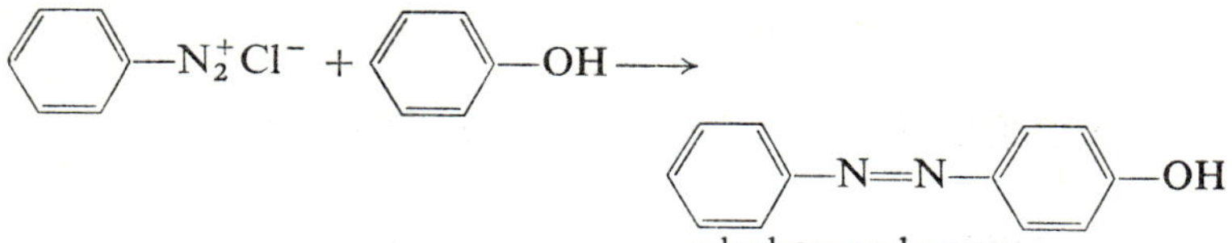

$$\text{C}_6\text{H}_5\text{—N}_2^+\text{Cl}^- + \text{C}_6\text{H}_5\text{—OH} \longrightarrow$$

$$\text{C}_6\text{H}_5\text{—N}{=}\text{N—C}_6\text{H}_4\text{—OH}$$

p-hydroxyazobenzene

(*iii*) Condenses with dimethylaniline:

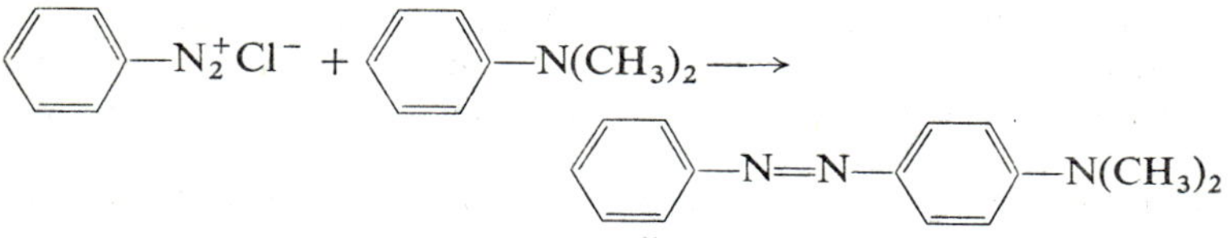

$$\text{C}_6\text{H}_5\text{—N}_2^+\text{Cl}^- + \text{C}_6\text{H}_5\text{—N(CH}_3)_2 \longrightarrow$$

$$\text{C}_6\text{H}_5\text{—N}{=}\text{N—C}_6\text{H}_4\text{—N(CH}_3)_2$$

p-dimethylaminoazobenzene

N.B. The group —N=N— is known as the *azo* group and such a group always gives colour to a molecule. The two dyes obtained in reactions (*ii*) and (*iii*) above are simple examples of ' azo ' or ' aniline ' dyes.

EXAMINATION QUESTION

(1) Describe exactly how you would prepare a solution of benzene diazonium chloride (diazobenzene chloride). Indicate how you would obtain from this solution (*a*) iodobenzene, (*b*) phenylhydrazine, (*c*) aminoazobenzene, (*d*) an azo-dye, (*e*) benzoic acid. (O. S.)

25

CARBOHYDRATES AND PROTEINS

CARBOHYDRATES, as the name suggests, consist of carbon, hydrogen and oxygen only, the hydrogen and oxygen being in the same proportions as in water. The general formula is therefore $C_x(H_2O)_y$. The name carbohydrate is unfortunate since not one of them can be prepared by the direct action of water on carbon, whereas the hydrates of inorganic chemistry can often be prepared by direct addition of water to a substance.

Carbohydrates are widely distributed in nature, particularly in the plant kingdom. The carbohydrate cellulose is the chief constituent of the structural material of many plants. Starch, another carbohydrate, is found in different parts of many plants, particularly as a food reserve stored in the roots and stems and also in many seeds. Fructose and glucose give the sweet taste to many fruits and are also present in nectar and therefore in honey.

Although carbohydrates constitute one of the most important foodstuffs of animals, providing heat and energy, they are not present to the same extent in the animal as in the plant body. Glycogen, a carbohydrate similar to starch, is a food reserve of many animals, being stored in the liver. Lactose, a carbohydrate never found in the vegetable kingdom, is present in the milk of mammals—up to 5 per cent in cow's milk and 7 per cent in human.

Carbohydrates are also important in modern industry. Paper, cotton, linen, synthetic products made from cellulose, such as rayon, viscose rayon, celluloid and guncotton, starch, gums, lacquers, —these are just a few examples of carbohydrates or their derivatives in common use.

The better known carbohydrates can be classified in three groups.

(1) *Monosaccharides**

e.g. glucose and fructose, isomers of molecular formula $C_6H_{12}O_6$.

(2) *Disaccharides*

e.g. sucrose, maltose, lactose, isomers of molecular formula $C_{12}H_{22}O_{11}$.

(3) *Polysaccharides*

e.g. cellulose, starch, glycogen, all of very high molecular weights. They are, as the name suggests, polymers (p. 24).

* Latin, *saccharon*=Sugar.

MONOSACCHARIDES

Monosaccharides constitute the building units of which poly-saccharides are composed. Thus on mild hydrolysis all disaccharides and polysaccharides are broken down to monosaccharides, which are themselves stable to hydrolysis. The best known members are the two isomers glucose and fructose of molecular formula $C_6H_{12}O_6$. They are hence referred to as *hexoses*. Both occur, as mentioned above, in fruits, in nectar and honey, and glucose is present also in the leaves and roots of many plants either free or as derivatives known as glucosides.

Glucose or dextrose, $C_6H_{12}O_6$

Glucose is *manufactured* by the hydrolysis of starch under the influence of a mineral acid, either sulphuric or hydrochloric, in very dilute solution. The hydrolysis is carried out under high pressure and at a high temperature:

$$(C_6H_{10}O_5)_n + nH_2O \longrightarrow n(C_6H_{12}O_6)$$
$$\text{starch} \qquad\qquad\qquad \text{glucose}$$

The syrupy solution obtained is decolorized by filtering over animal charcoal. The resulting solution may be used direct in the manufacture of jam and sweets or may be evaporated under reduced pressure and allowed to crystallize to obtain the solid.

Glucose is also formed as an intermediate product in the alcoholic fermentation of starch (p. 72):

$$C_{12}H_{22}O_{11} + H_2O \longrightarrow 2C_6H_{12}O_6$$
$$\text{maltose} \qquad\qquad\qquad \text{glucose}$$

It is also obtained along with fructose by the hydrolysis of sucrose under the influence of either (*a*) dilute hydrochloric or sulphuric acid, or (*b*) the enzyme invertase:

$$C_{12}H_{22}O_{11} + H_2O \xrightarrow[\text{or invertase}]{\text{mineral acid}} C_6H_{12}O_6 + C_6H_{12}O_6$$
$$\text{sucrose} \qquad\qquad\qquad\qquad\quad \text{fructose} \quad\;\; \text{glucose}$$

Properties and constitution of glucose

Glucose is a crystalline solid, possessing a sweet taste but not so sweet as that of fructose or sucrose. It is soluble in water but almost insoluble in absolute alcohol.

(1) On *reduction* of glucose with hydrogen iodide and phosphorus, n-hexane is obtained. This points to a straight chain of six carbon atoms.

(2) On treatment of glucose with acetic anhydride, five acetyl radicals are introduced, indicating the presence of five hydroxyl groups.

(3) Glucose adds on hydrogen cyanide (but not sodium bisulphite), and condenses with hydroxylamine and hydrazine. These reactions indicate the presence of the carbonyl group $\diagup C {=\!\!=} O$.

(4) Glucose reduces Fehling's solution, ammoniacal silver oxide and is also oxidized by sodium hypobromite. In each case the product contains the carboxyl group —COOH. This suggests that the carbonyl group is present as an aldehyde.

On this behaviour and on the further study of the nature of the derivatives obtained, a structural formula for glucose may be written in the form I below:

```
        CHO
         |                          OH
      *CH·OH                        |
         |                          CH
      *CH·OH          or         /      \
         |                     O          CHOH
      *CH·OH                    |           |
         |              HOH2C—CH          CHOH
      *CH·OH                     \        /
         |                          CH
      CH2OH                         |
                                    OH
         I                          II
```

This formula I contains four asymmetric carbon atoms (p. 334), marked with an asterisk. Thus, there should be sixteen possible optical isomers. All these have been identified. The naturally occurring glucose is one of these isomers; it is dextro-rotatory, hence the alternative name of *dextrose*.

Emil Fischer† (whose name is famous for his study of the structure of sugars) attempted to make an acetal derivative of glucose and obtained two isomeric products resembling acetals, and no longer containing the aldehyde group. The reaction is readily explainable on the theory of a ring structure of glucose as shown in II. If, however, glucose possesses this formula, the ring (in the case of glucose but not necessarily some of its derivatives) must be easily broken up, and this would appear to occur first before the chemical properties of glucose given above can be understood.

(5) The reaction of glucose with phenylhydrazine is more extensive than that with a simple aldehyde. It is a reaction which played

† Emil Fischer lived from 1852–1919. He was Professor of Organic Chemistry in the University of Berlin and received the Nobel prize in 1902, partly for his work on sugars.

an important part in the solution of the constitution of carbohydrates. The compound formed is called an *osazone*. Using
formula I, the reaction may be represented as below:

$$
\begin{array}{l}
\text{CHO} \\
| \\
\text{CHOH} \\
| \qquad + 3C_6H_5{\cdot}NH{\cdot}NH_2 \longrightarrow \\
(\text{CHOH})_3 \\
| \\
\text{CH}_2\text{OH}
\end{array}
\qquad
\begin{array}{l}
\text{CH}{=}\text{N}{\cdot}\text{NH}{\cdot}C_6H_5 \\
| \\
\text{C}{=}\text{N}{\cdot}\text{NH}{\cdot}C_6H_5 \\
| \qquad\quad + C_6H_5{\cdot}NH_2 + NH_3 \\
(\text{CHOH})_3 \qquad\qquad + 2H_2O \\
| \\
\text{CH}_2\text{OH} \\
\text{glucose phenylosazone}
\end{array}
$$

Fructose or laevulose, $C_6H_{12}O_6$

This isomer of glucose is frequently found mixed with glucose in
nature. The naturally occurring fructose is laevo-rotatory, hence the
alternative name of *laevulose*. Like glucose, it possesses a sweet taste.

It is obtained by the hydrolysis of sucrose (p. 277). It is also
obtained by the hydrolysis of *inulin*, very dilute sulphuric acid or
oxalic acid being the hydrolysing agent. Here the product is fructose
only. Inulin is a starch-like polysaccharide present in dahlia tubers,
Jerusalem artichokes, dandelion roots, etc.

$$(C_6H_{10}O_5)_n + nH_2O \xrightarrow{\text{dil. H}_2\text{SO}_4} nC_6H_{12}O_6$$
$$\text{inulin} \qquad\qquad\qquad \text{fructose}$$

Properties and constitution of fructose

Fructose is a crystalline solid more soluble in water than is glucose.
It is also soluble to some extent in alcohol.

(1) Fructose shows the properties of a hydroxy compound. Its
reaction with acetic anhydride shows the presence of five hydroxyl
groups in the molecule.

(2) Fructose shows properties of the carbonyl group. It adds on
hydrogen cyanide. With phenylhydrazine it forms a *phenylosazone*
identical with that obtained from glucose—a fact of importance in
establishing the constitutional formula of fructose.

(3) Fructose reduces Fehling's solution, and ammoniacal silver
oxide. It is not, however, oxidized by sodium hypobromite, indicating
absence of an aldehyde group. Its reducing power is explained
(*a*) by the presence in the molecule of the group —CO·CH(OH)—
a grouping which is known to produce reducing properties in
simpler compounds, or (*b*) by the fact that in presence of alkali
fructose forms an equilibrium mixture with glucose (and one of its

optical isomers, *mannose*), and in both Fehling's solution and ammoniacal silver oxide, an alkali is present.

As with glucose, there is evidence for both open and closed chain formulae; the open chain may be represented thus:

$$CH_2OH$$
$$|$$
$$CO$$
$$|$$
$$*CH \cdot OH$$
$$|$$
$$*CH \cdot OH$$
$$|$$
$$*CH \cdot OH$$
$$|$$
$$CH_2OH$$

asymmetrical carbon atoms again being marked with an asterisk. Fructose is one of the possible optical isomers.

Because of the supposed presence of a carbonyl group, fructose is often referred to as a *ketose*, or more strictly a *keto-hexose*, in contrast to glucose which is an *aldose*.

The optical activity of fructose in the laevo direction is greater than the dextro-rotatory power of glucose. A mixture of these two hexoses in equimolecular proportions is thus laevo-rotatory. Such a mixture is obtained by the hydrolysis of sucrose, which itself shows optical activity in the dextro direction. Hence as the hydrolysis of sucrose proceeds, the dextro-rotatory power gradually falls until a point is reached where the solution is optically inactive. Laevo-rotation then increases to a steady figure which marks completion of the hydrolysis. The process is thus often referred to as an *inversion*† and the mixture of equal amounts of glucose and fructose as *invert sugar*. It is sold under this name for use in confectionery:

$$C_{12}H_{22}O_{11} + H_2O \longrightarrow C_6H_{12}O_6 + C_6H_{12}O_6$$
sucrose glucose fructose
'invert sugar'

DISACCHARIDES

The molecular formula of the disaccharides is $C_{12}H_{22}O_{11}$. The characteristic distinguishing them from other carbohydrates is that

† By means of a polarimeter, the rate of reaction can readily be followed.

on hydrolysis, *one* molecule adds on one molecule of water to give *two* molecules of the formula $C_6H_{12}O_6$, i.e.:

$$C_{12}H_{22}O_{11} + H_2O \longrightarrow 2C_6H_{12}O_6$$

These may be two molecules of the same hexose or one each of two different ones.

The common disaccharides are *maltose, lactose* and *sucrose*.

Maltose is an intermediate product of the alcoholic fermentation of starch. It is hence commonly called *malt sugar*. *Lactose* is present in the milk of mammals and thus commonly called *milk sugar*. *Sucrose*, the best known of the disaccharides, is often called *cane sugar* or *beet sugar* because of its presence in the sugar cane and in sugar beet.

The molecules of all these disaccharides contain asymmetric carbon atoms and optical isomers of each of them (which are themselves structural isomers) are known.

Distinction between disaccharides

(1) *Hydrolysis:*—On hydrolysis, either by enzyme or dilute mineral acid, different hexoses are obtained.

$$\underset{\text{maltose}}{C_{12}H_{22}O_{11}} + H_2O \xrightarrow[\text{or acid}]{\text{maltase}} \underset{\text{glucose}}{2C_6H_{12}O_6}$$

$$\underset{\text{lactose}}{C_{12}H_{22}O_{11}} + H_2O \xrightarrow[\text{or acid}]{\text{lactase}} \underset{\text{glucose}}{C_6H_{12}O_6} + \underset{\text{galactose}}{C_6H_{12}O_6}$$

$$\underset{\text{sucrose}}{C_{12}H_{22}O_{11}} + H_2O \xrightarrow[\text{or acid}]{\text{invertase}} \underset{\text{glucose}}{C_6H_{12}O_6} + \underset{\text{fructose}}{C_6H_{12}O_6}$$

(2) *Reaction with Fehling's solution and also ammoniacal silver oxide:*—While maltose and lactose reduce these reagents, sucrose does not. If, however, sucrose is hydrolysed and the product heated with these reagents, then reduction does occur because of the reducing power of the glucose and fructose formed.

(3) *Reaction with phenylhydrazine:*—Maltose and lactose form a phenylosazone; sucrose does not.

Each of the disaccharide molecules consists of two hexose units joined to each other through an atom of oxygen. Thus adopting the ring structure, the molecule of maltose may be represented as:

The rings are not actually flat, but this diagram does show the spatial arrangements of attached groups. On hydrolysis the molecule splits at the central linkage, to give two identical molecules (of glucose).

The lactose molecule similarly consists of a unit of glucose joined through an oxygen atom to a unit of galactose, these two hexoses being set free on hydrolysis. In the same way, sucrose consists of a unit of glucose joined to one of fructose, the two units being separated as above when a water molecule is added on.

Sucrose, $C_{12}H_{22}O_{11}$

This, the most widely used of all the disaccharides, was isolated as early as 300 A.D., possibly because it is so easy to crystallize. The principal sources are the sugar cane and sugar beet. The original home of the sugar cane was probably in N. E. India. It was introduced into Egypt during the seventh century and by 1000 A.D. there was a flourishing sugar industry in that country. The main sugar plantations of the world are now in the West Indies where the cane was first introduced in the fifteenth century. The sucrose content varies from 10 to 15 per cent.

The presence of sucrose in sugar beet was first discovered by Marggraf in 1749 and extraction from sugar beet was first started in Europe towards the end of that century. Improved strains of the plant have since been raised and a sugar content up to 16 per cent is now obtainable.

The sugar canes are first crushed between rollers to squeeze out the juice which contains the sucrose. The sugar beet, after washing, are sliced and the sucrose extracted by heating the slices in water to a temperature of about 80°C. (The remaining pulp is a valuable cattle food, and contains all the protein matter originally present.)

In both cases, a brownish syrupy solution of sucrose is obtained. Lime is added and carbon dioxide passed through the mixture. The precipitated calcium carbonate carries down many impurities. (The alkaline character of the lime also prevents any hydrolysis of the sucrose to fructose and glucose.) After filtering off the insoluble impurities, a brownish solution is obtained. It is evaporated under reduced pressure and allowed to crystallize. The mother liquor is again subjected to the same treatment.

The raw sugar is brownish in colour. It is dissolved in water and the solution filtered over charcoal beds to decolorize it. The filtrate is then evaporated to recover the sucrose. The darker coloured sucrose obtained during this crystallization is sold as brown sugar.

The final mother liquor obtained from the first purification (known as *molasses*) contains about 50 per cent of fermentable sugar. It is used to make treacle, industrial alcohol, and rum, and also as a cattle food in the making of silage.

The chief use of sucrose is as a food. Since the development of the sugar plantations in the West Indies, it has almost completely displaced honey as a sweetening material. Its use as a fruit preservative and in jam-making depends upon the fact that the different micro-organisms causing decay are unable to develop in a concentrated solution of sucrose. A considerable amount of sucrose is hydrolysed to invert sugar (p. 276) which is used in confectionery.

Sucrose is a colourless crystalline substance readily soluble in water but only slightly soluble in alcohol. It melts at 160°C and solidifies to a transparent mass which is ' barley sugar '. If heated to 200°C, some water is lost to give a brownish substance known as *caramel* which is used to colour gravies, rum, etc.

POLYSACCHARIDES

The chief characteristic of these carbohydrates distinguishing them from monosaccharides and disaccharides is their high molecular weight (in most cases not known with certainty), which means that on hydrolysis, large numbers of molecules are obtained. The empirical formula is $C_6H_{10}O_5$ but the molecular weights of members of the group may vary from 30,000 to 14,000,000. In contrast to the monosaccharides and disaccharides, the polysaccharides are amorphous and insoluble in water and in organic solvents in general. Their purification is thus rendered more difficult. As a general rule they can only be brought into solution by hydrolysis to simpler monosaccharides and disaccharides.

The best known polysaccharides are *starch, inulin,* and *glycogen* (reserve foodstuffs in the plant or animal body), *cellulose* (a structural material of plants), and *chitin* (a nitrogen-containing compound having a similar function in many insects). The gums of many trees are polysaccharides.

Starch

This compound, also called *amylum*, has a molecular formula of $(C_6H_{10}O_5)_n$ where n is large but variable in starches from different sources. It is present in all green plants and constitutes a high percentage content of the seeds of wheat, barley, maize, rice and the tubers of potatoes, which are also its chief commercial sources. It is found in the plant cells as tiny granules, the shape and size of which vary in the different plant species. In hot water, the granules burst, the starch content going into colloidal solution. On adding alcohol the starch is precipitated to form the ' soluble starch ' which forms a colloidal solution again on adding water.

The molecule consists of a large number of glucose units united through oxygen atoms and when starch is hydrolysed, it is broken up into glucose molecules. Maltose is an intermediate product in the hydrolysis. The hydrolysis may be brought about by boiling with a dilute mineral acid. The enzymes diastase and ptyalin (in saliva) hydrolyse starch to maltose only:

$$\text{starch} \xrightarrow{\text{H}_2\text{O}} \text{maltose} \xrightarrow{\text{H}_2\text{O}} \text{glucose}$$
$$(C_6H_{10}O_5)_n \qquad C_{12}H_{22}O_{11} \qquad C_6H_{12}O_6$$

Dextrin (known also as ' British gum ') is a gum-like substance used in the manufacture of adhesives. It is made by heating dry starch to about 200°C. It is thus formed as the shiny substance when starched material is ironed.

The test for starch is the formation of an intense blue-black substance with iodine. This product is unstable at high temperature and if an aqueous mixture of iodine and starch is heated, the blue-black colour disappears to reappear on cooling. Dextrin gives a reddish brown colour with iodine. Starch does not reduce Fehling's solution or ammoniacal silver oxide and does not form an osazone.

Manufacture of starch

The main source of starch in Europe is potatoes; in the U.S.A., it is maize. The raw material is crushed (but not sufficiently to burst the granules) and washed with water through sieves to separate the cell tissues from the granules. Starch settles from the aqueous suspension on standing. The plant tissues are used as cattle food.

Uses of starch

Starch is used in the manufacture of alcohol and glucose. Starch paste is used as an adhesive, in laundry work and in sizing paper.

Cellulose $(C_6H_{10}O_5)_n$

Cellulose is the chief constituent of the cell walls of plants. The woody material of a plant contains up to 50 per cent cellulose and the fibres of cotton, hemp and flax up to 90 per cent. Pure cellulose can be obtained from cotton wool by first 'dewaxing', by washing with alcohol or ether and then treating with hot dilute caustic soda. After final washing with water, white amorphous cellulose is obtained. Good quality filter paper is almost pure cellulose. The commercial sources of cellulose are the fibres of cotton seeds (giving a good quality product), flax, hemp, wood (sawdust and shavings), straw, and grasses (particularly esparto grass).

Cellulose is insoluble in water, alcohol and ether. It will dissolve in concentrated acids, e.g. sulphuric acid, a concentrated solution of zinc chloride, a mixture of caustic soda and carbon disulphide, and in an ammoniacal solution of copper(II) hydroxide.

The commercial value of cellulose depends on the fact that it occurs in the form of fibres rendering it suitable for making into paper or spinning into cloth. In the latter use, in particular, the length and strength of the fibre are extremely important. The fibres of cotton seed hairs may be up to 5 cm long, those of flax and hemp being very much longer, while those in wood are very much shorter. Fibres much longer than those occurring in nature can be made artificially from cellulose.

Uses of Cellulose

(1) *In the manufacture of paper:*—Wood is pulped for the making of paper by boiling it with a solution of caustic soda and calcium bi-sulphite.

If paper is dipped into concentrated sulphuric acid for a few seconds, then washed and dried, the product resembles parchment and is called *parchment paper*. It is much stiffer and is not dis-integrated by water. Like all carbohydrates, cellulose contains alcohol groups and when dipped into the sulphuric acid, cellulose sulphate is formed.

(2) *In the manufacture of cotton cloth:*—The longer fibres of the cotton seed are spun and woven into cloth.

Towards the beginning of the last century, the English chemist and calico printer John Mercer attempted to filter a concentrated solution of caustic soda through cotton cloth. He found that the fibres swelled to the extent that filtration stopped. (Filtering of even dilute solutions of caustic soda through filter paper is slow for the same reason.) After washing and drying, Mercer found that the cloth was much more easily dyed, but that during the treatment with alkali much shrinkage had occurred.

Towards the end of the century, an attempt to prevent shrinkage was made by treatment with caustic soda while the cloth was under tension. After washing, the cloth was found to have a high lustre like silk and to be much stronger than the original. This is, in principle, the method of manufacturing 'mercerized' cotton.

(3) *In the manufacture of glucose:*—Glucose is obtained by the hydrolysis of cellulose by the prolonged action of sulphuric acid. In Germany and Sweden glucose is obtained from the cellulose in sawdust by this process.

(4) *In the manufacture of cellulose esters:*—Owing to the presence of alcohol groups, cellulose esters can be made. The two of greatest commercial importance are the *nitrates* and the *acetates*.

(*a*) The *cellulose nitrates* are obtained from cellulose by the action of sulphuric and nitric acids. The product has been wrongly called *nitrocellulose*, for it is not a nitro compound but a nitrate. If the nitration is carried out to completion the product is *gun cotton* used in making smokeless powder, by treating it with an organic solvent, e.g. acetone, to gelatinize it. When gun cotton is similarly treated with ' nitroglycerine ', *cordite* is obtained.

If the cellulose is nitrated only partially, the product is known as *pyroxylin*. Pyroxylin is a very valuable raw material, and dissolves readily in a mixture of alcohol and ether to give *collodion*. If camphor is added to the mixture, *celluloid* is the product. Motion picture films were made by adding camphor to a solution of pyroxylin in methyl alcohol and acetone. An *artificial leather* has been made by coating cotton fabric with a mixture of pyroxylin and castor oil. *Lacquers* used in the motor industry are made from cellulose nitrates by dissolving them in a mixture of different solvents. Pigments are then added.

(*b*) The *cellulose acetates* are of even greater importance than the nitrates. They are made by the action of acetic acid and acetic anhydride, in the presence of sulphuric acid, on cellulose. The cellulose dissolves to form a syrupy solution from which the acetates are precipitated by addition of water. They are used in the making of lacquers, aeroplane dope and artificial silk or rayon (see below).

(5) *Rayon:*—Synthetic fibres from cellulose are called rayon. There are three processes in use: the *viscose*, the *acetate* and the *cuprammonium* process. The principle in all three is the same, viz. a solution of cellulose is made and then forced through tiny jets, called spinnerets. The cellulose is precipitated from the stream of solution either by injecting it into some solution, or by allowing the solvent to evaporate in air—the latter process being a direct imitation of the spider spinning its web. In all cases, the films are cylindrical and possess smooth surfaces, giving the high lustre characteristic of silk.

(*a*) In the *viscose process*, cellulose is dissolved in a mixture of caustic soda and carbon disulphide. The orange solution so obtained is squirted through spinnerets into a bath of sodium bisulphite which hydrolyses the product and regenerates the cellulose. The filaments are twisted and may be woven with wool, cotton or linen as required. Eighty per cent of artificial silk is made by this process.

Cellophane is made by squirting the solution through slots. The product is treated with glycerine to make it flexible. Sausage skins are made in this way.

(*b*) In the *cuprammonium process* the cellulose is dissolved in an ammoniacal solution of copper(II) hydroxide*. As above the filaments are coagulated by sodium bisulphite.

(*c*) In the *acetate process*, the acetate of cellulose is dissolved in the volatile solvent acetone. The solution is forced through the spinnerets into warm air when the solvent evaporates and no treatment with a coagulating bath is needed.

PROTEINS

Proteins derive their name† from their great importance in animal tissues; unlike the carbohydrates they contain nitrogen as well as carbon, hydrogen and oxygen. There are two main classes of proteins—the first, the *fibrous* proteins, are insoluble in water, and are analogous to cellulose in plants, acting as structural material for animals. Examples, are *fibroin* (in silk), *collagen*, and *keratin* (in hair, horns and feathers). The second class are the *globular* proteins, having some degree of solubility in water or in acids or bases. Egg *albumen*, *casein* (in milk), and the proteins of the blood plasma are examples of this class.

Proteins are polymeric, with chain-like structures and molecular weights which can be thousands or even millions. A solution of a globular protein is therefore not a true solution but a colloidal one; and if the solution is boiled, or treated with acid or base, an irreversible change, called *denaturation*, is produced, leading to loss of solubility and change in chemical and biological character. The change when egg-white coagulates on boiling is a familiar example of this. More prolonged boiling of a protein with mineral acid produces a mixture of *amino-acids*. These have the general formula

$$\begin{array}{c} R \\ \diagdown \\ H_2N \diagup \end{array} CH\cdot C \begin{array}{c} \diagup O \\ \diagdown \\ OH \end{array}$$

The simplest amino-acid, with $R = H$, is *amino-acetic acid*, $H_2N\cdot CH_2\cdot COOH$, first obtained by Braconnot in 1820 by the hydrolysis of gelatine; the sweet taste caused it to be called *glycine*. Other amino acids include *alanine* $(R = CH_3)$ and *phenylalanine* $(R = C_6H_5\cdot CH_2)$.

* Known as Schweitzer's reagent, after the discoverer.
† Greek *proteios*, primary.

The formation of amino acids when a protein is hydrolysed is the key to the structure of these substances, for it shows that they all contain the *polypeptide* chain; they are in fact polyamides.

Each part of the chain like that enclosed within the dotted brackets can be seen to yield an amino-acid on hydrolysis, when the chain is broken up. The chain unit within the full brackets is called the *peptide link*, and is the characteristic group of proteins. There are some thirty known amino-acids originating from proteins, the R group showing wide variations. Some of these can be synthesized by the human body, but nine are essential to man—the *essential amino-acids*. In the laboratory, synthesis of polypeptide chains is possible, e.g. by condensation of two or more amino-acid molecules; but even a relatively simple protein like insulin (minimum molecular weight *c*. 6000) contains two polypeptide chains, one of which contains 21 amino-acid residues and the other 30. The long chains of proteins are often helical in form, for hydrogen bonding between the —NH and $>$C$=$O groups along the chain bends it into this kind of configuration (p. 294).

Amino-acids—glycine

Since amino-acetic acid has an acidic group, —COOH, and a basic group, —NH$_2$, it is amphoteric and can behave as an acid or a base, thus:

$$H_2N \cdot CH_2 \cdot COO^- \underset{\xleftarrow{\text{alkali}}}{\rightleftharpoons} \overset{+}{H_3}N \cdot CH_2 \cdot COO^- \underset{\xrightarrow{\text{acid}}}{\rightleftharpoons} \overset{+}{H_3}N \cdot CH_2 \cdot COOH$$

amino-acetate   a dipolar   an ammonium
anion   ion   cation

At some intermediate point, the ' dipolar ion ' shown is formed; here the hydrogen ion from the carboxyl group is not transferred to the solvent (as happens with an ordinary acid, p. 291) but is transferred internally to the —NH$_2$ group. This intermediate form is also called an *inner salt* or ' zwitterion ' (meaning ' between-ion '). The value of the pH at this intermediate point is called the *iso-electric point*, because at it the dipolar ion has no net charge.

Glycine is a white solid substance, soluble in water. It is prepared by hydrolysis of a protein (e.g. gelatine) or readily by the reaction of concentrated ammonia solution on monochloracetic acid:

$$ClCH_2 \cdot COOH + 2NH_3 \longrightarrow H_2N \cdot CH_2 \cdot COOH + NH_4Cl$$

The properties are, in general, those expected for $-NH_2$ and $-COOH$ groups, e.g. for $-NH_2$:

(1) Adds on acids to form salts, e.g.

$$H_2N \cdot CH_2 \cdot COOH + HCl \longrightarrow [\overset{+}{H_3N} \cdot CH_2 \cdot COOH]Cl^-$$

(2) Reacts with nitrous acid to give glycollic acid:

$$H_2N \cdot CH_2 \cdot COOH + HNO_2 \longrightarrow \underset{\substack{\text{glycollic} \\ \text{acid}}}{CH_2OH \cdot COOH} + N_2 + H_2O$$

(3) Reacts with acetyl chloride (or acetic anhydride) to give an acetyl derivative:

$$H_2N \cdot CH_2 \cdot COOH + CH_3 \cdot COCl$$
$$\longrightarrow CH_3 \cdot CO \cdot NH \cdot CH_2 \cdot COOH + HCl$$

With benzoyl chloride (p. 250) the product is hippuric acid (p. 249).

Like simple carboxylic acids, glycine forms esters in presence of an inorganic acid as catalyst (note the effect of the latter on the product) e.g.

$$H_2N \cdot CH_2 \cdot COOH + C_2H_5OH \xrightarrow{\text{HCl}} [\overset{+}{H_3N} \cdot CH_2 \cdot COOC_2H_5]Cl^-$$
$$+ H_2O$$

Test for proteins:—The biuret test (p. 174) is given by any compound containing a peptide link.

26

THE STRUCTURE AND REACTIVITY OF ORGANIC COMPOUNDS

In earlier chapters, organic reactions have been classified as either *free radical* or *ionic*. In the former, covalent bonds are broken so that one electron from the shared pair goes to each of the two fragments formed (homolysis), while in the ionic reaction both electrons go to one fragment (heterolysis). Reagents involved in heterolytic reactions are classified either as *nucleophiles* (which provide an electron pair for the formation of a new bond) or *electrophiles* (which accept a share in the electron pair, so forming a new bond). So far, however, we have given relatively little consideration to the nature of the covalent bond itself. In this chapter, therefore, some bond types are discussed more fully, and some further concepts relating structural properties and reactivity of organic compounds are developed.

The idea of an electron as a particle revolving in an orbit is not in itself useful in visualizing the covalent bond as a shared electron pair. In 1924 it was suggested, on theoretical grounds, that an electron might have some of the characteristics of a wave, e.g. wavelength and amplitude. This idea received support when it was found that electrons could undergo diffraction; indeed, electron diffraction, like X-ray diffraction, can be used to investigate the structure of matter, and electron ' waves ' are made use of in the electron microscope as are light waves in an ordinary microscope.

The application of wave theory to the electrons in atoms and molecules is called *wave mechanics*. There are two important results of this; first, that the position of an electron in space cannot be precisely located, i.e. it cannot be pin-pointed; but, second, that the ' amplitude ' of the electron wave (usually given the symbol ψ^2) can be used to indicate the *probability* of finding the electron at a given point. ψ (psi) is usually called the *wave function*. In wave mechanics, it is possible to write down a wave-equation, involving ψ and the energy of the electron, and to this equation there can be several solutions, i.e. ψ can have several values, to each of which there is a corresponding value of the energy. So now, instead of the Bohr idea of electrons in *orbits* of different energy, we speak of electrons having different ψ values, i.e. ' *orbitals* ' of different energy. There are then several possible orbitals each with its own energy level—and each orbital can ' accommodate ' two electrons of opposing spin.   Hence we can speak of a vacant orbital (containing no electrons), a half-filled orbital (containing one electron) and a filled orbital (containing two electrons).

In the simplest atom, hydrogen, the orbital, called a $1s$ orbital, is spherically symmetrical; ψ (and, therefore, ψ^2 also) diminishes rapidly with increasing distance from the nucleus.   Hence the best picture we can get of the hydrogen atom is:

The 'density' of the dots at any point represents the probability of finding the one electron there—in other words, the dots represent the electron density in a spherical 'cloud' around the nucleus. When two hydrogen atoms come together to form the hydrogen molecule, the two clouds overlap, and calculation shows that we now have a charge cloud concentrated between the two nuclei, thus:

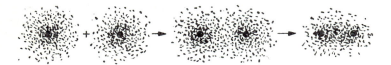

It is this overlap and consequent concentration of charge between the nuclei which gives the H—H bond its great strength (much greater than if it were an ionic bond $H^+ \cdots : H^-$ involving only electrostatic attraction).   Putting this in another way, we can say that when two orbitals of hydrogen atoms overlap, they unite to form a new *molecular orbital*, filled by two electrons of opposite spin.

Now in the atoms of carbon, nitrogen and oxygen, the outer (valency) electrons fill up the L 'shell' with $n = 2$ (p. xix) and here there are four orbitals of importance—one spherically symmetrical, called the $2s$ orbital and three others called $2p$ orbitals; each of these $2p$ orbitals has a 'double-pear' shape (as shown shaded on the left of *Figure 26.2*).   When any of these atoms form covalent bonds, some or all of these $2s$ and $2p$ orbitals can be mixed or *hybridized* to form new, *hybrid* orbitals, each of which has, effectively, a 'single-pear' shape, well suited for overlap with

the orbital of another atom. Thus, in the methane molecule, the
the four orbitals of the carbon atom, 2s, 2p, 2p, 2p can be mixed
to form four new hybrid orbitals (called sp^3 orbitals because they
are formed from one s and three p orbitals). These new orbitals
appear as in *Figure 26.1*, i.e. they form four 'lobes' which project
tetrahedrally. The four valency electrons of carbon go one into
each orbital, and overlap of these singly-occupied orbitals with the
four spherically-symmetrical 1s orbitals of the hydrogen atoms gives
the four covalent bonds of the methane molecule.

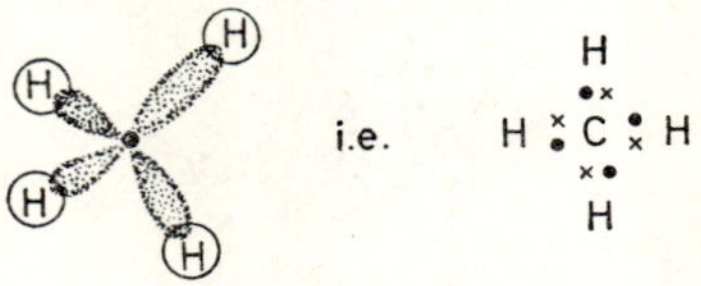

Figure 26.1.

In ethylene the situation is rather different. Here, each carbon
atom has one 2s and *two* 2p orbitals hybridized to form three sp^2
hybrid orbitals, which are trigonal planar (shown unshaded in
Figure 26.2). Two of these orbitals are used to form the two C—H
bonds on each carbon, and the third hybrid orbital overlaps the
similar orbital of the other carbon atom to give a C—C covalent
bond. The unhybridized 2p orbitals (shown shaded) overlap
'sideways-on' and we obtain a molecule as depicted on the right
of *Figure 26.2*—the overlap requiring the molecule to be *planar*.

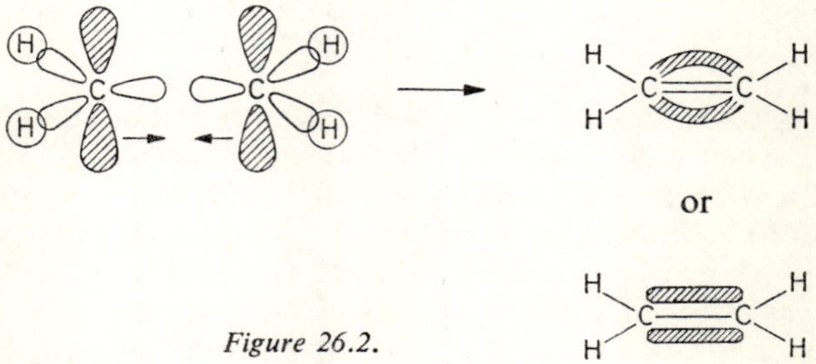

Figure 26.2.

There are *two* bonds between the carbon atoms; one, the direct
C—C bond, called a σ-*bond*, is formed by end-on overlap, the other,
formed by side-ways overlap and consisting of two sausage-like
electron clouds one above the other below the C—C line, is
called a π-*orbital* and the bond a π-*bond*. Because of this double bond,
the C—C bond in ethylene is shorter and mechanically stronger,
but more reactive, than in ethane.

In acetylene, the ' skeleton ' H—C—C—H is made up in a similar way to that of ethylene, but each carbon atom has two *sp* hybrid orbitals, one to overlap with hydrogen, the other to overlap with the other carbon atom, i.e.

There remain two ' double-pear ' *p* orbitals on each carbon, each singly occupied, and these are at right angles:

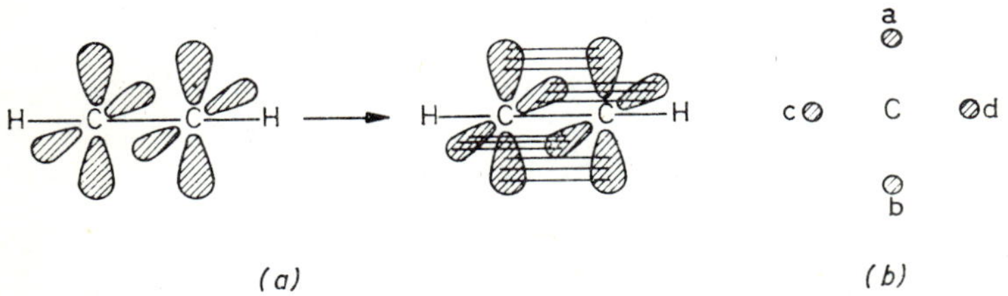

(a)   *(b)*

Figure 26.3

Hence as *Figure 26.3 (a)* shows, *two* π-bonds can be formed, each made up of two sausage-like clouds. An ' end-on ' view of the acetylene molecule is shown in *Figure 26.3 (b)*, a and b being one π-bond, c and d the other. Note that good overlap of the orbitals is achieved in a linear molecule here. The high concentration of electrons around the mid-point of the molecule is probably the reason for the ' acidity ' of acetylene.

We may now consider the valency behaviour of three other atoms which are important in organic chemistry, viz. nitrogen, oxygen and chlorine. The three-dimensional structure of the simplest compounds of these elements, with hydrogen, can be represented thus:

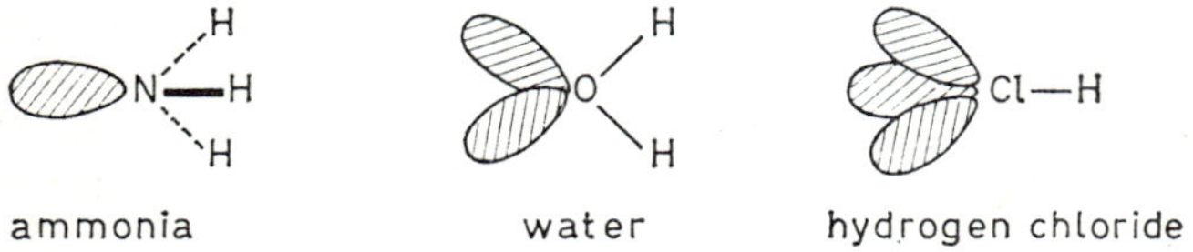

ammonia   water   hydrogen chloride

Here, each shaded ' pear ' represents a *doubly-occupied sp³* hybrid orbital, i.e. containing *two* electrons. The N—H, O—H and Cl—H bonds are covalent and are formed in a similar way to those in methane, so here we have (' lone pair ' orbitals + bonding orbitals) = 4 in each case. The hydrogen atoms in these compounds can be replaced by alkyl groups, with little alteration to structure, i.e.

1H replaced—primary amine   1H replaced—alcohol   alkyl halide
2H replaced—secondary amine   2H replaced—ether
3H replaced—tertiary amine

Replacements of H by R in this way have some effect (see below, p. 298); but the essential function of the doubly-filled orbitals —the *lone pairs*—remains the same. First, these lone pairs can be shared with another atom; the effect is strongest with nitrogen with one lone pair, weaker with oxygen (two lone pairs) and weaker still with chlorine; usually only one pair can be shared. Consider nitrogen; if some other atom has a *vacant* orbital, i.e. containing no electrons, then overlap between this and the filled orbital of nitrogen will produce a covalent bond just as if two half-filled orbitals overlapped. An atom with a vacant orbital is boron, and so with boron trifluoride and trimethylamine as examples, we have:

i.e. $F_3B \leftarrow N(CH_3)_3$

or $F_3\bar{B}\!-\!\overset{+}{N}(CH_3)_3$

i.e. addition takes place. The covalent bond so formed is called the *co-ordinate** or *dative* bond; it does not differ from the ordinary covalent bond except that both electrons come from the nitrogen (the donor atom) and none from the boron (the acceptor atom). The co-ordinate link is represented as shown, either by the arrow as in B←N, or by charges as in $\bar{B}\!-\!\overset{+}{N}$; these positive and negative charges are purely formal and merely indicate that the nitrogen has lost, and the boron has gained, a half-share in two electrons.

---

* This name arises from the belief that such bonds also occur in co-ordination compounds, in complex ions such as $[Co(NH_3)_6]^{3+}$, $[Cr(H_2O)_6]^{3+}$.

There is another important acceptor of electrons—the proton, H^+. Ammonia, for example, can add on to the proton thus:

$$
\begin{array}{c}
H \\
| \\
H—\overset{\displaystyle \cdot}{N}\! : \ +\ H^+ \\
| \\
H
\end{array}
\quad\longrightarrow\quad
\left[
\begin{array}{c}
H \\
| \\
H—N—H \\
| \\
H
\end{array}
\right]^{+}
$$

ammonium ion

The ammonium ion formed then has four identical N—H bonds, and the sign N→H is not used here. An amine can add on a proton in the same way, e.g.:

$$
\begin{array}{c}
CH_3 \\
| \\
H—\overset{\displaystyle \cdot}{N}\! : \ +\ H^+ \\
| \\
H
\end{array}
\quad\longrightarrow\quad
\left[
\begin{array}{c}
CH_3 \\
| \\
H—N—H \\
| \\
H
\end{array}
\right]^{+}
$$

methylamine   methylammonium ion

Again, water forms the *hydroxonium* or *oxonium* ion:

$$
\begin{array}{c}
H \\
| \\
H—\underset{\displaystyle \cdot\cdot}{O}\! : \ +\ H^+
\end{array}
\quad\longrightarrow\quad
\left[
\begin{array}{c}
H \\
| \\
H—O\! : \\
| \\
H
\end{array}
\right]^{+}
$$

This ion, H_3O^+, is the ion existing in aqueous solutions of acids, and is what is meant when we speak of ' hydrogen ions in solution '. Alcohols can also take up a proton, but have less tendency to do so than water; ethers have less tendency still.

The proton, H^+, is never found free under ordinary conditions; it has to be *abstracted* from a compound containing a covalently-bound hydrogen atom, say HX, by heterolytic fission. Hence the process in water is:

$$H_2O + HX \rightleftharpoons H_3O^+ + X^-$$

e.g.
$$H_2O + H : \overset{\cdot\cdot}{\underset{\cdot\cdot}{Cl}} : \ \rightleftharpoons H_3O^+ + : \overset{\cdot\cdot}{\underset{\cdot\cdot}{Cl}} : ^-$$

$$H_2O + HOOC\cdot CH_3 \rightleftharpoons H_3O^+ + CH_3\cdot COO^-$$
acetic acid

If the equilibrium is well over to the right, HX is a strong acid; if not, it is a weak acid. Since, however, alcohol and ether are poorer donor molecules than water, hydrogen chloride in either of these solvents is not as strong an acid as it is in water.   If ammonia is present in water to which hydrogen chloride is added, the reaction:

$$NH_3 + HCl \rightleftharpoons NH_4^+ + Cl^-$$

occurs in preference to the reaction:

$$H_2O + HCl \rightleftharpoons H_3O^+ + Cl^-$$

in water only, since ammonia is a better donor molecule than water. This power of sharing an electron pair with a proton is a characteristic of a *base*; hence ammonia is a stronger base than is water*. The position of equilibrium in the reaction:

$$R_3N + HX \rightleftharpoons R_3NH^+ + X^-$$

will depend upon the donor power of R_3N.   For example, trimethylamine is a stronger base than ammonia, i.e. abstracts protons better than ammonia, because of the methyl groups on it (p. 298). The position of equilibrium also depends upon the attraction of X for the hydrogen.   Thus the proton is more easily abstracted from $CCl_3 \cdot COO$—H, trichloracetic acid, than from $CH_3 \cdot COO$—H, acetic acid, because of the effect of the chlorine atoms in the 'X' part of the molecule (see later, p. 296).

The presence of lone pairs in the molecules that we have been considering is important in other properties.   The doubly-filled orbitals projecting from one side of the molecule in ammonia or water give the molecule *polarity*; this is discussed later, but here we may note that the ability of alcohol to form alcoholates with, for example, calcium salts (p. 60) is due to attraction of the negative end of the molecule to the calcium ion, i.e.:

R = H or an alkyl group

Intermolecular forces

We may now consider why organic compounds of similar nature differ in their physical properties.   For example, methane, CH_4, is a gas at ordinary temperature, while chloroform, $CHCl_3$, is a liquid, though in both we have carbon attached covalently to four other atoms.   The differences in properties arise through differences in the forces between the molecules (the higher the boiling point the greater must be the force) and the origin of such forces must now be considered.

* These statements refer to the Brønsted-Lowry definition of acids and bases.

Some molecules possess a *dipole moment*, i.e. behave as if they possessed a positive charge at one end and a negative charge at the other, electrically analogous to a magnet*. The origin of this dipole moment is considered later (p. 295); here we can note that the negative end of one dipole can attract the positive end of the other; hence *dipole–dipole attraction* is one kind of intermolecular force. As an example, methane, CH_4, has no dipole moment, but methyl chloride, CH_3Cl, has a considerable one; and methane has a b.p. of $-162°C$, whereas methyl chloride boils at $-24°C$. Nitromethane $CH_3 \cdot NO_2$ has a very large dipole moment and this accounts for its high boiling point of $101·5°C$.

Now some molecules, though not possessing a permanent dipole moment, can be *polarized*, i.e. opposite electrical charges can be produced on them by some neighbouring charge such as an ion or another molecule having a dipole moment; and then mutual attraction can occur here also. But further than this, any two molecules which are near together can mutually polarize one another, even though neither of them possesses a dipole moment originally. The extent to which this polarization can occur depends very much on *size*—bigger molecules being polarized more easily than small ones. Hence in the homologous series of the alkanes, C_nH_{2n+2}, as n increases the size increases, and so do the intermolecular forces, with consequent rise in boiling point (p. 9). The formation of branched chains gives more compact, less polarizable molecules with lower intermolecular forces and so lower boiling points are found. In the methyl halides, there is firstly dipole-dipole attraction which raises the boiling point above that of the corresponding alkane (methane) as we have seen; but the boiling point also rises sharply from methyl chloride through the bromide to the iodide because, as the size of the halogen atom increases, the whole molecule increases in size, and so mutual polarization forces increase. These are superimposed on the dipole attraction which does not change very much down this halide series. In general we find that bromo- or iodo- compounds are less volatile than chloro- compounds (and indeed may often be solids). We can see here also that the extent to which a molecule can be polarized— its *polarizability*—will be important in its reactions, for the polarization can be effected by the attacking reagent molecule.

Compounds which contain oxygen or nitrogen united with hydrogen (e.g. containing the —OH group or —NH_2 group) seem to possess much stronger intermolecular forces than all other compounds of similar size; and these intermolecular forces

* But not to be confused with a magnetic moment!

give rise to other phenomena besides an increase of boiling point. Thus, we often find *association*, e.g. acetic acid in benzene solution is associated as double molecules $(CH_3 \cdot COOH)_2$; *good solvent power*, especially for substances of similar nature, e.g. water, liquid ammonia, alcohol, acetic acid are all good solvents and often completely miscible, (acetic acid and water, alcohol and water where mixing releases heat, liquid ammonia and amines). A good example of the specific nature of these forces is given by the two isomers ethyl alcohol and dimethyl ether, both having the formula C_2H_6O (p. 76) and with similarly sized molecules. Ethyl alcohol, C_2H_5OH, has b.p. 78°C, and dimethyl ether $-24°C$ — a very large difference. Ethyl alcohol has good solvent power especially for other hydroxylic compounds. Dimethyl ether is also a solvent, but is much less effective as a solvent for hydroxylic compounds. Moreover, ethyl alcohol is soluble in water, being completely miscible with it, whereas the solubility of dimethyl ether is much less.

It seems therefore that there is some strong intermolecular force dependent upon the presence of hydrogen attached to oxygen or nitrogen. This force is called the *hydrogen bond*. We have seen that in ammonia and amines, the nitrogen atom has a ' lone pair ' of electrons in a position such that one end of the molecule carries a concentration of negative charge; there are two such pairs on the oxygen atom of water, alcohols, ethers and all compounds of the general formula $\overset{A}{}\overset{B}{} \cdot \underset{\cdot}{O} \cdot$ It is believed that ' hydrogen bonding ' arises because of the attraction between the small positively charged hydrogen* atom of one molecule and the negative charge due to the lone pair on another, e.g. with water or an alcohol:

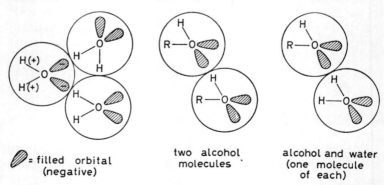

$\mathscr{D}$ = filled orbital
(negative)

two alcohol molecules

alcohol and water (one molecule of each)

*Since there is an excess of negative charge at the lone pair end of the molecule, there must be corresponding excess positive charge at the other and this is located on the hydrogen atom or atoms.

With ethers, $\begin{smallmatrix} R \\ \diagdown \\ \diagup \\ R \end{smallmatrix} O$, there are no oxygen-attached hydrogen atoms, and the positive charge is distributed over the alkyl groups, so that there is much smaller attraction between neighbouring molecules. Hence ether molecules have only weak interaction with one another, and moderately weak interaction with water, whereas the availability of both oxygen-attached hydrogen and oxygen lone pairs in alcohols, and in carboxylic acids like acetic acid makes such substances miscible. (In acetaldehyde, $\begin{smallmatrix} CH_3 \\ \diagdown \\ \diagup \\ H \end{smallmatrix} C{=}O$, the hydrogen again is not attached to oxygen directly and the inter-molecular forces are not strong.)

Polarity in σ-bonds. The inductive effect

As already noted, some molecules are known to possess a **dipole moment**. Hence the molecule behaves as if it had some positive charge at one end and an equal negative charge at the other; the magnitude of the dipole moment increases with the size of the charges and with the distance between them. A dipole moment can arise in several ways, but is usually due to lone pairs of electrons in the molecule or to unequal sharing of the electron pairs in covalent bonds.

Consider two atoms A and B. If A is a very electropositive element (i.e. having a great tendency to lose one or more electrons, e.g. $Na \rightarrow Na^+$, $Ca \rightarrow Ca^{2+}$) and B a strongly electronegative element (i.e. having a tendency to gain one or more electrons, e.g. $Cl \rightarrow Cl^-$) then the bond between A and B is likely to be ionic (p. xx), i.e. $A + B \rightarrow A^+B^-$, and the 'ion pair' A^+B^- will have a very large dipole moment. If, however, A is carbon and B hydrogen, then neither of these elements is particularly electropositive or electro-negative, and hence the C—H bond will be covalent, and the electron pair will be shared more or less equally. (Actually, only in molecules where A equals B, e.g. Cl—Cl, H—H, can we be sure that sharing is exactly equal.) The dipole moment can sometimes indicate the degree of inequality of sharing, or the *polarity* of the bond, provided that there is no other effect due to unshared electron pairs. The water molecule has a fairly large dipole moment, but this does not necessarily mean that the sharing of electrons is unequal in the two O—H bonds; it is the two unshared pairs on the oxygen atom (see p. 289) which give the molecule polarity.

The displacement of the shared electrons in one direction or another is sometimes called the *inductive effect*. This term applies only to electrons in σ-bonds (p. 288); π-bonds require different considerations and are discussed later, p. 300. The sign of the inductive effect is indicated thus:

	standard
C⟶O	C—H
oxygen more electronegative than carbon, i.e. electron-attracting	assumed no inductive effect

The inductive effect can be transmitted along a chain of atoms. The presence of a polar bond *induces* some polarization in adjacent bonds, but the effect falls off rapidly with increasing distance:

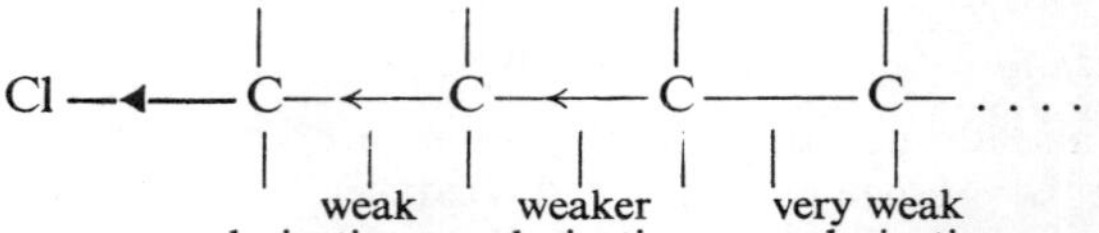

$$\text{Cl} \longrightarrow \overset{|}{\underset{|}{C}} \longleftarrow \overset{|}{\underset{|}{C}} \longleftarrow \overset{|}{\underset{|}{C}} \longrightarrow \overset{|}{\underset{|}{C}} \ldots$$

weak polarization   weaker polarization   very weak polarization

This type of induced polarization in σ-bonds has a noteworthy effect on the strengths of aliphatic carboxylic acids.

Consider the structures of acetic acid and its chlorination products:

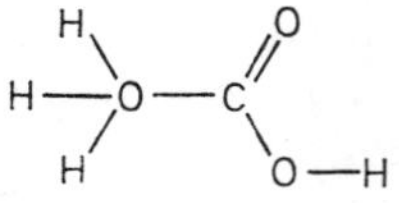

pk$_a$ = 4·76 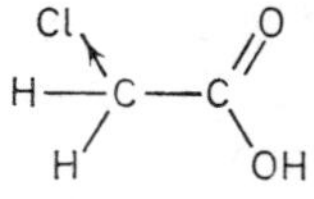   pk$_a$ = 2·86

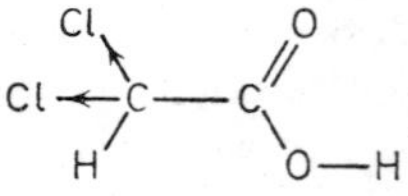

pk$_a$ = 1·29   and   pk$_a$ = 0·65

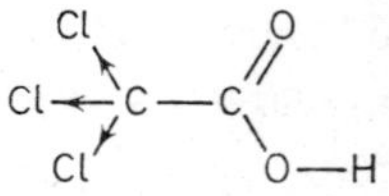

The observed pK$_a$ values* indicate that as the CH_3— group of acetic acid is increasingly substituted by chlorine atoms, the strength of the acid increases markedly. Clearly, the presence of polar

* Acid strengths can be expressed in terms of K$_a$ (the familiar dissociation constant of the acid) or as pK$_a$ values (pK$_a$ = $-\log_{10}$K$_a$; the *smaller* the value of pK$_a$ the *stronger* the acid).

C—Cl bonds increases the tendency of the O—H bond to undergo dissociation (heterolysis). Taking chloracetic acid as an example, the effect may be represented as follows:

$$\begin{array}{c}
Cl\\
\backslash\\
H-C\leftarrow C\\
/\qquad\\
H\qquad O\leftarrow H
\end{array}$$

The powerfully electron-attracting chlorine atom affects the sharing of the electrons in all the σ-bonds between itself and the acidic hydrogen atom, and although the polarization becomes weaker as the number of bonds increases, nevertheless there is still a sufficient effect on the O—H bond to make the separation of a proton significantly easier than in acetic acid*. With two C—Cl bonds the effect will obviously be stronger, and still more so with three C—Cl bonds. Thus, trichloracetic acid is an acid of considerable strength.

It is interesting to note that β-chloropropionic acid, Cl—CH₂—CH₂—CO₂H (where the polar C—Cl bond is further removed from the O—H bond) is a weaker acid than chloracetic acid, but is a stronger acid than either propionic or acetic acids. Formic acid, HCO_2H, is a stronger acid than acetic acid, and acetic acid is stronger than propionic acid. Thus, in propionic acid, it is inferred that the terminal methyl has a small electron-repelling effect:

$$\begin{array}{c}
CH_3\\
\nwarrow\\
C-C\\
/|\qquad\\
H\ H\qquad O-H
\end{array}$$

making the dissociation of the O—H bond slightly more difficult.

---

* Strengths of substituted acetic acids can also be discussed in terms of the effect of substituents in the methyl group on the stability of the anion formed, e.g. a polar C—Cl link will stabilize the anion, thus;

$$Cl\leftarrow CH_2 \leftarrow C\begin{array}{c}O\\ \diagdown\\ O-\end{array}$$

Electron repulsion by alkyl groups also appears to be important in aliphatic amines; if ammonia is compared with the amines, e.g.

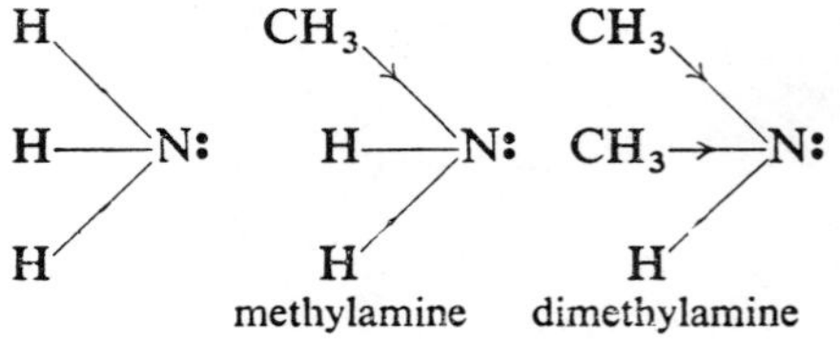

methylamine   dimethylamine

then the increasing repulsion effect as we go from ammonia to dimethylamine increases the proton-attracting power of the nitrogen atom (p. 291) and so the base strength increases. However, trimethylamine is a *weaker* base than dimethylamine (in aqueous solution) and it is clear, therefore, that the inductive effect of alkyl groups is not the *only* factor involved here.

Displacement reactions of halides

The presence of polar C-halogen bonds in alkyl halides has an important effect on reactivity; notably, the halogen atom can be displaced (as an anion) by certain nucleophiles, e.g. OH^-, CN^-, OR^- etc. (p. 48). Reactions of this kind have been studied extensively and it has become clear that more than one mechanism is possible for the displacement.

Primary halides, e.g. propyl halides, react by the kind of mechanism indicated in outline in Chapter 5. In more detail, the steps involved are shown diagrammatically below, using OH^- as the nucleophile:

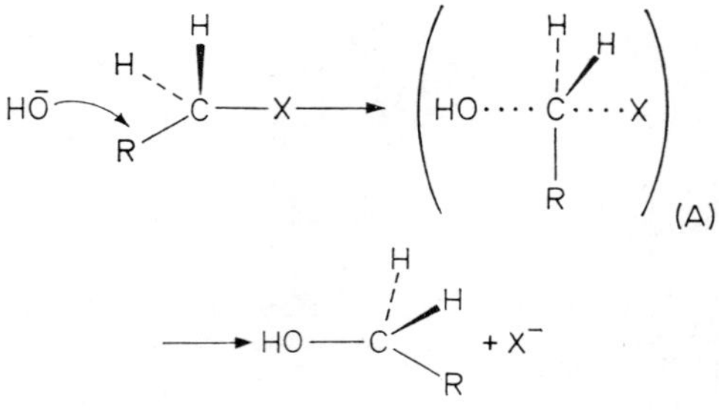

In this scheme, (A) represents a ' transition state ' which has only the most transitory existence (the transition state really represents the energy ' hump ' over which the reactants must pass).

The nucleophile attacks at the back side of the carbon atom carrying the halogen atom. As the nucleophile approaches the carbon atom, the C—X link begins to stretch, so that as the new bond is formed, the original C—X bond gradually breaks. In the transition state, both nucleophile and halogen are attached, partially, to the carbon atom. It should be noted that the other groups (H, H and R) attached to this carbon atom move, first away from one another [into the arrangement in (A)], and then subsequently close up again into the final arrangement. (The change is often likened to an umbrella blowing inside-out in a high wind.)

Part of the evidence for this kind of mechanism comes from studies of the *kinetics* of displacement reactions. With saturated primary halides, the reaction rates depend on the concentrations of *both* alkyl halide and nucleophile. On the other hand, with tertiary halides, e.g. *tert*-butyl compounds, the reaction rate has been found under similar conditions to depend only on the concentration of the halide and to be independent of the concentration of nucleophile, indicating that the nucleophile is not involved in the rate-determining step of the reaction. The mechanism is then thought to be as follows:

$$(CH_3)_3C \longrightarrow X \xrightarrow[\text{step}]{\text{slow}} (CH_3)_3C^+ + X^- \text{ (ions solvated)}$$
$$(+ \text{ solvent})$$

$$(CH_3)_3C^+ + OH^- \xrightarrow[\text{step}]{\text{fast}} (CH_3)_3COH$$

As can be seen, the slow (rate-determining) step is a reaction of the halide only. Its rate *will* depend on the halide concentration, but not on the concentration of hydroxide ions. The first step is, in fact, ionic fission (heterolysis) of the carbon-halogen bond, leading to a *carbonium* ion. This will only take place in certain polar types of solvent (e.g. those containing water, alcohols etc.) which can stabilize the intermediate ions by solvation processes; in less polar solvents, the mechanism of the reaction is different and is, in fact, analogous to the mechanism of the primary halide reaction.

Studies of this kind on alkyl halide reactions suggest that tertiary carbonium ions, R_3C^+, are more favoured* as intermediates than primary carbonium ions RCH_2^+. One reason why this may be so is that alkyl groups can act as electron-repelling groups (p. 298); thus the three methyl groups in a *tert*-butyl halide may help to

---

* 'More favoured' here means energetically more favoured, i.e. of lower energy, and therefore requiring less energy for their formation.

compensate the central carbon atom for its loss of electrons to the halogen atom:

$$CH_3 \diagdown$$
$$CH_3 \longrightarrow C \longrightarrow X \longrightarrow CH_3 \longrightarrow \overset{+}{C} \qquad + X^-$$
$$CH_3 \diagup$$

In a primary halide, there is only one alkyl group to act in this way; consequently, primary halides do not normally react by initial ionization—the carbonium ion mechanism is avoided altogether.

It should be noted that detailed studies of the elimination reactions of alkyl halides (leading to alkenes) also indicate that more than one mechanism is possible, and that one of these mechanisms involves a rate-determining ionization step, occurring most easily with tertiary halides.

π-Bond reactions of alkenes

The remaining sections of this chapter are concerned with systems containing π-bonds, the simplest of which is ethylene. The structure of ethylene has already been discussed in some detail (p. 298).

It has been seen that ethylene behaves as a nucleophile in addition reactions (Chapters 2 and 8); it is able to react in this way because its π-electrons are accessible, and therefore available for sharing with another nucleus (i.e. for bond formation). Thus, in the reaction with bromine, the close approach of a bromine molecule to the π-electron cloud of the ethylene molecule will induce polarization of the Br—Br bond:

The positively charged intermediate may be described in several ways, e.g.

Remembering that the bromine atom is bulky, it is not surprising that when addition is completed by reaction with a bromide anion, this anion approaches the system from the opposite side so that the overall process is a *trans* addition:

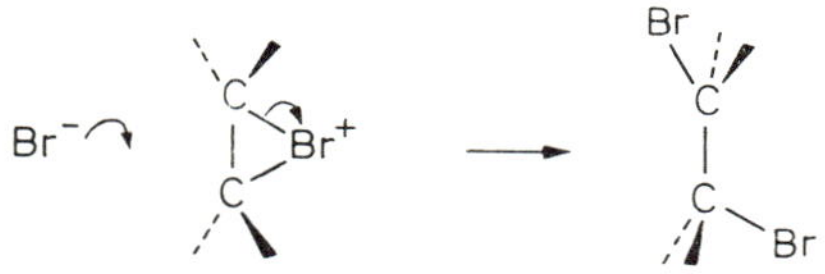

When the alkene is ethylene, the *trans* nature of the addition is not revealed in the product because the $BrCH_2-$ groups can rotate about the C—C bond. But in other, more complicated examples, clear evidence has been obtained that the two bromine atoms become attached at opposite sides of the original double bond.

The other 'ionic' addition reactions of alkenes discussed in Chapter 2 are also π-bond reactions. When the reaction involves the addition of two dissimilar fragments to the double bond (as, for example, in additions of HCl, H_2SO_4 or of bromine in aqueous solution) then, with unsymmetrical alkenes like propylene, two isomeric products are theoretically possible:

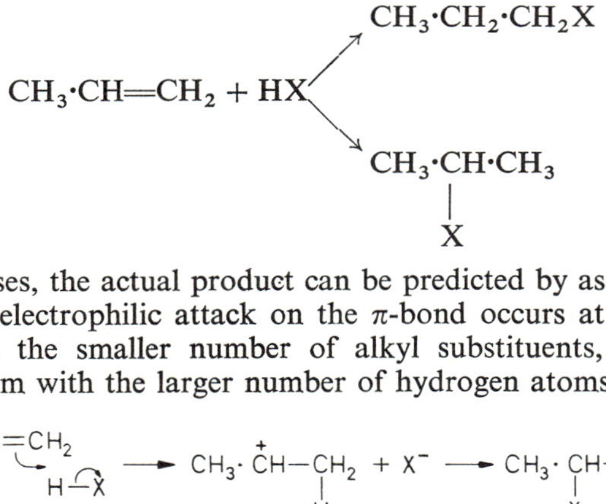

In such cases, the actual product can be predicted by assuming that the initial electrophilic attack on the π-bond occurs at the carbon atom with the smaller number of alkyl substituents, i.e. at the carbon atom with the larger number of hydrogen atoms:

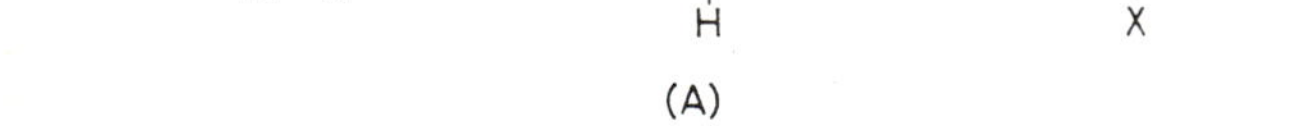

(A)

The alternative route would be:

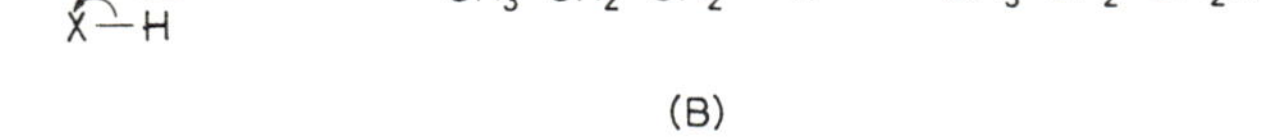

(B)

In general, this route is *not* favoured, probably because, as has already been suggested in this chapter (p. 299), the intermediate carbonium ion (B) is a very high energy species (carbonium ions of this type, with the positive charge on a terminal CH_2 group, have been referred to as ' hot ' carbonium ions!). Although extremely reactive, the secondary carbonium ion (A) is of lower energy than (B) and is thus the more favourable intermediate.

Electronic effects in π-bonds.   Electron delocalization

In simple alkenes like ethylene and propylene the π electrons may be assumed to be symmetrically distributed in the undisturbed molecule.   But in the carbonyl group, which also contains a σ-bond and a π-bond, the electron distribution is unequal, the oxygen having the greater share of the electrons.   The π electrons particularly, being more easily polarizable, tend to be closer to the oxygen atom than to the carbon atom.   In the earlier discussion of aldehydes, this situation was illustrated with diagrams I and II below:

$$\overset{\delta+}{>}C=\overset{\delta-}{O} \quad \text{or} \quad >C=O \longleftrightarrow \overset{+}{>}C-\overset{-}{O}$$

$$\qquad\qquad\qquad (a) \qquad\qquad (b)$$

$$\text{I} \qquad\qquad\qquad\qquad \text{II}$$

the double-headed arrow in II having special significance in meaning that the true state of the system represented is *between* the two forms shown.   In this example, the symbol ⟷ means that the carbonyl group is not properly represented by either II (*a*) or II (*b*) alone but that a state between the two extremes is nearer to the true state.

This way of looking at the bonds in ions or molecules has its basis in Resonance theory which, in turn, is based upon a particular wave-mechanical approach to valency known as Valence–Bond theory.   The forms written as II (*a*) and II (*b*) in our example are referred to as *canonical forms*, or, frequently, as *limiting forms*, and the true state is said to be a *hybrid state* or a *resonance hybrid*. Because the words ' resonance ' and ' hybrid ' are used in other senses in chemistry and physics, the term *mesomerism* is sometimes preferred to resonance, with mesomeric states corresponding to resonance hybrids.   However, resonance is a commonly used word in more advanced chemistry, and it is therefore essential to know what it means in this context.

The unequal distribution of π-electrons in the carbonyl group can also be represented, rather crudely, by diagrams of the type:

implying a higher electron density around the oxygen atom. Because it is virtually impossible to represent in one diagram all the orbitals, molecular and atomic, for a given molecule, chemists commonly use the valence–bond or resonance convention. Nevertheless, the alternative wave-mechanical approach to valency, known as Molecular Orbital theory, where all orbitals are taken into account, has certain advantages over valence–bond theory, particularly when the energy levels of the electrons in molecules have to be considered.

A good example of mesomerism (or resonance) is provided by the carboxylate anion (p. 124):

$$-C\overset{\displaystyle O}{\underset{\displaystyle O^-}{<}} \quad \longleftrightarrow \quad -C\overset{\displaystyle O^-}{\underset{\displaystyle O}{<}}$$

where the negative charge is shared between the oxygen atoms. One of the important postulates of resonance theory is that the hybrid state is of lower energy (and therefore more stable) than any one of the limiting forms we may write down. Thus, mesomerism is associated with stabilization. One important reason why carboxylic acids are more acidic than alcohols is that the carboxylate ion is *stabilized* by mesomerism (or resonance), whereas the alkoxide ion, RO^-, is not so stabilized; the negative charge is necessarily located on the oxygen atom, where it provides a centre of attraction for the proton, and hence the equilibrium $RO^- + H^+ \rightleftharpoons ROH$ is such that ROH molecules are very weakly acidic. In the carboxylate ion, the π electrons are said to be *delocalized;* in molecular orbital terminology they are placed in a molecular orbital extending over the *three* atoms (C,O,O) of the carboxylate ion rather than over the C and one O.

Benzene presents another example of stabilization by resonance. In valence–bond terms, benzene can be represented (simply) as the hybrid

$$\underset{\text{(benzene Kekulé structures)}}{} \quad \longleftrightarrow \quad $$

implying that the C—C bonds are neither single nor double bonds, nor are they alternately single or double, but are of intermediate *bond order*. Thus we should expect benzene to be somewhat different in character from simple alkenes, and we should expect it to be more stable than aliphatic analogues (e.g. dienes).

The planar six-membered ring of carbon atoms in benzene can be seen to involve sp^2 hybrid orbitals for each carbon atom, as in ethylene (p. 288); but here two hybrid orbitals overlap to form the two C—C σ-bonds of each carbon and one is used to form each C—H bond, as in (a):

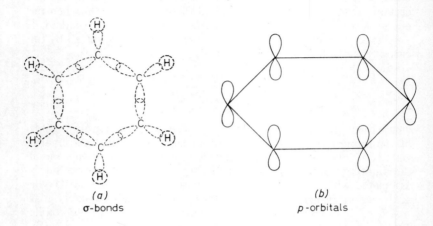

(a)
σ-bonds

(b)
p-orbitals

Each carbon atom is now left with one *p* orbital which projects at right angles to the plane of the ring [as in (b)], and there is one valency electron for each *p* orbital. These *p* orbitals overlap laterally, but do not form localized π-orbitals [as in (c)] but a π-orbital system extending round the ring as in (d), i.e. delocalized over the six carbon atoms.

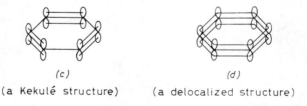

(c)
(a Kekulé structure)

(d)
(a delocalized structure)

The presence of the π electrons allows benzene to behave as a nucleophile (e.g. in substitution reactions, p. 208), but it is generally less reactive than are alkenes and does not readily undergo addition reactions (in which the stabilization due to electron delocalization would be lost).

A further illustration of the usefulness of the resonance concept is provided by the acidity of phenol. As we have seen, phenol is much more acidic than alcohols like ethyl alcohol, although it is a weaker acid than (e.g.) acetic acid. If we consider the structure of the anion of phenol (phenate ion) we can immediately imagine the two Kekulé forms as contributors to the hybrid:

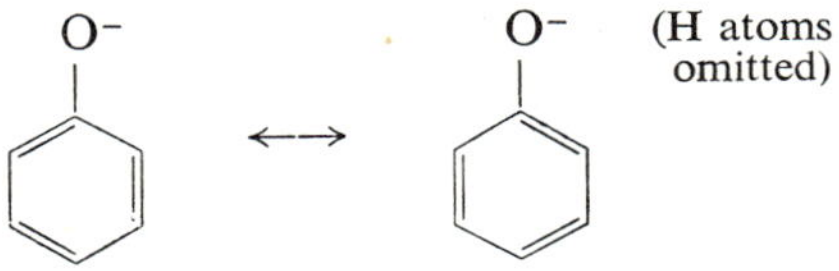

O⁻   O⁻   (H atoms
                        omitted)

⟷

But, in addition, we can visualize other forms in which the negative charge appears at various positions in the ring:

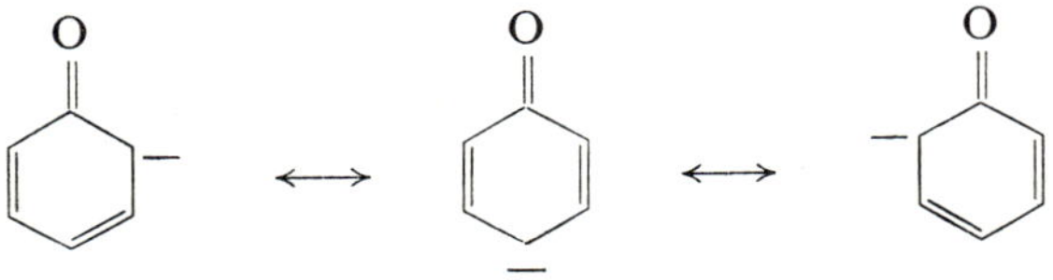

O   O   O

⟷   ⟷

The various canonical forms given above differ only in the position of electrons (resonance theory is not applicable where atoms change positions). We attempt, in this way, to represent delocalization of π electrons (and of the negative charge). Thus the phenate ion is more stable than we would expect on the basis of a single classical Kekulé-type structure, and it is not surprising that phenol is a stronger acid than ethyl alcohol.

EXAMINATION QUESTION

(1) What are the **four** principal types of reaction which are given by the hydroxyl group in methyl alcohol? Give **one** example of each type.

Compare the chemical behaviour of the hydroxyl group in methyl alcohol with that shown by the hydroxyl group in (a) phenol, (b) acetic acid.

From the facts which you have mentioned, what conclusions, expressed in terms of the electronic theory of valency, can you draw concerning the influence of neighbouring groups on the character of the hydroxyl group?

(J. M. B. A.)

27

PHYSICAL METHODS OF SEPARATION AND PURIFICATION

In marked contrast to inorganic chemistry, many of the reactions between organic compounds proceed to an equilibrium state only, and not to completion; furthermore, the main reaction is often accompanied by side reactions. In many organic preparations therefore, separation of the required product from other substances formed is not easy. However, since most organic compounds are covalent and exist as molecules held together by relatively weak forces (Chapter 26), they have usually low boiling points and are soluble in a more varied range of solvents, and hence more methods of separation and purification are available than in inorganic chemistry. The methods most commonly used are:

(1) *Crystallization*
(2) *Distillation*
    (*a*) Fractional
    (*b*) Steam
(3) *Solvent Extraction*
(4) *Chromatography*

(1) *Crystallization*

This method, available for solids, is that commonly used in inorganic chemistry and depends on the fact that in general the solubility of a solid solute increases with rise of temperature. Hence on cooling a hot saturated solution, the solute separates. The small amount of impurity present usually remains in the mother liquor.

For inorganic compounds, which are in general ionic, water is often the only suitable solvent, and on cooling a hot saturated solution crystallization is often slow. But for covalent organic compounds, there is a wide choice of solvents, e.g. alcohols, ethers, acetone, chloroform, benzene, toluene, etc.; and cooling will very frequently produce rapid crystallization of a pure product.

Crystallization can also be achieved in some cases by adding to a solution some other solvent (miscible with it) in which the solute is less soluble. Thus acetamide (p. 142) can be obtained as pure crystals by dissolving in warm methyl alcohol, then adding ether and allowing to stand. This method is also used to obtain a polymer of definite molecular weight from a mixture of polymers of different molecular weights.

(2) Distillation

Distillation is the conversion of a liquid to a vapour and the condensation of the vapour to liquid again by cooling.

(a) *Fractional distillation:*—Liquids which are miscible in all proportions can usually be separated by fractional distillation (i.e. where the distillate is collected in ' fractions ' coming over between known temperatures), if there is some appreciable difference in their boiling points.

Suppose a mixture of two liquids A and B (of different boiling points and miscible in all proportions) is heated to boiling point in a flask connected to a reflux condenser. A thermometer (the bulb of which dips into the liquid) will show that the mixture boils at a steady temperature.

If in this apparatus, the boiling points of a series of mixtures of A and B, of known compositions, are determined, and these plotted against composition, a graph such as that marked ' Liquid ' in *Figure 27.1* is obtained in most cases.

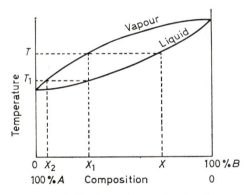

Figure 27.1. Boiling point graph of a mixture.

Now in a second series of experiments, a mixture of approximately equal amounts of A and B is distilled, using a distillation flask and a Liebig condenser, but in this case, the thermometer is so placed that the bulb is opposite the side tube of the distillation flask, to read the temperature of the distillate instead of the liquid. When distillation starts, fractions of the distillate are collected over ranges of each 5°C, and the mean temperature of distillation of each fraction noted. The composition of each fraction is now determined. If these temperatures are plotted against the corresponding composition, a graph of the type marked ' Vapour ' is obtained.

Suppose now an attempt is made by fractional distillation to separate the two liquids A and B in a mixture of composition represented by X. Vapour will start coming off when the temperature T is reached, and the composition of the distillate will be given by X_1. The first fraction coming over, it must be noted, does not consist of pure A but is merely a mixture richer in A than was the original one.

Suppose now the fraction of composition X_1 is redistilled. The distillate starts coming over at the temperature of T_1 with a composition given by X_2, to yield again a liquid which, although still a mixture of A and B, contains a higher proportion of A than did the first fraction of the first distillation. More than one distillation—in fact, an infinite number—will be needed to obtain pure A.

A much better separation can be achieved in one distillation by using one of the many different fractionating columns (*Figure 27.2*) instead of an ordinary distillation flask, for in such an apparatus, many distillations do in fact take place. The first vapour coming off the liquid condenses at the bottom of the fractionating column. As the liquid so formed flows down the column, it meets more vapour and is redistilled. The process is repeated again and again to amount almost to a series of distillations. By the time vapour reaches the top of the column to come over into the condenser, it is very rich in A.

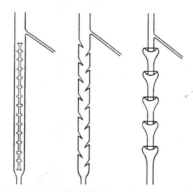

Figure 27.2. Some fractionating columns.

If in a distillation of a mixture, one or more of the constituents tends to be unstable near or at its boiling point under atmospheric pressure, or if the boiling points under such pressure are inconveniently high, the distillation may be carried out under a reduced pressure when the boiling points are lower. An apparatus in which such a distillation may be carried out is shown in *Figure 27.3*. If the pressure of the distillation is required to be known, a mercury manometer may be fitted between the receiver and the pump.

With a mixture of ethyl alcohol and water an equilibrium diagram different from *Figure 27.1.* is obtained. At one point on the graph,

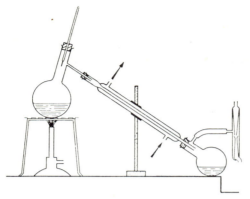

Figure 27.3. Distillation under reduced pressure.

at a temperature below the boiling point of either constituent (*Figure 27.4*), the vapour and liquid curves meet. This mixture, of composition 95·6 per cent alcohol and 4·4 per cent water, boils at a constant temperature and distils unchanged in composition. It is known as a *constant boiling* or *azeotropic mixture.* (It was at one time thought to be a compound, but was shown to be a mixture by the fact that the composition varies with the pressure of distillation.) Other pairs of liquids, e.g. ethyl alcohol and benzene, acetic acid and benzene, form similar azeotropes. Such mixtures result from the existence of intermolecular forces, prominent among which are hydrogen bonds, causing association between molecules (p. 294).

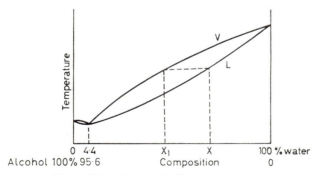

Figure 27.4. Constant boiling mixture graph.

Suppose a mixture of composition X is distilled. Then the first fraction collected will have a composition indicated by X_1, i.e. richer in the azeotrope. As the distillation proceeds, the residue in the flask will become richer in water, and finally pure water can be obtained from such a mixture—but never (by ordinary distillation) pure alcohol.

Suppose on the other hand, a mixture richer in alcohol than the azeotrope be distilled. Then as the distillation proceeds, the residue in the flask becomes richer in alcohol and finally, pure alcohol distils over. In this case, pure water could never be obtained by distillation.

N.B. Some pairs of liquids give azeotropes boiling at temperatures above the boiling points of both constituents, e.g. hydrochloric acid and water.

(b) *Steam distillation:*—A common preparation in which this method of separation is used is that of aniline, obtained by the reduction of nitrobenzene by iron and hydrochloric acid (p. 255). The aniline formed is separated along with water, from an aqueous mixture containing oxidation products of iron, excess slaked lime, etc., by passing steam through the mixture. The apparatus for this process (known as *steam distillation*) is shown in *Figure 27.5*.

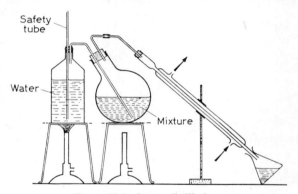

Figure 27.5. Steam distillation.

A mixture of two liquids which do not react together chemically nor dissolve in each other to any appreciable extent, boils at a temperature below the boiling point of both constituents. For example, a mixture of aniline and water boils at 99°C under a pressure of 760 mm. At this temperature, the vapour pressure of water (P_W) is 733 mm and therefore, by Dalton's law of Partial Pressures, that of aniline (P_A) will be $760-733=27$ mm. These partial pressures are proportional to the respective numbers of molecules of each constituent in a given volume of vapour, i.e.

$$\frac{P_W}{P_A} = \frac{\text{number of molecules of water vapour in a given volume of vapour}}{\text{number of molecules of aniline in an equal volume of vapour}}$$

These numbers are proportional to the respective **weights of** vapour (W_w and W_A) divided by their molecular weights (M_w and M_A) i.e.:

$$\frac{P_w}{P_A} = \frac{W_w}{M_w} \times \frac{M_A}{W_A}$$

The molecular weights of water vapour and aniline ($C_6H_5NH_2$) are 18 and 93 respectively. Substituting:

$$\frac{733}{27} = \frac{W_w}{18} \quad \frac{93}{W_A}$$

$$\frac{W_w}{W_A} = \frac{733 \times 18}{27 \times 93} = \frac{5}{1} \quad \text{(approx.)}$$

Thus from the above mixture, a distillate consisting of 1 part of aniline to 5 parts of water is obtained. This amount of aniline is surprisingly high in view of its relatively low vapour pressure, but is due to its relatively high molecular weight.

(3) Solvent Extraction

If a solute is added to two immiscible solvents A and B which are in contact with each other, then some of the solute dissolves in one solvent and some in the other. If the concentrations of the solute in the two solvents are determined, then it will be found that no matter how much solute is added or what the relative amounts of the two solvents are, provided that the temperature is the same in all cases:

$$\frac{\text{Concentration in } A}{\text{Concentration in } B} = \text{constant } (K)$$

The above relation is known as the *Distribution or Partition Law* and the constant K is called the *Distribution or Partition Coefficient*.

The Law is obeyed by all such systems when the condition of the solute in both solvents is the same. It does not obtain when there is dissociation or association in one solvent and not in the other.

It may frequently happen in preparative organic chemistry that the substance required is obtained in aqueous mixture. Although the bulk of it may not mix with water (making mechanical separation easy) some smaller amount may have dissolved in the water to give a true solution. Examples occur in the recovery of aniline and phenol by steam distillation. The smaller amount of phenol in the aqueous layer can be removed by shaking up with a variety of liquids, e.g. ether, chloroform, amyl alcohol—i.e. liquids which do not mix with water and in which phenol is more soluble than it is in water. If 200 ml of amyl alcohol, for example, are to be used, more phenol is extracted by shaking successively with two volumes each of 100 ml than by using the whole 200 ml in one extraction. The following calculation, based on the Distribution Law, illustrates this:

Phenol dissolves with its normal molecular weight in both water and amyl alcohol. An aqueous solution of phenol containing 3 g per litre is in equilibrium with an amyl alcohol solution containing 48 g per litre.

Suppose an aqueous solution of phenol contains 25 g per litre. Calculate what weight of phenol would be extracted from 500 ml of it by shaking it up (*a*) with 200 ml amyl alcohol and (*b*) with two successive amounts of 100 ml of amyl alcohol.

Let C_W and C_A be concentrations in g per litre of phenol in water and amyl alcohol respectively. Then:

$$K = \frac{C_A}{C_W} = \frac{48}{3} = 16$$

(*a*) *First extraction* with 200 ml amyl alcohol.

In 500 ml aqueous solution of phenol there are 12·5 g. Let the weight of phenol extracted by 200 ml amyl alcohol be *x* g.

Then weight left in 500 ml water = (12·5−*x*) g

$$\begin{aligned} \text{and} \quad C_A &= 5x \\ C_W &= 2(12\cdot5 - x) \\ 16 &= \frac{5x}{2(12\cdot5 - x)} \\ x &= 10\cdot8 \end{aligned}$$

Hence, weight of phenol extracted by 200 ml amyl alcohol
= 10·8 g

Weight of phenol left in 500 ml aq. solution
= (12·5 − 10·8) g
= 1·7 g

(*b*) *Second extraction*

Let *y* g be weight of phenol extracted by first 100 ml amyl alcohol.

$$\text{Now,} \quad C_A = 10y$$

The weight left in 500 ml water = (12·5−*y*) g

$$\begin{aligned} \text{and} \quad C_W &= 2(12\cdot5 - y) \\ 16 &= \frac{10y}{2(12\cdot5 - y)} \\ y &= 9\cdot53 \end{aligned}$$

Hence weight of phenol extracted = 9·53 g

Weight of phenol left in 500 ml aq. solution = (12·5 − 9·53) g
= 2·97 g

Let *z* g be weight of phenol extracted by second 100 ml amyl alcohol.
$$C_A = 10z$$

and (as above) $C_W = 2(2·97-z)$

$$16 = \frac{10z}{2(2·97-z)}$$

$$z = 2·27$$

Therefore, total weight of phenol extracted by method (b) is (9·53 + 2·27) = 11·8 g compared with 10·8 g in method (a).

(4) Chromatography

If a mixture of variously coloured organic substances is dissolved in a suitable solvent and poured slowly into the top of a vertical glass tube (e.g. a burette), packed with a powder which does not react chemically with the substances used, it is found that as the solution passes down the column, separation into differently coloured zones occurs. This process, called *chromatography*, was first used in 1906 by the Russian botanist Twsett, who separated a chlorophyll extract in petrol solution into the two chlorophylls ($a-$ and $\beta-$) and xanthophyll and carotene. Little notice was taken of this discovery until over twenty years later, since when the process has been increasingly used to separate not only coloured but colourless substances, both organic and inorganic. It is now used on an industrial as well as the laboratory scale.

Columns of material such as chalk, finely-powdered alumina or filter paper strips can be used. The separation zones can be further separated by running pure solvent down the column after the solution has been run through—a process dependent on desorption and re-adsorption further down the column of the less strongly adsorbed substances, thereby increasing the distance between the coloured zones or rings. This is the process of 'development' which makes it easier to extract the separate coloured zones from the column for subsequent analysis and identification.

Alternatively, separation after development can be effected by running a substance through which is more strongly taken up on the column than any of those to be separated. This is *displacement chromatography*. Separation can also be enhanced by using a mixture of solvents; this is *partition chromatography*. This is, in fact, the principle involved in the use of filter paper with an organic solvent, partition being effected between this solvent and the water adsorbed on the 'dry' filter paper (and constituting about 20 per cent of its weight)—the process being fundamentally the same as the partition described on p. 311.

The method is applicable to liquids, vapours and gases, and substances which are chemically very similar. For example, amino acids can be effectively separated and, although these substances are not coloured, the separate fractions can still be identified by suitable analytical methods.

EXAMINATION QUESTION

(1) Describe how you would separate the components of the following mixtures and obtain one component in a pure condition:

(a)   methyl alcohol and acetone;
(b)   benzoic acid and phenol;
(c)   benzoic acid and acetanilide;
(d)   aniline and benzene.

(O. and C. S.)

(N.B.   This question clearly involves chemical methods of separation, followed by purification using physical methods.)

28

DETERMINATION OF MOLECULAR FORMULAE

MOST of the tests used to analyse inorganic compounds are carried out in aqueous solution, in which, in general, the inorganic compound is dissociated.   These tests, which are thus tests for ions, are clearly inapplicable to organic compounds.

One test to distinguish between an inorganic and an organic compound is to heat a little of the substance on a piece of broken porcelain.   If organic, it burns away completely leaving no residue. The test is not always conclusive; for example, many ammonium salts and some mercury salts sublime and leave no residue when heated.   If sodium acetate or some other metallic salt of an organic acid is heated, the organic part burns away and leaves an inorganic residue which is often the carbonate or the oxide of the metal.

Suppose we are in the position of a research worker who has prepared or isolated some new compound.   In our report on its chemical nature, we should, of course, want to give its formula. To arrive at the molecular formula we should have to follow the procedure set out below.

(1) Determination of purity.
(2) Qualitative analysis, to find the elements present.
(3) Quantitative analysis, to determine its percentage composition.
(4) Determination of molecular weight.

(1) *Determination of purity*

It is necessary to find out whether a given substance is pure before any analysis, whether qualitative or quantitative, is attempted.   In the case of a *liquid*, this can be done by noting its behaviour when boiled.   Some of the liquid is put into a small flask and a thermometer inserted so that the bulb is dipping into the liquid. The flask is then heated until the liquid starts to boil. If pure, the temperature will remain constant while boiling continues.   If the liquid contains some impurity, then when boiling starts either impurity or the liquid dissolving it·will boil off.   In any case, the composition of the mixture will change and so will the boiling point.

This test depends upon the fact that the vapour pressure of a liquid at any temperature has a definite value, and this value is often changed by presence of impurity.   Hence determination of vapour pressure at various temperatures is also a test for purity.

The refractive index is another very good criterion of the purity of a liquid. The liquid is redistilled until a constant value of its refractive index is obtained.

There are several methods of testing the purity of a *solid:*

(*a*) A solution of the solid is made in some suitable organic solvent (preferably a volatile one). A little of the solution is smeared on a microscope slide by means of a glass rod. The crystals separating are then examined under the microscope. If the solid is pure, all the crystals will have the same shape; if not, some of a different shape will be seen. (If the impurity is one having the same crystalline shape, this test fails, but it is very unlikely that this would be the case.)

(*b*) The behaviour on melting is noted. This can be seen as follows: A thin-walled capillary tube is first made by heating a test-tube until it becomes soft, and then drawing it out quickly. The capillary tube so obtained is cut into lengths of about four inches and one end of each piece sealed. A little of the solid under investigation is now put into a capillary tube and by tapping gently it can be brought to the bottom of the tube. The capillary is then fixed by means of a rubber band to a thermometer, so that the solid is alongside the bulb. Thermometer and capillary tube are now clamped in a small beaker (or boiling tube which has been cut short, see *Figure 28.1*) containing a silicone oil, and fitted with a

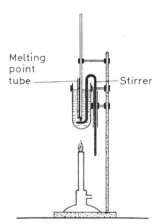

Melting
point
tube

Stirrer

Figure 28.1. Melting point apparatus.

stirrer made from a piece of glass rod. The temperature of the oil is raised slowly, stirring all the time. The first indication of melting will be a shrinking of the solid in the capillary tube. If the temperature interval between the beginning and the completion of melting is long (e.g. 2 degrees), then the solid is probably impure, for *pure solids melt sharply.*

Silicone D.C. 550 is a suitable oil with little change of viscosity with change of temperature. It is colourless and remains so even when heated to 300°C. It is quite stable, is involatile and is non-inflammable. Other liquids may be used, dependent on the temperature required, e.g. glycerol (for low temperature work), liquid paraffin, concentrated sulphuric acid (with which extreme caution is obviously necessary).

This is the method of determining the actual melting point of a solid. It may be used also as a means of identification. Suppose the compound, the melting point of which has been determined, is believed to be *A*. Some of it is then mixed with a little of the pure compound *A*, and the melting point of the mixture determined. If the melting point is unchanged, then the original compound is *A* —if depressed, it is not. The test is known as the ' mixed melting point ' test.

In modern research laboratories, spectroscopic methods (e.g. infrared and nuclear magnetic resonance spectroscopy) are regularly used to test purity and as a means of identification. These methods can be applied to gases, liquids and (usually) solids.

(2) *Qualitative analysis*

If it is known that the compound is organic, tests for the presence of carbon and hydrogen are unnecessary since all such compounds contain these elements.

(*a*) *Carbon and hydrogen:*—An intimate mixture of the compound with powdered copper(II) oxide is heated in an ignition tube. The carbon and hydrogen are oxidized to their respective oxides. Carbon dioxide is identified by the lime-water test, and the water is seen to condense in the cool mouth of the tube.

Both reagents must be dry. Since copper(II) oxide is hygroscopic, it is advisable to heat it strongly before use, to remove any water which may be present.

(*b*) *Nitrogen:*—Many, but not all, nitrogen-containing organic compounds evolve ammonia when heated with soda-lime.

Nitrogen can be detected in all organic compounds by the following test due to Lassaigne. A very small piece of sodium is placed in an ignition tube and then some of the organic compound added. When the tube is heated (CARE!), the sodium breaks up the organic compound and if nitrogen is present forms sodium cyanide.

An alternative method of making sodium cyanide is that of Middleton in which an intimate mixture of the organic compound, with zinc dust and sodium carbonate is heated strongly in an ignition tube.

In both methods, the tube while still hot is put into a little water in a mortar. The contents are then ground up and filtered. The filtrate contains the sodium cyanide.

Some iron(II) sulphate solution is then added and the mixture boiled, when sodium hexacyanoferrate(II) is formed:

$$6NaCN + FeSO_4 = Na_4Fe(CN)_6 + Na_2SO_4$$

The solution is acidified with dilute hydrochloric acid and aqueous iron(III) chloride added when there is a precipitate of Prussian blue.

(c) *Halogens:*—In general, halogen atoms are not present in organic compounds as ions and are therefore not detectable by the tests for inorganic halides.

A simple test for a halogen is to heat a little of the substance on a copper wire in a bunsen flame. The copper halide formed colours the flame green.

In the Lassaigne or Middleton test above, any halogen present is converted into the sodium halide. After extracting the residue with water and filtering, the solution is acidified with dilute nitric acid and aqueous silver nitrate is added. A white precipitate indicates the presence of chlorine, a yellowish one, bromine and a deep yellow, iodine.

Sodium cyanide will also give a white precipitate with silver nitrate. If nitrogen is present, the solution must first be boiled with dilute nitric acid to remove the cyanide before the silver nitrate is added:

$$NaCN + HNO_3 = NaNO_3 + HCN \uparrow$$

It is thus necessary to find whether nitrogen is present before testing for a halogen.

N.B. If sulphur is present in an organic compound, it would be converted into insoluble zinc sulphide during Middleton's test. The residue after filtration on treatment with hydrochloric acid would then evolve hydrogen sulphide.

There is no satisfactory test for oxygen. If after quantitative analysis, the percentages of the elements known to be present do not add up to 100 (making allowance for experimental error), the difference is concluded to be due to oxygen (unless, of course, there is reason to think that other elements, e.g. phosphorus, may be present).

(3) *Quantitative analysis*

(a) *Carbon and hydrogen:*—The method (due to Liebig) is, in principle, that of qualitative analysis to show the presence of these two elements.

A known weight of the organic compound is heated with copper oxide. The amount of water is determined by absorbing it in a weighed amount of calcium chloride and that of carbon dioxide by absorption in caustic potash. Great care is required for an accurate determination.

A quantity of copper(II) oxide (of the wire form type) is packed into a combustion tube about two and a half feet long, and is kept in position between two plugs of asbestos wool (*Figure 28.2*). A

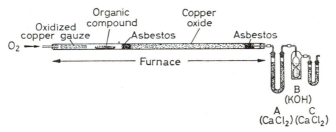

Figure 28.2. Determination of percentage carbon and hydrogen.

coil of freshly oxidized copper gauze is placed at one end of the combustion tube as shown. The tube is now heated strongly over a furnace while a stream of oxygen (freed from carbon dioxide and water) is passed through. This preliminary heating removes water vapour from the tube and oxidizes any organic matter which might be present.

When cool, a boat containing a weighed amount of the organic compound is placed in the combustion tube in the position shown, and the weighed calcium chloride tubes A and C, and the caustic potash absorption vessel B are connected.

Heating of the copper oxide can now start. The burners farthest away from the boat are lit first. When the copper oxide is hot, the burners under the gauze and then under the boat are lit. A gentle stream of oxygen is now passed through the tube. The vapour of the organic compound is driven into the copper oxide which oxidizes it. The water formed is absorbed by the weighed calcium chloride tube A, and the carbon dioxide by the caustic potash in B. If the oxygen passing through should remove any water from B, this is trapped in C. Increase in weight of B and C thus gives the weight of carbon dioxide.

Since 18 g of water contain 2 g of hydrogen, and 44 g of carbon dioxide contain 12 g of carbon, the weights of these two elements and hence their percentages can be calculated.

One of the great disadvantages of Liebig's method is that relatively large amounts of the organic compound are needed, although with more sensitive balances the amount can be reduced.

The method was improved upon by Pregl in 1911 and made available for the analysis of minute amounts (e.g. of some of the first hormones isolated) with the use of a micro-balance. Pregl's method is now used on a semi-micro scale. The combustion tube is filled as originally—a known weight of the compound mixed with copper oxide and then a large bulk of copper oxide. Oxygen from an aspirator (at a pressure just above atmospheric) is passed through the tube, being first purified by (a) solid caustic potash to remove carbon dioxide, then (b) dehydrite (magnesium perchlorate trihydrate) to remove water vapour and (c) carbasorb (caustic soda on pumice) to remove final traces of carbon dioxide. The combustion tube is heated electrically.

Beyond the combustion tube is some lead chromate (to trap compounds of sulphur), lead dioxide (to remove oxides of nitrogen) and silver wool (to remove halogen elements). The section of apparatus containing these is maintained at a definite temperature by a boiler containing *cymene*, b.p. 177°C.

The weights of water and carbon dioxide from the combustion of the compound are obtained as in the original method.

(*b*) *Nitrogen:*—Two methods are in use.

(*i*) Dumas' method:—Here a known weight of the organic compound is mixed intimately with copper oxide and the mixture placed in a combustion tube (*Figure 28.3*), containing also a piece of

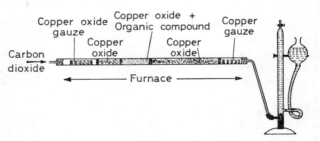

Figure 28.3. Determination of percentage nitrogen.

copper gauze. (The copper gauze is freshly reduced by heating it and, when hot, dipping it into methyl alcohol.) The combustion tube is connected to a nitrometer filled with a concentrated solution of caustic potash (1 part potash to 1 part water). A stream of carbon dioxide is now passed through the apparatus to remove all air. When all the gas passing through dissolves completely in the caustic potash it is known there is no air present. By raising the reservoir containing caustic potash, the nitrometer can be filled with alkali.

The combustion tube is now heated, when the organic compound is oxidized by the copper oxide. Any nitrogen present is set free and collects in the nitrometer where its volume can be measured. A gentle stream of carbon dioxide will ensure no nitrogen is left in the combustion tube. In case any nitrogen should come off as oxides of nitrogen, these are reduced to nitrogen by the copper gauze.

From the volume of nitrogen, measured at the laboratory temperature and pressure, the weight of nitrogen and hence its percentage in the organic compound can be calculated.

(*ii*) Kjeldahl's method:—A known weight of the organic compound is heated with a small amount of concentrated sulphuric acid. Charring is first noticed but after a time the liquid becomes colourless. The nitrogen present has now been converted into ammonium sulphate, in which the amount of ammonia is determined. Addition of a little mercury(II) iodide (or oxide) assists the conversion to an ammonium salt.

An excess amount of caustic soda is added cautiously at first, from a tap funnel, and the mixture heated (*Figure 28.4*). The

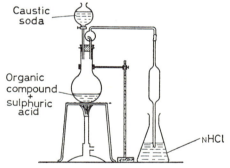

Caustic soda

Organic compound + sulphuric acid

NHCl

Figure 28.4. Kjeldahl's method of estimating percentage nitrogen.

ammonia liberated is absorbed in the known accurately measured volume of normal hydrochloric acid (in excess of that required) in the conical flask. When evolution of ammonia is complete, the contents of the conical flask are transferred to a graduated flask and diluted with water. Portions of this diluted solution are then titrated against standard alkali to determine the amount of acid unused and hence the amount required to absorb the ammonia. (This method of estimating the amount of ammonia in an ammonium salt is described fully in most books of volumetric analysis.) From the following relation, the weight of nitrogen can be calculated:

$$1 \text{ litre NHCl} \equiv 17 \text{ g NH}_3 \equiv 14 \text{ g N}$$

While the method of Kjeldahl is much easier to carry out than that of Dumas, it is not so reliable, for in some organic compounds, not all the nitrogen is converted into ammonium sulphate by the sulphuric acid. The method is used extensively for the estimation of nitrogen in fertilizers and foods, in which the nitrogen is readily convertible into ammonia.

(4) *Determination of molecular weight*

From the results of quantitative analysis, the percentage composition of the elements present and hence the empirical formula can be calculated. If oxygen is present in the compound, its percentage is found by difference.

The relationship between the empirical formula and the molecular formula is determined by finding the molecular weight. The following are standard methods by which this can be done. They are described fully in most books of physical chemistry.

(*a*) By the determination of the Vapour Density (H = 1) by the methods of *Victor Meyer, Dumas* or *Hofmann*. These are methods suitable for volatile liquids. Then:

Vapour Density × 2 = Molecular Weight

(*b*) By the determination of the lowering of freezing point, raising of boiling point or lowering of vapour pressure of some suitable solvent by the addition of a known weight of the organic compound to a known volume of the solvent.

(*c*) By the measurement of osmotic pressure. This method of determining molecular weights, used many years ago, has been revived recently with better osmometers, to find the very high molecular weights of polymers.

(*d*) By viscosity measurements. The solution of a polymer in a solvent increases the viscosity of the solvent. The viscosity is measured by determining the time a given volume of liquid takes to flow through a capillary tube. The increase in viscosity depends upon the molecular weight of the polymer, which can hence be determined.

Example

A given compound contains carbon, hydrogen, oxygen and nitrogen only. On combustion, 1·124 g of the compound gave 1·678 g of carbon dioxide and 0·857 g of water. In a Kjeldahl estimation, the ammonia obtained from the same weight of compound was passed into 50 ml of NHCl. The resulting solution was diluted to 250 ml and 25 ml of it were titrated with 0·1NNaOH to find the amount of unused acid. The average titration reading of the caustic soda was 30·96 ml.

The molecular weight of the given compound is 59.

Find (*a*) the empirical formula, (*b*) the molecular formula and identify the compound.

To find the weight of carbon:

$$C \longrightarrow CO_2$$
$$12 \text{ g} \qquad 44 \text{ g}$$

i.e. $3 \longrightarrow 11$

Hence, the weight of carbon in 1·678 g of carbon dioxide and therefore in 1·124 g of the original compound

$$= \frac{3}{11} \times 1\text{·}678 \text{ g}$$
$$= 0\text{·}457 \text{ g}$$

To find the weight of hydrogen:

$$H_2 \longrightarrow H_2O$$
$$2 \text{ g} \qquad 18 \text{ g}$$

i.e. $1 \longrightarrow 9$

Hence weight of hydrogen in 1·124 g of compound

$$= \frac{1}{9} \times 0\text{·}857 \text{ g}$$
$$= 0\text{·}095 \text{ g}$$

To find the weight of nitrogen:

30·96 ml 0·1NNaOH $\equiv$ 3·096 ml NHCl

This is the volume of unused acid in the 25 ml titrated.
Total volume of unused acid = 30·96 ml
Volume neutralized by the ammonia

$$= (50 - 30\text{·}96) \text{ ml}$$
$$= 19\text{·}04 \text{ ml}$$

The following relationship holds:

$$N \longrightarrow NH_3 \equiv HCl$$
$$14 \text{ g} \qquad\qquad 1 \text{ litre } N$$

Weight of nitrogen in 1·124 g compound

$$= \frac{19\text{·}04}{1000} \times 14 \text{ g}$$
$$= 0\text{·}267 \text{ g}$$

To find the weight of oxygen:

N.B. Since there is no direct method of estimating the oxygen content, it is obtained by difference.

Weight of oxygen in 1·124 g compound

$$= \{1\text{·}124 - (0\text{·}457 + 0\text{·}095 + 0\text{·}267)\} \text{ g}$$
$$= 0\text{·}305 \text{ g}$$

To calculate the empirical formula:

	C	H	O	N
Combining weights	457	95	305	267
Ratio of atoms (divide by respective atomic weight)	$\frac{457}{12}$	$\frac{95}{1}$	$\frac{305}{16}$	$\frac{267}{14}$
	38·1	95	19·1	19·1

(Divide by smallest to get $\dfrac{38 \cdot 1}{19 \cdot 1}$ $\dfrac{95}{19 \cdot 1}$ $\dfrac{19 \cdot 1}{19 \cdot 1}$ $\dfrac{19 \cdot 1}{19 \cdot 1}$
simple ratio)

i.e.   2   5   1   1

∴ Empirical formula $= C_2H_5ON$
Empirical formula weight $= 59$
∴ Molecular formula $= C_2H_5ON$

Identification of compound:—

The compound cannot be aromatic since there are less than six carbon atoms in the molecule.

Aliphatic compounds containing only one atom of oxygen and one of nitrogen are: cyanates, isocyanates and amides.

Taking away the functional cyanate and isocyanate group of —OCN, this leaves the unknown group CH_5. Hence the compound is not a cyanate or an isocyanate.

Subtracting the functional amide group of —$CONH_2$, we are left with the radical CH_3. Hence the compound is acetamide, $CH_3 \cdot CONH_2$.

(N.B. In this worked example, the separate steps are indicated by sub-headings and where thought helpful, explanatory details are given. This has been done for simplicity.)

EXAMINATION QUESTION

(1) A compound of carbon, hydrogen and oxygen is found by analysis to contain 48·7 per cent C and 8·1 per cent H and has a vapour density of 37. Derive its molecular formula and suggest possible structural formulae, giving names. Describe simple tests which would enable you to identify the various isomers. (L. A.)

29

DETERMINATION OF STRUCTURAL FORMULAE

OWING to the common occurrence in organic chemistry of the phenomenon of isomerism, it is necessary in order to understand the chemical behaviour of an organic compound not only to know the number and kind of atoms in the molecule but also the manner of their arrangement, i.e. to know its constitutional formula and structure. Little progress was made in organic chemistry until means of finding such formulae were known. The steps in the process have already been mentioned in the Introduction. The first contribution came from Frankland in 1852 in his *Theory of Valency*, to be followed six years later by the establishment of the tetravalency of carbon. *The tetravalency of carbon is the fundamental principle on which the structural formulae of organic compounds are based.*

It has been mentioned that as early as 1832, both Wöhler and Liebig recognized the existence of groups of atoms which were residues of compounds and which were then called *radicals*. In chemical reactions these radicals behaved as single atoms. Examples are the methyl, CH_3—, and the ethyl, C_2H_5—, radicals. Here is the *second principle* on which structural formulae are founded, for it is assumed that *when an atom or group is replaced, the incoming atom or group takes up the position occupied by the atom or group displaced.* The assumption is justified if the change is effected at temperatures below 200°C (although exceptions are known). Above this temperature, reshuffling or rearrangement of atoms is more likely.

Bearing these two principles in mind, the structural formulae of simple organic compounds may be worked out by one or more of the following methods of approach:

(1) The chemical properties of the compound are investigated with the object of recognizing the presence of certain groups. The method may be referred to as the *analytical method.*

(2) The compound may be *synthesized* from compounds the constitutional formulae of which are already known. Here, in particular, it is assumed that in the synthesis, the groups already present in the reactants remain intact.

(3) By using *physical methods of analysis*, i.e. examining the molecular structure by methods such as spectroscopy which do not involve any chemical changes in the compound. Such methods are now widely used for structure determination.

Some examples of the first two methods have already been given (pp. 76, 86). A further example is discussed below.

From the results of quantitative analysis and molecular weight determination, the molecular formula of acetic acid is found to be $C_2H_4O_2$. One of the four hydrogen atoms can be shown to be acidic by titration with alkali or by analysis of, for example, sodium acetate $(C_2H_3O_2)^-Na^+$. Phosphorus pentachloride replaces one hydrogen atom and one oxygen atom in acetic acid by a chlorine atom; the product, C_2H_3OCl (acetyl chloride) reacts with water to give back acetic acid and hydrogen chloride:

$$C_2H_4O_2 \xrightarrow{PCl_5} C_2H_3OCl \xrightarrow{H_2O} C_2H_4O_2 + HCl$$

These reactions suggest that acetic acid contains the grouping —O—H which is responsible for its acidity (note that in alcohols the hydroxyl group is replaceable by chlorine with phosphorus pentachloride and that alcohols form salts, RO^-Na^+).

By the action of chlorine on acetic acid, three hydrogen atoms can be replaced by chlorine atoms. The product, $C_2HO_2Cl_3$ is still acidic (in fact, strongly acidic) forming salts such as $(C_2O_2Cl_3)^-Na^+$. Thus we can write acetic acid as $(H_3)C_2O(OH)$. Remembering that acetic acid can be 'decarboxylated' to give methane (p. 128), the inference is clearly that it contains a CH_3— group; this might be linked to oxygen, which must be linked to carbon:

$$\begin{array}{c} \text{H} \qquad | \\ \text{HC—O—C—} \\ \text{H} \qquad | \end{array}$$

The only conceivable structure of this type would be $CH_3 \cdot O \cdot C\begin{smallmatrix}{} \diagup \text{H} \\ \diagdown \text{O} \end{smallmatrix}$, which does not contain a hydroxyl group (this structure is actually that of methyl formate).

The conventional structure for acetic acid, $CH_3 \cdot C\begin{smallmatrix}{} \diagup \text{O} \\ \diagdown \text{OH} \end{smallmatrix}$, is clearly the only satisfactory one, since it enables us to understand all the reactions given above. The correctness of the structure is indicated by syntheses of acetic acid, for example by the routes:

$$\begin{array}{c} CH_3I \longrightarrow CH_3 \cdot CN \longrightarrow CH_3 \cdot CO_2H \\ \big\downarrow \qquad\qquad\qquad \big\uparrow \\ \big\longrightarrow CH_3 \cdot MgI \xrightarrow{CO_2} CH_3 \cdot C\begin{smallmatrix}{} \diagup \text{O} \\ \diagdown \text{OMgI}\end{smallmatrix} \end{array}$$

The structure is also supported by (i) the formation of acetic acid by oxidation of ethyl alcohol:

$$CH_3 \cdot CH_2 \cdot OH \longrightarrow CH_3 \cdot C\!\!\begin{array}{c} \nearrow O \\ \searrow OH \end{array}$$

(ii) the formation of ethane by electrolysis of salts of acetic acid (Kolbe synthesis):

$$2CH_3 \cdot CO_2^- \longrightarrow 2CH_3\cdot + 2CO_2 + 2e^-$$
$$\downarrow$$
$$H_3C\!\!-\!\!CH_3$$

The above evidence for the structure of acetic acid was not obtained without a great amount of work involving many experiments. The use of physical methods has greatly shortened the time required to establish the structure of a compound, and some of these will now be considered using acetic acid as an example.

When light in the infra-red region of the spectrum (i.e. having longer wavelengths than visible light) is passed through the vapour or liquid form of an organic compound, light of certain wavelengths is absorbed. The light that is absorbed gives the molecules energy which appears as vibration or rotation of the atoms relative to one another. Now it is found that any one particular group,

e.g. the $\rangle C\!\!=\!\!O$ group in a ketone, or the $-C\!\!\begin{array}{c} \nearrow O \\ \searrow OH \end{array}$ group in a carboxylic acid, gives rise to certain characteristic absorptions in the infra-red, and that these vary little from compound to compound. Thus, examination of the infra-red spectrum of large and complicated molecules can often enable the chemist to pick out the presence of characteristic groups.

The infra-red absorption spectrum of any compound can be compared to the finger-prints of a man; no two compounds have exactly the same spectrum, but since the spectra of many thousands of compounds have been reported, an unknown compound can often be quickly identified from its ' finger-print ' which can be compared with those recorded. A modern infra-red spectrometer can record the spectrum of a compound in a few minutes.

Another group of physical methods of determining structure depends upon the fact that atoms or ions in a compound can cause the diffraction of a beam of X-rays, or of electrons or of neutrons in the same way that a diffraction grating can cause diffraction of light.

Hence passage of a beam of, say, X-rays through a crystal gives a pattern of spots on a photographic plate, of varying intensities. By a rather lengthy and often tedious process, the details of this pattern can be translated into a detailed structure of the crystal, and the positions of many of the atoms can be exactly located. This process can now be shortened somewhat by using a computer. If it is then found, for example, that two carbon atoms are located the same short distance apart as they are when bound together by a single bond, C—C, then it can be assumed that there is such a bond between them in the crystal under examination; and in this way, the structure of the molecules composing the crystal can be determined. However, X-ray diffraction can be used effectively only on solid crystalline compounds; electron diffraction can be used for liquids and gases, but does not give such exact results. Moreover, X-rays and electron beams do not show up hydrogen atoms; the position of these has to be inferred from that of other atoms in the crystal.

In recent years, use has been made of spectroscopic methods of investigating structure in which the wavelengths used are much greater than those used in ordinary spectroscopy involving a light source and a prism. In one of these methods, a compound in gaseous or liquid form is placed between the poles of a powerful electromagnet, the field of which can be varied. Now the nuclei of the hydrogen atoms (the protons) in this compound themselves behave as if they were tiny spinning magnets. It is found that, if energy of suitable wavelengths is supplied to the compound and the magnetic field varied, there is, at one particular value of the field, a sudden uptake of energy. Now the field strength at which this 'resonance'* occurs depends upon the way in which the hydrogen atoms in the compound are combined. Thus, in acetic acid, there are two 'resonance peaks' in the magnetic field. One corresponds to the uptake of energy in the three protons of the hydrogen atoms in the CH_3 group; the other smaller peak corresponds to the one hydrogen in the COOH group. In ethyl alcohol, there are three peaks, one for the CH_3 hydrogens, one for the CH_2, and one for the OH hydrogen. This kind of spectroscopy, called *nuclear magnetic resonance spectroscopy*, can therefore tell us a great deal about the way in which hydrogen atoms (and some other atoms also) are bound in organic compounds. Hence isomers of the same compound which often differ only slightly in structure (some examples are given in the following chapter) will often give different absorption 'peaks' because one or more of the hydrogen atoms are differently situated in each isomer.

* The use of the word resonance has no connection with resonance as applied to different chemical structures (p. 302).

EXAMINATION QUESTIONS

(1) Assuming the molecular formulae C_2H_6O, C_2H_4O, $C_4H_{10}O$ for ethyl alcohol, acetaldehyde and diethyl ether respectively, give the experimental evidence for the structural formulae of each of these compounds. (O. A.)

(2) Give an outline of the evidence upon which the formula $CH_3 \cdot CO \cdot OH$ has been assigned to acetic acid.
Describe how (a) acetic anhydride, (b) methane, (c) ethane may be prepared from acetic acid. (A. A.)

(3) The organic compound A contains 66·4 per cent of carbon, 5·54 per cent of hydrogen and 28·06 per cent of chlorine. When digested with caustic soda solution it produces an alcohol B, which upon treatment with acetic anhydride gives an ester of vapour density 75. Use these facts to name and formulate A.
Suggest methods of preparation (i) of A and B, (ii) of an isomer of A, (iii) of benzene, starting with either A or B. (A. S.)

(4) Give evidence in support of three of the following formulae:
(a) $(C_2H_5)_2O$ for diethyl ether; (b) $CH_3 \cdot COOH$ for acetic acid;
(c) $(COOH)_2$ for oxalic acid; (d) $C_2H_5 \cdot CN$ for ethyl cyanide.
(O. and C. A.)

(5) A substance X has the molecular formula $C_5H_{10}O_3$; when treated with a solution of sodium hydroxide it yields ethyl alcohol and a salt $C_3H_5NaO_3$. The acid $C_3H_6O_3$ obtained from this sodium salt yields methyl iodide when boiled with a concentrated solution of hydriodic acid. What is the constitutional formula of X? Suggest, in outline, a synthesis of X from acetic acid. (L. S.)

(6) By means of formulae and equations outline experiments which would enable you to determine the structures of the following compounds: (i) a compound C_4H_6, (ii) a substance C_4H_8O which reacts with hydroxyl-amine to give a derivative C_4H_9NO, (iii) a basic substance C_3H_9N.

30

ISOMERISM

Isomerism* is the occurrence of two or more compounds with the same molecular formulae, but having one or more different physical or chemical properties. It is due to different linkings or different positions of atoms or groups in space. Since, in general, there are more atoms in the molecules of organic than in those of inorganic compounds, it is understandable that the phenomenon of isomerism is more common in organic than it is in inorganic chemistry.

There are two main types of isomerism:

(1) Structural isomerism

(2) Stereoisomerism, which includes optical and geometrical isomerism.

(1) *Structural isomerism*

Structural isomerism results from the different ways in which atoms are linked within the molecules. These differences can be shown by writing the structural formulae in one plane, and numerous examples have been given already. The following are representative and it is appropriate, since the term was first coined at the beginning of the last century to explain the difference between ammonium cyanate and urea, that this pair should be the first example to be given:

ammonium
cyanate:
$$\left[\begin{array}{c} H \\ | \\ H-N-H \\ | \\ H \end{array} \right]^{+} OCN^{-} \text{ and urea:}$$

dimethyl ether: $CH_3 \cdot O \cdot CH_3$ and ethyl alcohol: $CH_3 \cdot CH_2OH$;
ethyl acetate: $CH_3 \cdot COOC_2H_5$ and methyl propionate: $C_2H_5 \cdot COOCH_3$;

* From the Greek, *isos* = equal, and *meros* = part.

α-hydroxypropionic acid (lactic acid): $CH_3 \cdot CH(OH) \cdot COOH$ and
β-hydroxypropionic acid: $CH_2OH \cdot CH_2 \cdot COOH$;

These few examples serve to show that structural isomerism may occur not only between members of different homologous series, in which case both chemical and physical properties are different, but among members of the same series where the main differences may be in physical characteristics only.

It is not easy to convert dimethyl ether into ethyl alcohol or vice versa, and this is generally true of structural isomers.   There are, however, exceptional cases in which the passage from one to the other is not only easy but difficult to avoid.   This was why Wöhler obtained urea when trying to prepare ammonium cyanate, for if either of these two compounds is dissolved in water, a dynamic equilibrium between the two isomers is established—an equilibrium in which, however, the concentration of urea is much higher than that of the cyanate:

$$NH_4OCN \rightleftharpoons CO(NH_2)_2$$

Wöhler obtained urea because this isomer is less soluble in water than ammonium cyanate.   As crystals of it separated out, more urea was formed in an attempt to re-establish equilibrium, until the change to urea was complete.

Structural isomerism where each isomer can be changed rapidly into the other, i.e. where two isomers are in dynamic equilibrium, is known as *tautomerism*.   The classical example on which most research work has been done is the ethyl ester of acetoacetic acid (which is the compound obtained when one of the methyl hydrogen atoms of acetic acid is replaced by the acetyl group).   The formula for the acid is $CH_3 \cdot CO \cdot CH_2 \cdot COOH$ and for its ethyl ester, $CH_3CO \cdot CH_2 \cdot COOC_2H_5$.   In aqueous solution, the ester which, as written above, contains a ketonic group, is in equilibrium with an isomeric form containing the hydroxyl group, thus:

$$CH_3 \cdot C \overset{\diagup O}{\underset{}{\diagdown}} CH_2 \, COOC_2H_5 \rightleftharpoons CH_3 \cdot C \overset{\diagup OH}{=\!=} CH \cdot COOC_2H_5$$

keto- isomer   enol- isomer

This type of tautomerism (and other similar examples are known) is appropriately known as *keto-enol isomerism*. The enol form, containing the ethylenic double bond, shows the characteristic properties of an unsaturated compound, decolorizing bromine water and aqueous potassium permanganate, for example. If either of these two reagents is added to the equilibrium mixture until the colour is seen, the colour will be found to disappear as soon as some of the enol isomer is formed in an attempt to re-establish equilibrium.

(2) *Stereoisomerism*

Stereoisomerism is not due to different linking of atoms or groups within the molecules, but to different positions taken up in *space* by the same atoms or groups similarly joined together. Two kinds of stereoisomerism are known:

(*a*) Optical isomerism

(*b*) Geometric isomerism

(*a*) *Optical isomerism:*—At the beginning of the nineteenth century, it was found that certain quartz crystals possessed the power of rotating polarized light (i.e. light which can be considered as vibrating in one plane only). The phenomenon was believed to be due to some peculiar arrangement within the crystals. But in 1815, the French physicist Biot showed that certain liquids possessed similar power, e.g. turpentine, an aqueous solution of sugar. In these cases, at any rate, the cause of the phenomenon could only be a molecular one.

In 1769, Scheele had prepared tartaric acid from the crystalline deposit argol, found in wine vats. A solution of this rotated the plane of polarized light to the right, or as we say, is *dextro-rotatory*. In 1831, Berzelius, experimenting with the mother liquor left after the separation of tartaric acid crystals above obtained another acid which he called *racemic acid* (from *racemes=a bunch of grapes*). This acid, although identical in chemical composition to tartaric acid, was optically inactive.

In 1780, Scheele had obtained from sour milk an acid which he called *lactic acid*. This had no effect on polarized light, but in 1808, Berzelius obtained from muscle a compound of identical composition with similar chemical properties, but which was dextro-rotatory. Lactic acid has the structural formula

$$CH_3 \cdot CH(OH) \cdot COOH$$

and tartaric acid $HOOC \cdot CH(OH) \cdot CH(OH) \cdot COOH$.

The problem of this new type of isomerism was taken up in 1848 by the great French scientist Pasteur (famous for his work on fermentation and on disease-causing micro-organisms). A few years previously, Mitscherlich (discoverer of the Law of Isomorphism) had made crystals of the sodium ammonium salt of tartaric and racemic acids, and had noticed that all the crystals did not have the same shape. Pasteur prepared crystals of the sodium ammonium salt of racemic acid. He found that these crystals possessed a characteristic facet which in some crystals pointed in one direction and in others the opposite one. The two kinds of crystals differed from each other as an object differs from its mirror image, or a left hand glove from a right hand one, i.e. the two varieties could not be superimposed. (Such crystals are said to be *enantiomorphous*.) Both kinds of crystals were asymmetric, i.e. there was no one plane through which they could be cut to get two identical halves.

The next step in the solution of the problem came in 1874 by the independent publications of van't Hoff and le Bel, who put forward the *tetrahedral valency theory of carbon*, in which a carbon atom was to be pictured at the centre of a tetrahedron with its four valencies directed towards the corners. (Some ways in which a tetrahedral molecule may be drawn in one plane have been given, for methane, on p. xxi.) Now in the molecules of both lactic and tartaric acids there are carbon atoms surrounded by *four different* atoms or groups, one such carbon atom in each molecule of lactic acid and two in each one of tartaric. This is seen clearly if the formulae are written as below:

$$CH_3 \overset{\displaystyle \overset{H}{\underset{|}{\,}}}{\underset{\displaystyle \underset{COOH}{|}}{\overset{|}{C}}}\!\!{}_*\!\!-OH$$

lactic acid

$$\overset{\displaystyle \overset{H}{\underset{|}{\,}}}{HO-\underset{\displaystyle \underset{H}{|}}{\overset{|}{C}}\!\!{}_*\!\!-COOH}$$
$$HO-\overset{|}{\underset{|}{C}}\!\!{}_*\!\!-COOH$$

tartaric acid

The carbon atoms referred to are indicated by an asterisk. A carbon atom surrounded by four dissimilar atoms or groups is called an *asymmetric carbon atom*, and it is found that in all cases where such an atom is present, the compound is capable of showing optical activity. If diagrams are drawn as I and II representing such molecules in space (*or better still, if models are actually constructed*), it will be seen that two different molecules are possible:

I   II

From whatever point the models are examined, they are always different. They compare with each other as do right and left hand gloves, or as an asymmetric object does to its mirror image. One type of molecule possesses dextro-rotatory and the other laevo-rotatory power. A mixture of equimolecular proportions of both is optically inactive.   Such a physical mixture of the two isomers is known as the *racemic mixture*. The lactic acid obtained by Scheele from sour milk and the ' racemic acid ' from the mother liquors by Berzelius were the racemic mixtures.   The lactic acid from muscle and the tartaric acid first obtained from argol by Scheele were the separate active isomers.

Tartaric acid is a more complicated compound than lactic acid since in each of its molecules there are two asymmetric carbon atoms.   Four forms have been isolated; the *laevo-isomer*, the *dextro-isomer* (in these cases, both asymmetric carbon atoms are contributing to the optical power in the same direction), the *racemic* form (a physical mixture of the two in equimolecular proportions and therefore optically inactive) and a fourth one called *meso-tartaric* acid.   Meso-tartaric acid is one isomer only but one in which the rotatory powers of the two halves of the molecule are oppositely directed.   It has a plane of symmetry and is therefore inactive.

In the glucose molecule (p. 274) written as:

$CH_2 \cdot OH \cdot \overset{*}{C}HOH \cdot \overset{*}{C}HOH \cdot \overset{*}{C}HOH \cdot \overset{*}{C}HOH \cdot CHO$, there are four asymmetric carbon atoms.   A much greater number of optical isomers is possible here.   More than one isomer is dextro-rotatory (or laevo-rotatory) but these differ in rotatory power. In one such isomer, all the optically active parts of the molecule tend to turn the plane of polarized light in the same direction. In others, one or more parts may be working in opposite directions to reduce the rotatory power. The optical isomerism of not only glucose but of most other carbohydrates is very complicated. A reference to the optical activity of the naturally-occurring sucrose and its hydrolysis to the optically active isomers of glucose and fructose has been mentioned on p. 276, and also the term *inversion* commonly used to describe the process.

It must not, however, be thought that the phenomenon of optical activity occurs with carbon compounds only, for it is found in inorganic compounds also. For example, wherever an element M is bound tetrahedrally to four different elements or groups M*abcd*, then optical activity is possible. Thus M may be another element such as silicon or lead in Group IV of the Periodic Classification, or it may be an element forming a complex ion, e.g. nickel, giving $(Ni abcd)^{2+}$.

When a compound capable of showing optical isomerism is synthesized in the laboratory, the racemic mixture is always obtained. Consider the synthesis of lactic acid from acetaldehyde by the formation of the cyanhydrin (p. 194). This may be represented as below (although the mechanism can be seen more clearly if space models are constructed):

$$
\begin{array}{cc}
\text{CH}_3-\overset{\text{H}}{\underset{\overset{|}{\text{CN}^-}}{\text{C}}}\overset{\frown}{=}\text{O} & \text{CH}_3-\overset{\text{H}}{\underset{}{\text{C}}}\overset{\frown}{=}\text{O} \quad \text{CN}^- \\[2mm]
\text{(frontal attack)} & \text{(rear attack)} \\[4mm]
\downarrow & \downarrow \\[4mm]
\underset{\text{H}_3\text{C}----\text{CN}}{\overset{\text{H}}{\text{C}}-\text{OH}} & \underset{\text{H}_3\text{C}-----\text{OH}}{\overset{\text{H}}{\text{C}}-\text{CN}} \\[2mm]
\text{III} & \text{IV}
\end{array}
$$

The two different molecules, III and IV, result from the different way in which the double bond breaks. Since, in a laboratory synthesis, millions of molecules are of necessity used, by the law of chance, 50 per cent will break in one direction and the remainder in the other, to give a mixture of equimolecular proportions, i.e. the racemic mixture of the product.

There are several methods (all due to Pasteur) of resolving such a mixture into the two (or more) optically active isomers:

(1) By crystallization and picking out the differently shaped crystals. This method is not always possible for not all optically isomeric compounds give two types of crystal as does tartaric acid.

(2) If the compound is an acid (or an alcohol), its ester with either the laevo- or the dextro- isomer of some optically active alcohol (or acid) is made. It is usually found that the two resulting esters— called *diastereoisomers*—(viz. *dl* and *ll*) possess some other different physical property, e.g. solubility or boiling point, making separation by physical methods possible.

(3) If the optically active compound is one on which some species of mould, for example, can feed, then if the mould is introduced it is usually found that one isomer is ' eaten ' before the other. If the optical power is measured at intervals, it will be found to increase to a maximum, at which point one isomer will have been removed completely. It may be mentioned here that if one eats racemic glucose, only the dextro- is absorbed, the laevo- being excreted unchanged.

Following from these methods of resolution, it has been suggested that the occurrence of the separate optical isomers of certain compounds in the bodies of plants and animals (instead of their racemic mixtures) is due to their syntheses in those organisms under the influence of optically active enzymes.

(*b*) *Geometric isomerism:*—A second type of stereoisomerism occurs among certain compounds containing a double bond. By restricting free rotation within a molecule, such a bond usually fixes in space the positions of those atoms or groups linked to it.

Two dibasic acids of molecular formula $C_4H_4O_4$ are known. Both show the characteristic properties of a double bond compound and would seem to have the structural formula:

$$HOOC \cdot CH = CH \cdot COOH$$

The two acids, called respectively *maleic* and *fumaric* acids, differ in one important chemical property, viz. that maleic acid gives an internal acid anhydride while fumaric acid does not. To explain this, the formulae given below have been suggested. The formula of the maleic anhydride is also given:

$$\text{H—C—COOH} \\ | \\ \text{H—C—COOH}$$

maleic acid

$$\text{H—C—C}\diagup^{O} \\ || \qquad \diagdown_{O} \\ \text{H—C—C}\diagdown \\ \qquad \diagdown_{O}$$

maleic anhydride

$$\text{H—C—COOH} \\ || \\ \text{HOOC—C—H}$$

fumaric acid

Both acids on reduction yield the same succinic acid, of constitution $\text{HOOC·CH}_2\text{·CH}_2\text{·COOH}$, in which, since the two carbon atoms of the two methylene groups are united by a single bond, free rotation within the molecule is possible and the positions of the two carboxyl groups are no longer fixed.

Other similar cases are known. The maleic isomer is usually referred to as the *cis*-isomer and that of the fumaric as the *trans*-form. General methods of changing one isomer to the other are known. For example, a mixture of sulphur dioxide and hydrogen sulphide will convert maleic to fumaric acid. If fumaric acid is strongly heated, it gives maleic anhydride.

A similar isomerism occurs among the aldoximes, the double bond between the carbon and nitrogen atoms fixing the positions of the attached groups in the same way. The structures of the two isomers of benzaldoxime are written:

$$\text{C}_6\text{H}_5\text{—C—H} \\ || \\ \text{N—OH}$$
cis-benzaldoxime
m.p. 35°C

$$\text{C}_6\text{H}_5\text{—C—H} \\ || \\ \text{HO—N}$$
trans-benzaldoxime
m.p. 125°C

Ring isomerism

From what has been said previously about the benzene ring (e.g. p. 201), it will be readily realized that this ring is planar in benzene, i.e. flat. But in cyclohexane C_6H_{12} (p. 201), we have only C—H and C—C bonds, as in an alkane, and the bonds around each carbon atom would be expected to be tetrahedral. This could be achieved by making the ring non-planar; and in fact two possible forms of the ring which make the bonds tetrahedral are the chair

I   II

form (I) and the boat form (II). These forms are known as *conformations*. In cyclohexane itself, the two forms are so readily interconvertible that neither can be isolated, although there is evidence that the chair form predominates at room temperature. However, whether the rings exist as 'chairs' or 'boats', *cis-trans* isomerism is possible with disubstituted cyclohexane derivatives (and other saturated ring compounds). Thus, assuming for simplicity that the rings are flat, isomers of the types shown below are quite distinct compounds; they will remain distinct so long as no bonds are broken:

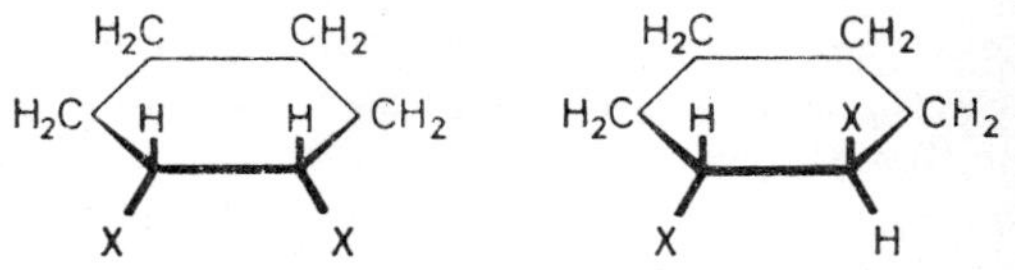

If one hydrogen on each carbon atom of cyclohexane is replaced by chlorine to give $C_6H_6Cl_6$ (also called benzene hexachloride) there are nine possible isomers, and six of these have been identified and named by Greek letters α, β, γ, etc. Only one—the γ isomer—possesses marked insecticidal activity, and is marketed as 'Gammexane'.

EXAMINATION QUESTIONS

(1) Briefly explain the reasons for the common occurrence of structural isomerism among the paraffin hydrocarbons and their halogen derivatives.

How many isomeric paraffins of formula C_5H_{12} can be expected, how many alkyl chlorides $C_5H_{11}Cl$, and how many olefins C_5H_{10}? Show (as concisely as you can) the basis of your predictions, and for each of these three molecular formulae indicate whether there is any possibility of additional isomers other than the types already mentioned.

Name three isomers of formula C_3H_9N and state what tests you would make to distinguish between them. (J. M. B. S.)

(2) Illustrate the stereoisomerism of the lactic acids and the tartaric acids with the aid of suitable formulae and brief explanatory notes. Explain clearly any conventions which you find to be necessary in drawing the formulae effectively.

Explain very briefly what is meant by a *racemic mixture* and outline the principles of a chemical method for the separation of the components of a typical example of such a mixture.

Two isomeric, optically active carboxylic esters A and B of molecular formulae $C_6H_{12}O_2$ yield, respectively, sodium acetate and methyl alcohol as one of the products of hydrolysis by alkali. Deduce the structural formulae of A and B. (W. S.)

INDEX

Calcium acetylide, 31, 34
Calcium formate, 131
Calor gas, 14
Cannizzaro's reaction, 105, 240, 246
Canonical forms, 302
Caramel, 279
Carbamic acid, 172
Carbamide, see Urea
Carbasorb, 320
Carbohydrates, 272–283
Carbolic acid, see Phenol
Carbon black, 41
Carbon tetrachloride, 10, 54
Carbonic acid, 171
Carbonium ion, 21, 299, 301
Carbonyl chloride, 171
Carbonyl group, 89, 124, 302
Carboxyl group, 120, 126
Carboxylic acids, aliphatic, 120–132
    derivatives, 136–144
    nomenclature, 120
    summary 132–133
    synthesis of, 190
Carboxylic acids, aromatic, 249–252
Carbylamines, see Isonitriles
Casein, 283
Cellophane, 283
Celluloid, 282
Cellulose, 24, 280–282
    uses, 281–283
Cellulose acetates, 282
Cellulose nitrates, 282
Cetyl alcohol, 151
Chevreul, 150
Chitin, 279
Chloral 101, 107
    summary, 111
Chloral hydrate, 107
Chlorobenzene, 225–227
    preparation, 225
    properties, 225–226
    uses, 227
Chloroform, 53, 101
    preparation, 101
    properties, 53
    summary, 56
Chlorotoluenes, 213, 227
Chromatography, 313
Chromium dibenzene, 210
Chromophore, 269
Cis and trans isomerism, 35
Clostridium acetobutylicum, 114
Coal berginization, 203
Coal carbonization, 203
Collagen, 283
Collodion, 282
Conformations, 338
Conjugated double bond, 29
Constant boiling mixture, 75
Co-ordinate link, xxii, 288
Copper(I) acetylide, 34
Cordite, 117, 282

Couper, xvii, xviii
Covalent bond, xxi
Cracking, 9, 43–44
Cream of tartar, 180, 183
Cresols, 212, 216, 234
Crystallization, 306
Cumene, 235
Cyanhydrin formation, 96–97, 335
Cyanhydrins, 96
Cyanides, 156–157
Cyclohexane, 201, 205, 337
Cycloparaffins, 28
Cyclopropane, 28

D.D.T., 227
Dalton, xv, 15
Dative bond, 290
Davy, xv, 54
Decarboxylation, 128, 205, 250
Dehydrite, 320
Dehydrogenation, 64
Delocalization, 124–125, 201, 303
Denaturation, 283
Detergents, 154
Dextrin, 280
Dextrose, see Glucose
Diastereoisomers, 336
Diazomethane, 27
Diazonium salts, 265–270
Diazotization, 265
Dibasic acid and derivatives, 171–180
    summary, 180–183
1,1-Dibromoethane, 52, 53
1,2-Dibromoethane, 21–22, 47, 52, 77, 98
1,2-Dibromoethylene, 35
    cis and trans, 36
1,1-Dichloroethane, 33, 52, 103
1,2-Dichloroethane, 18, 30, 52
1,2-Dichloropropane, 18
N-Diethyl acetamide, 165
Diethyl ammonium iodide, 162
Diethyl ether, 83–88
    preparation, 83–85
    properties, 86
    structural formula, 86–88
    uses, 86
Diethyl nitrosamine, 166
Diethyl oxalate, 176
Diethyl sulphate, 62
Diethylamine, 166
p-Dimethylaminoazobenzene, 269
Dimethylaniline, 254
m-Dinitrobenzene, 220
2,4-dinitrophenylhydrazine, 100
Di-olefins, 29
Diphenylamine, 254
Dipolar ion, 284
Dipole moment, 293
Dipole-dipole attraction, 293
Dipropargyl, 199

Trinitrotoluene, 221
Trioxan, 104
Twisett, 313

Urea, xxii, 172–174
  manufacture, 173
  preparation, 173
  properties, 173–174
  summary, 181
  uses, 174
Urea xxii, nitrate, 173

Valence-bond theory, 302
Vaseline, 42
Vinegar, 121, 122–123
Vinyl chloride, 26, 33

Vinyl group, 32
Viscose, 282

Waxes, 150–151
Westron, 37
Williamson, 49, 85, 188, 237
Wines, 73
Wöhler, xv, xvi, xvii, 172, 325
Wort, 73
Wurtz, 8, 51, 102, 161, 185

X-ray diffraction, 327–328
Xylenes, 203, 214

Zwitterion, 284
Zymase, 73